CAMBRIDGE LIBRARY COLLECTION

Books of enduring scholarly value

Mathematical Sciences

From its pre-historic roots in simple counting to the algorithms powering modern desktop computers, from the genius of Archimedes to the genius of Einstein, advances in mathematical understanding and numerical techniques have been directly responsible for creating the modern world as we know it. This series will provide a library of the most influential publications and writers on mathematics in its broadest sense. As such, it will show not only the deep roots from which modern science and technology have grown, but also the astonishing breadth of application of mathematical techniques in the humanities and social sciences, and in everyday life.

The Algebra of Invariants

Invariant theory is a subject within abstract algebra that studies polynomial functions which do not change under transformations from a linear group. John Hilton Grace (1873–1958) was a research mathematician specialising in algebra and geometry. He was elected a Fellow of the Royal Society in 1908. His co-author Dr Alfred Young (1873–1940) was also a research mathematician before being ordained in 1908; in 1934 he too was elected a Fellow of the Royal Society. Abstract algebra was one of the new fields of study within mathematics which developed out of geometry during the nineteenth century. It became a major area of research in the late nineteenth and early twentieth centuries. First published in 1903, this book introduced the work on invariant theory of the German mathematicians Alfred Clebsch and Paul Gordan into British mathematics. It was considered the standard work on the subject.

Cambridge University Press has long been a pioneer in the reissuing of out-of-print titles from its own backlist, producing digital reprints of books that are still sought after by scholars and students but could not be reprinted economically using traditional technology. The Cambridge Library Collection extends this activity to a wider range of books which are still of importance to researchers and professionals, either for the source material they contain, or as landmarks in the history of their academic discipline.

Drawing from the world-renowned collections in the Cambridge University Library, and guided by the advice of experts in each subject area, Cambridge University Press is using state-of-the-art scanning machines in its own Printing House to capture the content of each book selected for inclusion. The files are processed to give a consistently clear, crisp image, and the books finished to the high quality standard for which the Press is recognised around the world. The latest print-on-demand technology ensures that the books will remain available indefinitely, and that orders for single or multiple copies can quickly be supplied.

The Cambridge Library Collection will bring back to life books of enduring scholarly value (including out-of-copyright works originally issued by other publishers) across a wide range of disciplines in the humanities and social sciences and in science and technology.

The Algebra of Invariants

JOHN HILTON GRACE
ALFRED YOUNG

CAMBRIDGE
UNIVERSITY PRESS

CAMBRIDGE UNIVERSITY PRESS

Cambridge, New York, Melbourne, Madrid, Cape Town, Singapore,
São Paolo, Delhi, Dubai, Tokyo, Mexico City

Published in the United States of America by Cambridge University Press, New York

www.cambridge.org
Information on this title: www.cambridge.org/9781108013093

This edition first published 1903
This digitally printed version 2010

ISBN 978-1-108-01309-3 Paperback

THE ALGEBRA

OF

INVARIANTS

London: C. J. CLAY AND SONS,
CAMBRIDGE UNIVERSITY PRESS WAREHOUSE,
AVE MARIA LANE.
Glasgow: 50, WELLINGTON STREET.

Leipzig: F. A. BROCKHAUS.
New York: THE MACMILLAN COMPANY.
Bombay and Calcutta: MACMILLAN AND CO., LTD.

THE ALGEBRA
OF
INVARIANTS

BY

J. H. GRACE, M.A.
FELLOW OF PETERHOUSE

AND

A. YOUNG, M.A.
LECTURER IN MATHEMATICS AT SELWYN COLLEGE,
LATE SCHOLAR OF CLARE COLLEGE

CAMBRIDGE:
AT THE UNIVERSITY PRESS.
1903

Cambridge:
PRINTED BY J. AND C. F. CLAY,
AT THE UNIVERSITY PRESS.

PREFACE.

THE object of this book is to provide an English introduction to the symbolical method in the theory of Invariants. It was started as an attempt to meet the need expressed by Elliott in the preface to *The Algebra of Quantics*—'a whole book which shall present to the English reader in his own language a worthy exposition of the method of the great German masters remains a desideratum.' Since then the need has been partly met by the article 'Algebra' by MacMahon in the Supplement to the *Encyclopædia Britannica*. The subject has been treated from the commencement in order that readers unacquainted with Elliott's treatise or any presentation of the elements may be able to understand the argument. Such readers should bear in mind that this treatise is only concerned with one part of a very extensive subject. The modern theory of Partitions will be found in the first part of the article by MacMahon mentioned above.

The first six chapters—a great portion of which, we hope, will be found easy reading—may be said to lead step by step to Gordan's wonderful proof of the finiteness of the system for a single binary form. The sixth chapter is, in fact, devoted to an exposition of Gordan's third proof, but here, as throughout the book, we have allowed ourselves a free hand in dealing with the memoirs and treatises quoted. For example, we have made much use of Jordan's great memoirs on Invariants in proving Gordan's theorem: in a later chapter on Types of Covariants the development of Jordan's method has led us to some results which we believe to be important as well as novel, notably to an exact formula for the maximum order of an irreducible covariant of a system of binary forms.

The remainder of the book is mainly of geometrical interest: much space is devoted to Apolarity and Rational Curves, and the treatment of ternary forms is from the geometrical rather than the analytical point of view. The only complete system of ternary forms given is that for two Quadratics: it may be felt that more should have been said on this subject, but we think that with the methods known up to the present the treatment of ternary forms is too tedious for a text-book.

The number of references to Mathematical Journals etc. will perhaps be found unusually small: for this there is no need to apologise since the admirable *Bericht über den gegenwärtigen Stand der Invariantentheorie** of Meyer gives references up to the last few years and in a more complete fashion than is desirable in a book which makes no pretensions to being exhaustive.

We wish to thank Dr H. F. Baker for help given to us in our early reading and Professor Forsyth for encouragement while writing. For reading of proof-sheets we are indebted to Mr J. E. Wright, B.A., of Trinity College, Mr P. W. Wood, B.A., of Emmanuel College, and in a still greater degree to the late Mr A. P. Thompson, B.A., of Pembroke College, whose enthusiasm for Mathematics and research was most helpful and whose early death is deplored alike by his teachers and his fellow-workers. Our thanks are also due to the officials of the University Press for great help received during the course of printing.

J. H. GRACE.

A. YOUNG.

* *Jahresbericht der Deutschen Mathematiker Vereinigung*, Vol. I., 1892. French translation by Fehr; Gauthier-Villars, Paris, 1897. Italian translation by Vivanti; Pellerano, Naples, 1899. Article, *Invariantentheorie* in the *Encyclopädie der mathematischen Wissenschaften.*

August 18, 1903.

CONTENTS.

CHAPTER I.

INTRODUCTION. SYMBOLICAL NOTATION.

1. IF in the expression

$$a_0x_1^2 + 2a_1x_1x_2 + a_2x_2^2,$$

we write

$$x_1 = \xi_1X_1 + \eta_1X_2,$$
$$x_2 = \xi_2X_1 + \eta_2X_2,$$

we obtain a new expression, viz.

$$A_0X_1^2 + 2A_1X_1X_2 + A_2X_2^2,$$

where

$$A_0 = a_0\xi_1^2 + 2a_1\xi_1\xi_2 + a_2\xi_2^2,$$
$$A_1 = a_0\xi_1\eta_1 + a_1(\xi_1\eta_2 + \xi_2\eta_1) + a_2\xi_2\eta_2,$$
$$A_2 = a_0\eta_1^2 + 2a_1\eta_1\eta_2 + a_2\eta_2^2.$$

It is easy to verify the identity

$$A_0A_2 - A_1^2 = (a_0a_2 - a_1^2)(\xi_1\eta_2 - \xi_2\eta_1)^2,$$

which shews that the function $A_0A_2 - A_1^2$ of the coefficients of the transformed expression differs from the same function $a_0a_2 - a_1^2$ of the coefficients of the original expression by a factor involving only the coefficients contained in the transformation.

2. In the present work we shall give an account of the theory and structure of functions of the coefficients possessing properties analogous to that described above; but before proceeding to generalities we shall give some further examples.

If we transform the two expressions

$$a_0x_1^2 + 2a_1x_1x_2 + a_2x_2^2,$$
$$a_0'x_1^2 + 2a_1'x_1x_2 + a_2'x_2^2,$$

in the same way as before, and they become

$$A_0X_1^2 + 2A_1X_1X_2 + A_2X_2^2,$$

$$A_0'X_1^2 + 2A_1'X_1X_2 + A_2'X_2^2,$$

then it is easy to verify the identity

$$A_0A_2' - 2A_1A_1' + A_0'A_2 = (a_0a_2' - 2a_1a_1' + a_0'a_2)(\xi_1\eta_2 - \xi_2\eta_1)^2.$$

Thus we have here a function of the coefficients of two expressions such that the new value differs from the original value by a factor depending only on the transformation employed

3. As a third example, if the cubic expression

$$a_0x_1^3 + 3a_1x_1^2x_2 + 3a_2x_1x_2^2 + a_3x_2^3$$

become $$A_0X_1^3 + 3A_1X_1^2X_2 + 3A_2X_1X_2^2 + A_3X_2^3,$$

when we put $$x_1 = \xi_1X_1 + \eta_1X_2,$$

$$x_2 = \xi_2X_1 + \eta_2X_2,$$

then we have

$$(A_0A_2 - A_1^2)X_1^2 + (A_0A_3 - A_1A_2)X_1X_2 + (A_1A_3 - A_2^2)X_2^2$$
$$= \{(a_0a_2 - a_1^2)x_1^2 + (a_0a_3 - a_1a_2)x_1x_2 + (a_1a_3 - a_2^2)x_2^2\}(\xi_1\eta_2 - \xi_2\eta_1)^2.$$

This identity indicates a property quite similar to that illustrated in the two previous examples, but the function, which is unaltered except for the factor $(\xi_1\eta_2 - \xi_2\eta_1)^2$, now involves the variables as well as the coefficients of the expression from which it is formed.

The result we have written down may be verified directly, but more easily as follows:

Denoting the original expression by f and the transformed expression by F we have to prove that

$$\frac{\partial^2F}{\partial X_1^2}\frac{\partial^2F}{\partial X_2^2} - \left(\frac{\partial^2F}{\partial X_1\partial X_2}\right)^2 = \left\{\frac{\partial^2f}{\partial x_1^2}\frac{\partial^2f}{\partial x_2^2} - \left(\frac{\partial^2f}{\partial x_1\partial x_2}\right)^2\right\}(\xi_1\eta_2 - \xi_2\eta_1)^2.$$

Now $$\frac{\partial F}{\partial X_1} = \frac{\partial F}{\partial x_1}\frac{\partial x_1}{\partial X_1} + \frac{\partial F}{\partial x_2}\frac{\partial x_2}{\partial X_1}$$

$$= \xi_1\frac{\partial F}{\partial x_1} + \xi_2\frac{\partial F}{\partial x_2},$$

and in like manner

$$\frac{\partial^2 F}{\partial X_1^2} = \xi_1^2 \frac{\partial^2 F}{\partial x_1^2} + 2\xi_1\xi_2 \frac{\partial^2 F}{\partial x_1 \partial x_2} + \xi_2^2 \frac{\partial^2 F}{\partial x_2^2}$$

$$\frac{\partial^2 F}{\partial X_1 \partial X_2} = \xi_1\eta_1 \frac{\partial^2 F}{\partial x_1^2} + (\xi_1\eta_2 + \xi_2\eta_1) \frac{\partial^2 F}{\partial x_1 \partial x_2} + \xi_2\eta_2 \frac{\partial^2 F}{\partial x_2^2}$$

$$\frac{\partial^2 F}{\partial X_2^2} = \eta_1^2 \frac{\partial^2 F}{\partial x_1^2} + 2\eta_1\eta_2 \frac{\partial^2 F}{\partial x_1 \partial x_2} + \eta_2^2 \frac{\partial^2 F}{\partial x_2^2}.$$

But these equations are exactly the same as those which express A_0, A_1, A_2 in terms of a_0, a_1, a_2 (§ 1), hence

$$\frac{\partial^2 F}{\partial X_1^2}\frac{\partial^2 F}{\partial X_2^2} - \left(\frac{\partial^2 F}{\partial X_1 \partial X_2}\right)^2 = (\xi_1\eta_2 - \xi_2\eta_1)^2 \left\{\frac{\partial^2 f}{\partial x_1^2}\frac{\partial^2 f}{\partial x_2^2} - \left(\frac{\partial^2 f}{\partial x_1 \partial x_2}\right)^2\right\}.$$

The expression $\dfrac{\partial^2 f}{\partial x_1^2}\dfrac{\partial^2 f}{\partial x_2^2} - \left(\dfrac{\partial^2 f}{\partial x_1 \partial x_2}\right)^2$ is called the Hessian of f.

4. Let us now explain the phraseology in common use when dealing with questions such as arise in our subject.

Quantics. A rational integral homogeneous algebraic function of any number of variables $x_1, x_2, \ldots x_p$, is called a quantic.

The degree in the variables is called the *order* of the quantic, and according as the number of variables is two, three, four ...... we call the quantic binary, ternary, quaternary ......

Thus a binary quantic of order n is a rational integral homogeneous algebraic function of two variables which is of the nth degree in those variables.

Such a quantic might be written

$$a_0 x_1^n + a_1 x_1^{n-1} x_2 + a_2 x_1^{n-2} x_2^2 + \ldots + a_n x_2^n,$$

but we shall find it invariably more convenient to write it

$$a_0 x_1^n + \binom{n}{1} a_1 x_1^{n-1} x_2 + \binom{n}{2} a_2 x_1^{n-2} x_2^2 + \ldots + a_n x_2^n,$$

i.e. with binomial coefficients prefixed to the various a's.

The former of these expressions is now commonly written

$$(a_0, a_1, a_2, \ldots, a_n \,\tilde{\between}\, x_1, x_2)^n,$$

and the latter $\quad (a_0, a_1, a_2, \ldots, a_n \between x_1, x_2)^n$,

a very convenient notation introduced by Cayley.

The mere consideration of the transformation of the binary form

$$a_0x_1^2 + 2a_1x_1x_2 + a_2x_2^2$$

will be sufficient to convince the reader of the advantage of the introduction of binomial coefficients.

Passing now to the case of any number of variables, we call the quantic a p-ary q-ic when it is homogeneous and of degree q in p variables.

Thus the most general ternary quadratic is written

$$a_{200}x_1^2 + a_{020}x_2^2 + a_{002}x_3^2 + 2a_{110}x_1x_2 + 2a_{101}x_1x_3 + 2a_{011}x_2x_3,$$

and in general the ternary n-ic is written

$$\Sigma \frac{n!}{p!\,q!\,r!} a_{pqr}x_1^p x_2^q x_3^r,$$

where the summation is extended to all values of p, q, r satisfying the equality

$$p + q + r = n.$$

It will be noticed that here we have prefixed multinomial coefficients to the a's.

5. Linear Transformations. The equations

$$x_1 = \xi_1X_1 + \eta_1X_2$$
$$x_2 = \xi_2X_1 + \eta_2X_2$$

are said to constitute a linear transformation from the variables x_1x_2 to the variables X_1X_2—it is of course implied that the coefficients on the right do not involve either set of variables.

The determinant

$$D = \begin{vmatrix} \xi_1 & \eta_1 \\ \xi_2 & \eta_2 \end{vmatrix}$$

is called the determinant of the transformation.

If D vanishes it is evident that x_1 and x_2 are virtually identical, for their ratio is constant, and hence, as the variables are always supposed to be independent, we shall throughout only deal with transformations which have a non-vanishing determinant.

On solving for X_1X_2 we find

$$X_1 = (\eta_2x_1 - \eta_1x_2)/D$$
$$X_2 = (-\xi_2x_1 + \xi_1x_2)/D$$

so that the passage back from the new variables to the old is effected by a linear transformation. This is called the inverse of the original transformation; it is evident at once that its determinant is equal to $\frac{1}{D}$.

6. Let us now regard a linear transformation as an operator, which acting on x_1, x_2 changes them to X_1, X_2, and let us consider the effect of two such operators acting successively.

If the coefficients of the first are

$$\xi_1, \eta_1; \xi_2, \eta_2,$$

and those of the second

$$\xi_1', \eta_1'; \xi_2', \eta_2',$$

then we have

$$\left.\begin{aligned} x_1 &= \xi_1 X_1 + \eta_1 X_2 \\ x_2 &= \xi_2 X_1 + \eta_2 X_2 \end{aligned}\right\},$$

$$\left.\begin{aligned} X_1 &= \xi_1' X_1' + \eta_1' X_2' \\ X_2 &= \xi_2' X_1' + \eta_2' X_2' \end{aligned}\right\},$$

and the effect of the two operators acting successively is to change from the variables x_1, x_2 to X_1', X_2'.

Now on elimination of X_1, X_2 we find

$$x_1 = (\xi_1\xi_1' + \eta_1\xi_2') X_1' + (\xi_1\eta_1' + \eta_1\eta_2') X_2',$$

$$x_2 = (\xi_2\xi_1' + \eta_2\xi_2') X_1' + (\xi_2\eta_1' + \eta_2\eta_2') X_2'.$$

And accordingly we can pass directly from the original to the final variables by means of a single linear transformation which we shall call Σ.

If we call the two preceding operators S and S' we may write

$$\Sigma = SS'$$

and Σ is called the product or the resultant of S and S'.

It must be carefully noticed that the order of the factors S and S' is essential in considering their product. In our example we supposed that S acted first and then S'. If S' had acted first and then S we should have

$$\Sigma' = S'S$$

and it is manifest that Σ and Σ' are not in general the same.

Since the resultant of two or any number of linear transformations is another such transformation, the whole set of linear transformations obtained by varying the coefficients is said to form a group—a continuous group because the coefficients ξ and η may be supposed to vary continuously.

The determinant of Σ is equal to the product of the determinants of S and S', as follows from the multiplication theorem for determinants.

The product of a transformation and its inverse is a transformation which does not affect the variables, *i.e.* it is

$$\left.\begin{aligned} x_1 &= X_1 \\ x_2 &= X_2 \end{aligned}\right\}$$

which is called the identical operator. The determinant of this is unity, and, as we have pointed out, the product of the determinants of a transformation and its inverse is also unity.

7. The idea of a linear transformation admits of immediate extension to any number of variables $x_1, x_2, \ldots x_p$ and now the transformation consists of n equations

$$x_r = \xi_{r1} X_1 + \xi_{r2} X_2 + \ldots + \xi_{rp} X_p, \qquad r = 1, 2, \ldots p.$$

The determinant D formed with the ξ's for elements is called the determinant of the transformation, and inasmuch as when D vanishes there is a linear homogeneous relation between the x's, we exclude as before all transformations having a vanishing determinant.

If $D \neq 0$ we can solve for the X's in terms of the x's and, as can be easily seen, each X is a linear function of $x_1, x_2, \ldots x_p$, so that we have

$$X_r = \eta_{r1} x_1 + \eta_{r2} x_2 + \ldots + \eta_{rp} x_p,$$

a linear transformation which is the inverse of the preceding one.

As in the case of two variables, the resultant of two linear transformations S and T is a third linear transformation

$$\Sigma = ST,$$

and on examining the coefficients in Σ it will be seen at once by the multiplication theorem that the determinant of Σ is the product of the determinants of S and T.

8. In the earlier portion of this work we shall deal almost entirely with binary forms, and although we shall be constantly considering linear transformations and their effects, yet the fact that they form a group will not be explicitly used. Our only object, in introducing these elementary properties of groups, is to point out that the connection between invariants and groups is intimate and universal—in other words, that every group has its accompanying invariants and, conversely, every set of invariants belongs to a group.

9. **Invariants of Binary Forms.** If a binary form f be changed by a linear transformation into a new form F, and a function I of the coefficients of F be equal to the same function of the coefficients of f multiplied by a factor depending solely on the transformation, then I is called an invariant of the binary form f.

Thus for example in § 1 the identity

$$(A_0A_2 - A_1^2) = (a_0a_2 - a_1^2)(\xi_1\eta_2 - \xi_2\eta_1)^2$$

shews that $a_0a_2 - a_1^2$ is an invariant of the binary quadratic

$$a_0x_1^2 + 2a_1x_1x_2 + a_2x_2^2.$$

An exactly similar definition applies to a joint invariant of several binary forms, *e.g.*

$$a_0a_2' - 2a_1a_1' + a_0'a_2$$

is an invariant of the two binary forms

$$a_0x_1^2 + 2a_1x_1x_2 + a_2x_2^2,$$

and

$$a_0'x_1^2 + 2a_1'x_1x_2 + a_2'x_2^2.$$

10. For the present we shall confine our attention to invariants which are rational integral functions of the coefficients. It is easy to see that there is no further loss of generality if we suppose the invariants to be homogeneous in each set of coefficients that they contain.

Thus for example if I be an invariant of a single binary form f which is not homogeneous in the coefficients a we can write I in the form

$$I_1 + I_2 + \dots + I_s,$$

where each element in this sum is homogeneous.

Now by definition we have

$$I(A) = M \times I(a),$$

and therefore

$$I_1(A) + I_2(A) + \ldots + I_s(A) = M\{I_1(a) + I_2(a) + \ldots + I_s(a)\}.$$

But the A's are linear functions of the a's and M is independent of both, and therefore the only part on the left-hand side which is of the same degree as $I_1(a)$ on the right-hand side is $I_1(A)$;

$$\therefore \quad I_1(A) = M I_1(a),$$

that is to say I_1 is an invariant. Hence a non-homogeneous invariant is the sum of several homogeneous invariants.

This result can be at once extended to any number of binary forms.

As an example

$$a_0 a_2 - a_1^2 + a_0 a_2' - 2a_1 a_1' + a_2 a_0'$$

is an invariant of the two binary quadratics

$$(a_0, a_1, a_2 \between x_1, x_2)^2 \text{ and } (a_0', a_1', a_2' \between x_1, x_2)^2,$$

but it is the sum of two expressions

$$a_0 a_2 - a_1^2,$$

and $$a_0 a_2' - 2a_1 a_1' + a_2 a_0'$$

each of which is homogeneous in the two sets of coefficients.

11. Covariants of Binary Forms. If a binary form f is changed into a form F by a linear transformation, and a function C of the coefficients of F and the new variables X_1, X_2 be equal to the same function of the coefficients of f and the old variables x_1, x_2 multiplied by a factor depending only on the transformation, then C is called a covariant of the binary form.

Thus from what we have seen

$$\frac{\partial^2 f}{\partial x_1^2}\frac{\partial^2 f}{\partial x_2^2} - \left(\frac{\partial^2 f}{\partial x_1 \partial x_2}\right)^2$$

is a covariant of the binary cubic f and in fact of any binary form.

An exactly similar definition applies to a joint covariant of several binary forms—as an example the reader will have no difficulty in shewing that the Jacobian

$$\frac{\partial f}{\partial x_1}\frac{\partial \phi}{\partial x_2} - \frac{\partial f}{\partial x_2}\frac{\partial \phi}{\partial x_1}$$

of any two forms f and ϕ is a covariant of those forms, the multiplier being $(\xi_1\eta_2 - \xi_2\eta_1)$ the determinant of the transformation.

We shall confine our attention to covariants which are rational integral functions both of the coefficients and the variables, and, as in the case of invariants, there is no difficulty in seeing that there is no further loss of generality in supposing such covariants to be homogeneous in the variables and in each set of coefficients involved. In fact if a covariant be not homogeneous it is the sum of several parts each of which is a covariant and homogeneous.

12. Degree and Order of a Covariant. The degree of a covariant of a single form is its degree in the coefficients of that form—the order is the degree in the variables.

The covariant $\dfrac{\partial^2 f}{\partial x_1^2}\dfrac{\partial^2 f}{\partial x_2^2} - \left(\dfrac{\partial^2 f}{\partial x_1 \partial x_2}\right)^2$ of a binary form of order n is of degree two and order $2n - 4$.

A covariant of several binary forms has a definite partial degree in each set of coefficients involved and the order is as before the degree in the variables.

The Jacobian of f and ϕ is of degree one in the coefficients of each of the two forms, and its order is the sum of the orders of f and ϕ diminished by two.

13. Symbolical Notation. In our investigations we shall find it of the utmost value to write the binary quantic

$$a_0 x_1^n + n a_1 x_1^{n-1} x_2 + \ldots + \binom{n}{r} a_r x_1^{n-r} x_2^r + \ldots + a_n x_2^n$$

in the symbolical form

$$(\alpha_1 x_1 + \alpha_2 x_2)^n,$$

so that $\alpha_1^n = a_0,\ \alpha_1^{n-1}\alpha_2 = a_1, \ldots \alpha_1^{n-r}\alpha_2^r = a_r, \ldots \alpha_2^n = a_n.$

This representation is startling at first sight, but consider how the use of it would introduce errors into calculation. They would arise because relations of the type

$$a_0 a_2 = \alpha_1^{2n-2}\alpha_2^2 = a_1^2$$

between the coefficients prevent our binary form from being a general one. Now in representing a function of the coefficients symbolically we allow no symbol such as α to occur more than n times in any one term, so that the possibility of relations giving rise to

$$a_0 a_2 = a_1^2$$

is entirely precluded. In fact to obtain this relation there must be $2n$ α's multiplied together in the representation of the function $a_0 a_2$ or a_1^2, whereas, when we allow no more than n α's to occur in any one term, the $(n+1)$ expressions

$$\alpha_1^n,\ \alpha_1^{n-1}\alpha_2, \dots \alpha_1^{n-r}\alpha_2^r, \dots \alpha_2^n$$

are independent quantities, *i.e.* with these restrictions on the use of our symbols the $(n+1)$ coefficients of the original quantic are not necessarily connected by any relation, and therefore the most general quantic can be represented in the form indicated.

Accordingly in addition to the symbol α we introduce a number of equivalent symbols β, γ, ... so that

$$f = (\alpha_1 x_1 + \alpha_2 x_2)^n = (\beta_1 x_1 + \beta_2 x_2)^n = (\gamma_1 x_1 + \gamma_2 x_2)^n = \dots$$

or as it will invariably be written

$$f = \alpha_x^n = \beta_x^n = \gamma_x^n \dots$$

The symbolical equivalent of $a_0 a_2$ is not

$$\alpha_1^{2n-2}\alpha_2^2,$$

because here there are more than n α's multiplied together.

To represent $a_0 a_2$ we must use two different symbols α, β and then

$$a_0 a_2 = \alpha_1^n \beta_1^{n-2}\beta_2^2,$$

which is of course equivalent to

$$\beta_1^n \alpha_1^{n-2}\alpha_2^2,$$

whereas in the same symbols a_1^2 is represented by $\alpha_1^{n-1}\alpha_2\beta_1^{n-1}\beta_2$.

In general to represent an expression of degree m in the coefficients, we have to use m different symbols of the type α, β, γ, ...

We have said that not more than n α's must be multiplied together in a given term—on the other hand if the expression has an actual as well as a symbolical significance not less than n of these symbols must occur together because only the expressions

$$\alpha_1^n,\ \alpha_1^{n-1}\alpha_2, \dots \alpha_2^n$$

have an actual meaning.

14. A function of the coefficients can generally be represented symbolically in different ways as we have seen in the case of $a_0 a_2$ for example, which is equivalent to both

$$\alpha_1^n \beta_1^{n-2} \beta_2^2 \text{ and } \beta_1^n \alpha_1^{n-2} \alpha_2^2.$$

There is one method of determining the symbolical representation which is very convenient because it often leads to the expression most suitable for our purpose.

Suppose, in fact, that P is a homogeneous function of the mth degree in $a_0, a_1, \dots a_n$, then

$$P_1 = \left(b_0 \frac{\partial}{\partial a_0} + b_1 \frac{\partial}{\partial a_1} + \dots + b_n \frac{\partial}{\partial a_n}\right) P,$$

is only of degree $m-1$ in $a_0, a_1, \dots a_n$. If in P_1 we replace each b by the corresponding a we obtain mP, as follows from Euler's Theorem relating to homogeneous functions.

In like manner if in

$$P_2 = \left(c_0 \frac{\partial}{\partial a_0} + c_1 \frac{\partial}{\partial a_1} + \dots + c_n \frac{\partial}{\partial a_n}\right)\left(b_0 \frac{\partial}{\partial a_0} + b_1 \frac{\partial}{\partial a_1} + \dots + b_n \frac{\partial}{\partial a_n}\right) P,$$

we replace each c and each b by the corresponding a we get $m(m-1)P$ and P_2 is of degree $m-2$ in the a's.

Proceeding in this way we can find an expression P_{m-1} which is linear in each of m sets of symbols

$$a, b, c, \dots k,$$

and which becomes equal to $P \times m!$ when each $b, c, \dots k$ is replaced by the corresponding a.

Now having formed the expression P_{m-1} we replace each a by the symbol α, each b by the symbol β, each c by the symbol γ and so on. Since the expression is linear in each set of letters, each symbol will occur exactly n times in every term, and then, regarding the symbols as referring to the same quantic, we have the required symbolical expression.

Thus for example

$$\begin{aligned} a_0 a_2 - a_1^2 &= \tfrac{1}{2}\left(b_0 \frac{\partial}{\partial a_0} + b_1 \frac{\partial}{\partial a_1} + b_2 \frac{\partial}{\partial a_2}\right)(a_0 a_2 - a_1^2)_{b=a} \\ &= \tfrac{1}{2}(b_0 a_2 + b_2 a_0 - 2a_1 b_1)_{b=a} \\ &= \tfrac{1}{2}(\beta_1^2 \alpha_2^2 + \beta_2^2 \alpha_1^2 - 2\alpha_1 \alpha_2 \beta_1 \beta_2)\, \alpha_1^{n-2} \beta_1^{n-2} \\ &= \tfrac{1}{2}(\alpha_1 \beta_2 - \alpha_2 \beta_1)^2\, \alpha_1^{n-2} \beta_1^{n-2}, \end{aligned}$$

and the convenience of this expression in terms of α, β will be abundantly evident in the sequel.

Ex. (i). For the binary quartic shew that

$$a_0a_4 - 4a_1a_3 + 3a_2^2 = \tfrac{1}{2}(a_1\beta_2 - a_2\beta_1)^4,$$

and

$$\begin{vmatrix} a_0 & a_1 & a_2 \\ a_1 & a_2 & a_3 \\ a_2 & a_3 & a_4 \end{vmatrix} = \tfrac{1}{6}(a_1\beta_2 - a_2\beta_1)^2 (\beta_1\gamma_2 - \beta_2\gamma_1)^2 (\gamma_1 a_2 - \gamma_2 a_1)^2.$$

Ex. (ii). By the same method shew that for any binary form

$$\frac{\partial^2 f}{\partial x_1^2}\frac{\partial^2 f}{\partial x_2^2} - \left(\frac{\partial^2 f}{\partial x_1 \partial x_2}\right)^2 = 2\binom{n}{2}^2 (a_1\beta_2 - a_2\beta_1)^2 a_x^{n-2}\beta_x^{n-2}.$$

Ex. (iii). Shew that for a binary form of odd order $(a_1\beta_2 - a_2\beta_1)^n$ is zero and write down its value for a form of even order in terms of the coefficients.

15. Polar Forms. The expression

$$\frac{(n-r)!}{n!}\left(y_1\frac{\partial}{\partial x_1} + y_2\frac{\partial}{\partial x_2}\right)^r f,$$

where

$$f = \alpha_x^n = \beta_x^n = \text{etc.}$$

is a binary form of order n, is called the rth polar of f with respect to y.

The operator $\left(y_1\frac{\partial}{\partial x_1} + y_2\frac{\partial}{\partial x_2}\right)$, which is frequently written $\left(y\frac{\partial}{\partial x}\right)$, is called a polarizing operator and the expression

$$\left(y_1\frac{\partial}{\partial x_1} + y_2\frac{\partial}{\partial x_2}\right)^r f$$

is said to be derived from f by polarizing r times with respect to y. The numerical factor $\frac{(n-r)!}{n!}$ is only introduced for convenience.

These polar forms admit of very simple representation in our symbols, for

$$\left(y_1\frac{\partial}{\partial x_1} + y_2\frac{\partial}{\partial x_2}\right)\alpha_x^n = n\alpha_x^{n-1}\alpha_y,$$

$$\left(y_1\frac{\partial}{\partial x_1} + y_2\frac{\partial}{\partial x_2}\right)^2\alpha_x^n = n(n-1)\,\alpha_x^{n-2}\alpha_y^2,$$

and so on.

Hence the rth polar of f with respect to y is

$$\frac{(n-r)!}{n!}\,n(n-1)\ldots(n-r+1)\,\alpha_x^{n-r}\alpha_y^r,$$

that is

$$\alpha_x^{n-r}\alpha_y^r.$$

The differential coefficients of f with respect to the variables are particular cases of polar forms.

For if $y_1 = 1$, $y_2 = 0$, the rth polar is

$$\frac{\partial^r f}{\partial x_1{}^r}$$

and if $y_1 = 0$, $y_2 = 1$, the rth polar is

$$\frac{\partial^r f}{\partial x_2{}^r}.$$

In general we have

$$\frac{\partial^{p+q} f}{\partial x_1{}^p \partial x_2{}^q} = \frac{n!}{(n-p-q)!} \alpha_x{}^{n-p-q} \alpha_1{}^p \alpha_2{}^q.$$

The form $\dfrac{(n-p-q)!}{n!} \left(y\dfrac{\partial}{\partial x}\right)^p \left(z\dfrac{\partial}{\partial x}\right)^q f$ is called a mixed polar with respect to y and z; its symbolical expression is

$$\alpha_x{}^{n-p-q} \alpha_y{}^p \alpha_z{}^q.$$

16. Effect of a Linear Transformation.

If we write

$$x_1 = \xi_1 X_1 + \eta_1 X_2$$
$$x_2 = \xi_2 X_1 + \eta_2 X_2,$$

then α_x becomes

$$(\alpha_1 \xi_1 + \alpha_2 \xi_2) X_1 + (\alpha_1 \eta_1 + \alpha_2 \eta_2) X_2$$

or

$$\alpha_\xi X_1 + \alpha_\eta X_2,$$

and hence the binary form $\alpha_x{}^n$ becomes

$$(\alpha_\xi X_1 + \alpha_\eta X_2)^n$$

or

$$\alpha_\xi{}^n X_1{}^n + n\alpha_\xi{}^{n-1} \alpha_\eta X_1{}^{n-1} X_2 + \ldots + \alpha_\eta{}^n X_2{}^n.$$

Accordingly in the transformed expression the coefficient of $X_1{}^n$ is found by replacing x by ξ in the original form, and the coefficients of $X_1{}^{n-1} X_2$, $X_1{}^{n-2} X_2{}^2$ ... are found by polarizing the coefficient of $X_1{}^n$ with respect to η once, twice .... Of course suitable numerical multipliers must be introduced.

The reader will easily illustrate this result by reference to the transformation of a binary quadratic in § 1.

17. The form $\alpha_x^n = \beta_x^n = \gamma_x^n \ldots$ becomes on transformation

$$(\alpha_\xi X_1 + \alpha_\eta X_2)^n = (\beta_\xi X_1 + \beta_\eta X_2)^n = (\gamma_\xi X_1 + \gamma_\eta X_2)^n = \ldots.$$

Now we have

$$\alpha_\xi \beta_\eta - \alpha_\eta \beta_\xi = \begin{vmatrix} \alpha_1 \xi_1 + \alpha_2 \xi_2, & \alpha_1 \eta_1 + \alpha_2 \eta_2 \\ \beta_1 \xi_1 + \beta_2 \xi_2, & \beta_1 \eta_1 + \beta_2 \eta_2 \end{vmatrix} = \begin{vmatrix} \alpha_1, & \alpha_2 \\ \beta_1, & \beta_2 \end{vmatrix} \times \begin{vmatrix} \xi_1, & \xi_2 \\ \eta_1, & \eta_2 \end{vmatrix}$$

$$= (\alpha_1 \beta_2 - \alpha_2 \beta_1)(\xi_1 \eta_2 - \xi_2 \eta_1),$$

a result of fundamental importance.

We shall denote the expression $(\alpha_1 \beta_2 - \alpha_2 \beta_1)$, which we call a symbolical determinantal factor, by $(\alpha\beta)$, so that $(\alpha\beta) = -(\beta\alpha)$, and the above relation may be written

$$\alpha_\xi \beta_\eta - \alpha_\eta \beta_\xi = (\alpha\beta)(\xi\eta).$$

To illustrate these remarks let us prove that

$$a_0 a_4 - 4a_1 a_3 + 3a_2^2$$

is an invariant of the binary quartic

$$f = \alpha_x^4 = \beta_x^4 \ldots.$$

We have

$$(a_0 a_4 - 4a_1 a_3 + 3a_2^2) = \tfrac{1}{2}\left(\Sigma b \frac{\partial}{\partial a}\right)(a_0 a_4 - 4a_1 a_3 + 3a_2^2)_{b=a} = \tfrac{1}{2}(\alpha\beta)^4.$$

Thus, if the coefficients of the new form be denoted by capital letters as usual, we have

$$(A_0 A_4 - 4A_1 A_3 + 3A_2^2) = \tfrac{1}{2}(\alpha_\xi \beta_\eta - \alpha_\eta \beta_\xi)^4$$

as follows from the symbolical expression given above.

But since $\quad \alpha_\xi \beta_\eta - \alpha_\eta \beta_\xi = (\alpha\beta)(\xi\eta),$

$$A_0 A_4 - 4A_1 A_3 + 3A_2^2 = (\xi\eta)^4 (a_0 a_4 - 4a_1 a_3 + 3a_2^2),$$

which shews that $a_0 a_4 - 4a_1 a_3 + 3a_2^2$ is an invariant and that the multiplying factor is the fourth power of the determinant of transformation.

18. Symbolical expressions representing Invariants. If the symbolical equivalent of an expression I, homogeneous and of degree i in the coefficients of the binary form

$$f = \alpha_x^n = \beta_x^n = \ldots,$$

be an aggregate of terms each of which is a product of factors of the type $(\alpha\beta)$, then I is an invariant of the quantic.

For let $I = \Sigma T$, where T is the product of w factors of the type $(\alpha\beta)$, then the total degree of T in the symbols is $2w$ and it is also $n \times i$, for there must be i sets of symbols, each set occurring to degree n; therefore $ni = 2w$, so that w is the same for every term in the aggregate representing I.

If I' be the same function of the coefficients of the transformed expression, then
$$I' = \Sigma T'$$
where T' is found from T by replacing α_1 by α_ξ, α_2 by α_η and so on.

But since
$$(\alpha_\xi \beta_\eta - \alpha_\eta \beta_\xi) = (\alpha\beta)(\xi\eta)$$
it follows at once that
$$T' = (\xi\eta)^w T,$$
and therefore $I' = (\xi\eta)^w I$ since w is the same for every term.

Hence I is an invariant.

Exactly the same result is true for any number of binary forms if we suppose that I is homogeneous in each set of coefficients, for it is easily seen that the number of determinantal factors must be the same in every term, it being in fact
$$\tfrac{1}{2}(n_1 i_1 + n_2 i_2 + \ldots)$$
when $n_1, n_2, \ldots$ are the orders of the forms and $i_1, i_2, \ldots$ the respective degrees of I in the coefficients of the forms.

The rest of the proof then depends only on the fact that, whatever α and β are, we have
$$\alpha_\xi \beta_\eta - \alpha_\eta \beta_\xi = (\alpha\beta)(\xi\eta).$$

Thus I is an invariant and the multiplying factor is now
$$(\xi\eta)^{\frac{1}{2}(n_1 i_1 + n_2 i_2 + \ldots)}.$$

19. This simple theorem enables us to construct as many invariants as we please—we have only to write down a product of factors $(\alpha\beta)$ and take care that the symbol α occurs in n of these factors where n is the order of the form to which the symbol α belongs. If this condition be not satisfied the invariant property still holds but the expression has only a symbolical meaning. On the other hand, if every symbol occur to the right degree but the expression be not reducible to the form above, it is an actual function of the coefficients which is not an invariant.

As an example we have an invariant of the second degree $(\alpha\beta)^n$ for a binary form of order n. This vanishes identically when n is odd, as can be seen by expressing it in terms of the coefficients; or thus, since α, β are equivalent symbols

$$(\alpha\beta)^n = (\beta\alpha)^n$$

and hence $$(\alpha\beta)^n = (-1)^n (\alpha\beta)^n,$$

giving the result at once.

Again, for the binary cubic we have the invariant

$$(\alpha\beta)^2 (\alpha\gamma) (\beta\delta) (\gamma\delta)^2$$

and for the binary quartic the invariants

$$(\alpha\beta)^4, \quad (\alpha\beta)^2 (\beta\gamma)^2 (\gamma\alpha)^2, \quad (\alpha\beta)^3 (\alpha\gamma) (\beta\delta) (\gamma\delta)^3.$$

In every case it will be observed that the multiplying factor is a power of $(\xi\eta)$.

As an example of invariants of several binary forms we may mention $(\alpha\beta)^n$, an invariant of the two different binary forms $\alpha_x{}^n$ and $\beta_x{}^n$. For quadratics this is the well-known invariant of § 2.

Again $(\alpha\beta)$ is an invariant of the two linear forms α_x and β_x and in this case α, β are actual coefficients as well as symbols. Then $(\alpha\beta)(\alpha\gamma)$ is an invariant of the quadratic $\alpha_x{}^2$ and the linear forms β_x, γ_x.

20. Covariants. A similar method exists for constructing covariants.

Commencing with an example let us prove that the Hessian

$$H = \frac{\partial^2 f}{\partial x_1{}^2} \frac{\partial^2 f}{\partial x_2{}^2} - \left(\frac{\partial^2 f}{\partial x_1 \partial x_2}\right)^2$$

is a covariant of the binary form $f = \alpha_x{}^n = \beta_x{}^n = \ldots$

Since H is of the second degree in the coefficients

$$H = \tfrac{1}{2}\left(\Sigma b \frac{\partial}{\partial a}\right) \left\{\frac{\partial^2 f}{\partial x_1{}^2} \frac{\partial^2 f}{\partial x_2{}^2} - \left(\frac{\partial^2 f}{\partial x_1 \partial x_2}\right)^2\right\}_{b=a}$$

$$= \tfrac{1}{2}\left(\frac{\partial^2 f}{\partial x_1{}^2} \frac{\partial^2 f'}{\partial x_2{}^2} + \frac{\partial^2 f'}{\partial x_1{}^2} \frac{\partial^2 f}{\partial x_2{}^2} - 2 \frac{\partial^2 f}{\partial x_1 \partial x_2} \frac{\partial^2 f'}{\partial x_1 \partial x_2}\right)_{b=a}$$

where $$f' = (b_0 b_1 \ldots b_n \between x_1 x_2)^n.$$

Replacing the a's by α's and the b's by β's as usual, we have

$$H = \tfrac{1}{2} n^2 (n-1)^2 \{\alpha_1{}^2 \alpha_x{}^{n-2} \beta_2{}^2 \beta_x{}^{n-2} + \alpha_2{}^2 \alpha_x{}^{n-2} \beta_1{}^2 \beta_x{}^{n-2} - 2\alpha_1 \alpha_2 \alpha_x{}^{n-2} \beta_1 \beta_2 \beta_x{}^{n-2}\}$$
$$= \tfrac{1}{2} n^2 (n-1)^2 (\alpha\beta)^2 \alpha_x{}^{n-2} \beta_x{}^{n-2},$$

as can be immediately verified by expressing this in terms of the coefficients.

The transformed quantic is

$$(\alpha_\xi X_1 + \alpha_\eta X_2)^n,$$

and the corresponding expression derived from this is

$$(\alpha_\xi \beta_\eta - \alpha_\eta \beta_\xi)^2 (\alpha_\xi X_1 + \alpha_\eta X_2)^{n-2} (\beta_\xi X_1 + \beta_\eta X_2)^{n-2}$$
$$= (\xi\eta)^2 (\alpha\beta)^2 \alpha_x{}^{n-2} \beta_x{}^{n-2},$$

which shews that the expression is a covariant and that the multiplying factor is $(\xi\eta)^2$.

In general, if an expression C, of degree i in the coefficients of f and of order m in the variables, can be symbolically represented as an aggregate of terms, each of which is the product of a number of factors of the type $(\alpha\beta)$ and a number of the type α_x, then C is a covariant of f.

In fact let $C = \Sigma\Gamma$, where Γ is such a product.

The number of factors with suffix x in Γ must be m, the order of C, and if w be the number of the type $(\alpha\beta)$ we have

$$2w + m = ni,$$

for each of these represents the degree of C in the symbols. Hence w is the same for every term.

If C' be the corresponding expression derived from the transformed quantic, then

$$C' = \Sigma\Gamma',$$

where Γ' is derived from Γ by replacing α_1 by α_ξ, α_2 by α_η and so on, and α_x by $(\alpha_\xi X_1 + \alpha_\eta X_2)$.

Thus since $\qquad (\alpha_\xi \beta_\eta - \alpha_\eta \beta_\xi) = (\alpha\beta)(\xi\eta)$

and $\qquad \alpha_x = \alpha_\xi X_1 + \alpha_\eta X_2$

we have $\qquad \Gamma' = (\xi\eta)^w \Gamma,$

$$\therefore C' = (\xi\eta)^w C,$$

that is to say C is a covariant.

Exactly the same method applies to a covariant of any number of binary forms, but now the symbols $\alpha, \beta, \ldots$ may refer to different forms and, of course, a symbol such as α must occur in the symbolical expression to the requisite degree.

We can thus easily construct any number of covariants of one or more forms, *e.g.* for a binary form of order n

$$(\alpha\beta)^r \alpha_x^{n-r} \beta_x^{n-r}$$

is a covariant for any integral value of r, but it vanishes when r is odd.

Again, if α_x^m, β_x^n are two different quantics,

$$(\alpha\beta)^r \alpha_x^{m-r} \beta_x^{n-r}$$

is a covariant; if $r=1$ it is the Jacobian.

As further examples we have the covariants

$$(\alpha\beta)(\alpha\gamma)\alpha_x^3\beta_x^4\gamma_x^4, \quad (\alpha\beta)^2(\alpha\gamma)\alpha_x^2\beta_x^3\gamma_x^4,$$

$$(\alpha\beta)^2(\beta\gamma)^2(\gamma\alpha)^2\alpha_x\beta_x\gamma_x, \quad (\alpha\beta)(\beta\gamma)(\gamma\alpha)(\alpha\delta)(\beta\delta)(\gamma\delta)\alpha_x^2\beta_x^2\gamma_x^2\delta_x^2$$

of the binary quintic

$$\alpha_x^5 = \beta_x^5 = \gamma_x^5 = \delta_x^5 = \ldots$$

As an exercise the reader may prove that the last one vanishes identically.

21. We have seen how useful the symbolical methods are in constructing invariants and covariants. In the next chapter we shall prove that they constitute an ideal calculus when we shew that every invariant and covariant can be represented as a sum of symbolical products of factors of the types $(\alpha\beta)$ and α_x. Meanwhile anticipating this result we shall indicate the methods of transforming symbolical expressions. These depend on two principles:

(i) Interchange of equivalent symbols,

(ii) Identities in symbolical expressions.

According to (i) if a symbolical expression have an actual meaning and contain two equivalent symbols then its value is not altered by interchanging those symbols. We have already used this method in proving that the invariant

$$(\alpha\beta)^n \text{ of the quantic } \alpha_x^n = \beta_x^n$$

vanishes when n is odd. As another easy example we have

$$(\alpha\beta)(\beta\gamma)(\gamma\alpha)=0$$

for the quadratic $\alpha_x^2=\beta_x^2=\gamma_x^2$,

or for the two different quadratics $\alpha_x^2=\beta_x^2$ and γ_x^2.

More generally the covariant

$$(\alpha\beta)(\beta\gamma)(\gamma\alpha)\,\alpha_x^{n-2}\beta_x^{n-2}\gamma_x^{n-2}$$

is always zero unless the three forms α_x^n, β_x^n and γ_x^n are all different.

22. Fundamental Identities. We have identically

$$(\beta\gamma)\,\alpha_x+(\gamma\alpha)\,\beta_x+(\alpha\beta)\,\gamma_x=0 \quad\text{..............(I)},$$

as can easily be verified.

From this identity many others may be deduced.

For example, replacing x_1 by δ_2 and x_2 by $-\delta_1$ we have

$$(\beta\gamma)(\alpha\delta)+(\gamma\alpha)(\beta\delta)+(\alpha\beta)(\gamma\delta)=0 \quad\text{.........(II)},$$

a result useful in transforming invariants.

Again from (i)

$$(\beta\gamma)\,\alpha_x=(\beta\alpha)\,\gamma_x-(\gamma\alpha)\,\beta_x$$

and hence by squaring

$$2\,(\alpha\beta)(\alpha\gamma)\,\beta_x\gamma_x=(\alpha\beta)^2\gamma_x^2+(\alpha\gamma)^2\beta_x^2-(\beta\gamma)^2\alpha_x^2\text{...(III)}.$$

As identities less generally used we may mention

$$(\beta\gamma)^3\alpha_x^3+(\gamma\alpha)^3\beta_x^3+(\alpha\beta)^3\gamma_x^3=3\,(\beta\gamma)(\gamma\alpha)(\alpha\beta)\,\alpha_x\beta_x\gamma_x$$

$$(\beta\gamma)^4\alpha_x^4+(\gamma\alpha)^4\beta_x^4+(\alpha\beta)^4\gamma_x^4$$
$$=2\,\{(\alpha\beta)^2(\alpha\gamma)^2\beta_x^2\gamma_x^2+(\beta\gamma)^2(\beta\alpha)^2\gamma_x^2\alpha_x^2+(\gamma\alpha)^2(\gamma\beta)^2\alpha_x^2\beta_x^2\}.$$

Ex. (i). For the quadratic

$$\alpha_x^2=\beta_x^2=\gamma_x^2=f$$

$$(\alpha\beta)(\alpha\gamma)\,\beta_x\gamma_x=\tfrac{1}{2}\{(\alpha\beta)^2\gamma_x^2+(\alpha\gamma)^2\beta_x^2-(\beta\gamma)^2\alpha_x^2\}$$
$$=\tfrac{1}{2}f\,.\,(\alpha\beta)^2,$$

since the symbols α, β, γ are equivalent.

Ex. (ii). If $f_1=a_x^2=\beta_x^2$ and $f_2=a'^2_x=\beta'^2_x$ be two different quadratics, to express the square of the Jacobian $J=(aa')\,a_x a'_x$ in terms of f and f'.

We have
$$J^2=(aa')\,a_x a'_x\,(\beta\beta')\,\beta_x\beta'_x$$
or since
$$(\beta\beta')\,a'_x=(\beta a')\,\beta'_x-(\beta' a')\,\beta_x$$
$$J^2=(aa')\,(\beta a')\,a_x\beta_x\beta'^2_x-(aa')\,(\beta' a')\,a_x\beta'_x\,.\,\beta_x^2$$
$$=\beta'^2_x\tfrac{1}{2}\{(aa')^2\,\beta_x^2+(\beta a')^2\,a_x^2-(a\beta)^2\,a'^2_x\}$$
$$-\beta_x^2\tfrac{1}{2}\{(aa')^2\,\beta'^2_x+(\beta' a')^2\,a_x^2-(a\beta')^2\,a'^2_x\}\text{ by (III),}$$
or if
$$(a\beta)^2=I_{11},\quad (aa')^2=(a\beta')^2=\ldots=I_{12},\quad (a'\beta')^2=I_{22},$$
we have
$$2J^2=f_2\{I_{12}f_1+I_{12}f_1-I_{11}f_2\}$$
$$-f_1\{I_{12}f_2+I_{22}f_1-I_{12}f_2\}$$
$$=-\{I_{22}f_1^2+I_{11}f_2^2-2I_{12}f_1f_2\}.$$

Ex. (iii). Prove that for the binary quartic
$$f=a_x^4=\beta_x^4=\gamma_x^4$$
$$(a\beta)^2\,(a\gamma)^2\,\beta_x^2\gamma_x^2=\tfrac{1}{2}f\,.\,(a\beta)^4$$
$$(a\beta)\,(a\gamma)\,a_x^2\beta_x^3\gamma_x^3=\tfrac{1}{2}f\,.\,(a\beta)^2\,a_x^2\beta_x^2.$$

CHAPTER II.

THE FUNDAMENTAL THEOREM.

23. It will be remarked that in every example of invariants and covariants, discussed in the preceding chapter, the symbolical expression for such a function involved only factors of the types $(\alpha\beta)$ and α_x, and further that the multiplier alluded to in the definition was always a power of the determinant of the transformation. We are now going to establish the general truth of these properties.

As a matter of history, we may observe that the original definition of an invariant stated that the multiplier was of the form mentioned; but following the logical, rather than the historical order, we shall first prove that the multiplier must be a power of the determinant and then proceed to prove the proposition relating to the symbolical forms for invariants and covariants.

24. Suppose that I is an invariant or covariant of a single binary form f—after what has been said, § 10, we may assume that I is homogeneous in the coefficients of f.

Let the linear transformation

$$x_1 = \xi_1 x_1' + \eta_1 x_2'$$

$$x_2 = \xi_2 x_1' + \eta_2 x_2'$$

change f into f' and let I' be formed from f' in the same way that I is formed from f; then, by definition,

$$I' = F(\xi_1, \eta_1, \xi_2, \eta_2) \times I$$

and we have to shew that F is simply a power of $(\xi_1\eta_2 - \xi_2\eta_1)$.

Now let a second transformation

$$x_1' = \xi_1' x_1'' + \eta_1' x_2''$$
$$x_2' = \xi_2' x_1'' + \eta_2' x_2''$$

change f' into f'', and let I'' be formed in the same way from f'', so that

$$I'' = F(\xi_1', \eta_1', \xi_2', \eta_2') \times I'.$$

Hence we have

$$I'' = F(\xi_1, \eta_1, \xi_2, \eta_2) \times F(\xi_1', \eta_1', \xi_2', \eta_2') \times I.$$

But we can pass from the variables x_1, x_2 to the variables x_1'', x_2'' by the single transformation

$$x_1 = (\xi_1\xi_1' + \eta_1\xi_2') x_1'' + (\xi_1\eta_1' + \eta_1\eta_2') x_2''$$
$$x_2 = (\xi_2\xi_1' + \eta_2\xi_2') x_1'' + (\xi_2\eta_1' + \eta_2\eta_2') x_2'';$$

therefore

$$I'' = F\{(\xi_1\xi_1' + \eta_1\xi_2'), (\xi_1\eta_1' + \eta_1\eta_2'), (\xi_2\xi_1' + \eta_2\xi_2'), (\xi_2\eta_1' + \eta_2\eta_2')\} \times I.$$

Consequently F must satisfy the functional equation

$$F\{(\xi_1\xi_1' + \eta_1\xi_2'), (\xi_1\eta_1' + \eta_1\eta_2'), (\xi_2\xi_1' + \eta_2\xi_2'), (\xi_2\eta_1' + \eta_2\eta_2')\}$$
$$= F(\xi_1, \eta_1, \xi_2, \eta_2) \times F(\xi_1', \eta_1', \xi_2', \eta_2').$$

The solution of this equation is not difficult. In the first place we remark that since $\xi_1 = 1$, $\eta_1 = 0$, $\xi_2 = 0$, $\eta_2 = 1$ gives the identical transformation,

$$F(1, 0, 0, 1) = 1.$$

Again putting $\xi_1 = \kappa$, $\eta_1 = 0$, $\xi_2 = 0$, $\eta_2 = \kappa$ each new coefficient is equal to the corresponding original coefficient multiplied by the same power of κ, in this case the multiplier is clearly a power of κ, *i.e.*

$$F(\kappa, 0, 0, \kappa) = \kappa^r.$$

Since

$$F(\xi_1, \eta_1, \xi_2, \eta_2) \times F(\kappa, 0, 0, \kappa) = F(\kappa\xi_1, \kappa\eta_1, \kappa\xi_2, \kappa\eta_2),$$

we have

$$F(\kappa\xi_1, \kappa\eta_1, \kappa\xi_2, \kappa\eta_2) = \kappa^r F(\xi_1, \eta_1, \xi_2, \eta_2),$$

therefore F is homogeneous and of degree r in the four variables $\xi_1, \eta_1, \xi_2, \eta_2$.

Finally let us choose $\xi_1', \eta_1', \xi_2', \eta_2'$ so that

$$\xi_1\xi_1' + \eta_1\xi_2' = 1, \quad \xi_1\eta_1' + \eta_1\eta_2' = 0,$$
$$\xi_2\xi_1' + \eta_2\xi_2' = 0, \quad \xi_2\eta_1' + \eta_2\eta_2' = 1,$$

which relations give

$$\left.\begin{aligned}\xi_1' &= \frac{\eta_2}{D}, & \xi_2' &= -\frac{\xi_2}{D}\\ \eta_1' &= -\frac{\eta_1}{D}, & \eta_2' &= \frac{\xi_1}{D}\end{aligned}\right\}(D = \xi_1\eta_2 - \xi_2\eta_1),$$

then we have

$$F(\xi_1, \eta_1, \xi_2, \eta_2) \times F(\xi_1', \eta_1', \xi_2', \eta_2') = F(1, 0, 0, 1)$$

or $$F(\xi_1, \eta_1, \xi_2, \eta_2) \times F\left(\frac{\eta_2}{D}, -\frac{\eta_1}{D}, -\frac{\xi_2}{D}, \frac{\xi_1}{D}\right) = 1.$$

Consequently since F is homogeneous and of degree r

$$F(\xi_1, \eta_1, \xi_2, \eta_2) \times F(\eta_2, -\eta_1, -\xi_2, \xi_1) = D^r.$$

But inasmuch as D is obviously irreducible—*i.e.* it cannot be resolved into factors—and F is clearly an integral function, this equation shews at once that both

$$F(\xi_1, \eta_1, \xi_2, \eta_2) \text{ and } F(\eta_2, -\eta_1, -\xi_2, \xi_1)$$

are integral powers of D.

Hence the theorem is established.

25. Assuming the truth of the proposition just proved, the proof of the fundamental theorem that invariants and covariants can be completely represented by factors of the types $(\alpha\beta)$ and α_x is very simple in principle. The actual work requires two lemmas of great importance in the present subject, and we shall give them separately. They are both concerned with properties of the differential operator

$$\Omega = \frac{\partial^2}{\partial x_1 \partial y_2} - \frac{\partial^2}{\partial x_2 \partial y_1}.$$

26. *Lemma I.* If n be a positive integer

$$\left(\frac{\partial^2}{\partial x_1 \partial y_2} - \frac{\partial^2}{\partial x_2 \partial y_1}\right)(x_1y_2 - x_2y_1)^n = n(n+1)(x_1y_2 - x_2y_1)^{n-1}.$$

In fact $$\frac{\partial}{\partial x_1}(x_1y_2 - x_2y_1)^n = ny_2(x_1y_2 - x_2y_1)^{n-1},$$

$$\therefore \frac{\partial^2}{\partial x_1 \partial y_2}(x_1y_2 - x_2y_1)^n = n(x_1y_2 - x_2y_1)^{n-1} + n(n-1)(x_1y_2 - x_2y_1)^{n-2}x_1y_2.$$

Similarly

$$\frac{\partial^2}{\partial x_2 \partial y_1}(x_1y_2 - x_2y_1)^n = -n(x_1y_2 - x_2y_1)^{n-1} + n(n-1)(x_1y_2 - x_2y_1)^{n-2}x_2y_1.$$

Consequently

$$\begin{aligned}\Omega(x_1y_2 - x_2y_1)^n &= \{n(n-1) + 2n\}(x_1y_2 - x_2y_1)^{n-1} \\ &= n(n+1)(x_1y_2 - x_2y_1)^{n-1},\end{aligned}$$

which establishes the lemma.

If we operate again with Ω we find

$$\Omega^2(x_1y_2 - x_2y_1)^n = n(n+1)(n-1)n(x_1y_2 - x_2y_1)^{n-2}$$

and in general

$$\begin{aligned}&\Omega^r(x_1y_2 - x_2y_1)^n \\ &= (n+1)n^2(n-1)^2 \ldots (n-r+2)^2(n-r+1)(x_1y_2 - x_2y_1)^{n-r}\end{aligned}$$

or $\Omega^r(xy)^n = (n+1)n^2(n-1)^2 \ldots (n-r+2)^2(n-r+1)(xy)^{n-r}.$

Finally

$$\Omega^n(x_1y_2 - x_2y_1)^n = (n+1)(n!)^2$$

a constant which is not zero—for our immediate purpose this is the important result, and it can be at once verified by expanding

$$\left(\frac{\partial^2}{\partial x_1 \partial y_2} - \frac{\partial^2}{\partial x_2 \partial y_1}\right)^n, \ (x_1y_2 - x_2y_1)^n$$

by the Binomial Theorem.

27. *Lemma II.* If the operator Ω be applied r times to the product of m factors of the type α_x by n factors of the type β_y, then each term in the resulting expression contains r determinantal factors $(\alpha\beta)$, $(m-r)$ factors α_x and $(n-r)$ factors β_y.

To ensure perfect generality we consider

$$\Omega^r P \, . \, Q$$

where $$P = \alpha_x^{(1)} \alpha_x^{(2)} \ldots \alpha_x^{(m)}$$

and $$Q = \beta_y^{(1)} \beta_y^{(2)} \ldots \beta_y^{(n)},$$

the α's and the β's being all different.

Now $$\frac{\partial^2}{\partial x_1 \partial y_2} P . Q = \Sigma\, \alpha_1^{(r)} \beta_2^{(s)} \frac{P}{\alpha_x^{(r)}} \frac{Q}{\beta_y^{(s)}},$$

where the summation extends so that r takes all the values 1, 2, ... m and s takes all the values 1, 2, ... n.

Similarly $$\frac{\partial^2}{\partial x_2 \partial y_1} P . Q = \Sigma\, \alpha_2^{(r)} \beta_1^{(s)} \frac{P}{\alpha_x^{(r)}} \frac{Q}{\beta_y^{(s)}}.$$

Hence on subtraction

$$\Omega P . Q = \Sigma\, (\alpha^{(r)} \beta^{(s)}) \frac{P}{\alpha_x^{(r)}} \frac{Q}{\beta_y^{(s)}},$$

which establishes the lemma for $r = 1$.

But since the operator Ω has no effect on a factor of the type $(\alpha^{(r)} \beta^{(s)})$ the theorem holds for $r = 2$; in fact

$$\Omega^2 P . Q = \Sigma\, (\alpha^{(r)} \beta^{(s)})\, \Omega \left\{\frac{P}{\alpha_x^{(r)}} \frac{Q}{\beta_y^{(s)}}\right\}$$

and performing the operation on the right we have the result.

Proceeding in this way we see that at each step a new factor of the type $(\alpha\beta)$ appears in each term while one factor of each of the types α_x and β_y disappears—this completely establishes our lemma.

Ex. (i). Prove that $\Omega^r a_x^m b_y^n = \dfrac{m!}{(m-r)!} \dfrac{n!}{(n-r)!} (ab)^r a_x^{m-r} b_y^{n-r}$.

Ex. (ii). With the notation of the text prove that $\Omega^r P . Q$ contains *every* term of type there written $r!$ times and that the number of different terms is $\dfrac{m!}{(m-r)!} \dfrac{n!}{(n-r)!} \dfrac{1}{r!}$. (Use induction.)

28. Fundamental Theorem. Suppose now that

$$F(a_0, a_1, \ldots a_n)$$

is an invariant of the binary form

$$(a_0, a_1, \ldots a_n \between x_1, x_2)^n = \alpha_x^n = \beta_x^n = \ldots,$$

then after the linear transformation

$$\left.\begin{aligned} x_1 &= \xi_1 X_1 + \eta_1 X_2 \\ x_2 &= \xi_2 X_1 + \eta_2 X_2 \end{aligned}\right\}$$

we have seen that α_1 becomes α_ξ and α_2 becomes α_η; so that if the new form be

$$(A_0, A_1, \ldots A_n \! \lbrace\!\! X_1, X_2)^n$$

we have

$$A_r = \alpha_\xi{}^{n-r} \alpha_\eta{}^r.$$

By definition

$$F(A_0, A_1, \ldots A_n) = (\xi_1 \eta_2 - \xi_2 \eta_1)^w F(a_0, a_1, \ldots a_n);$$

accordingly if the A's are replaced by their symbolical expressions F becomes the sum of a number of terms, say $\Sigma P \cdot Q$, where P contains only factors of the type α_ξ, and Q only those of the type α_η. As the degree in ξ and η must be w we infer that there are just w factors in P and w in Q.

If we operate on both sides with Ω^w, $P \cdot Q$ becomes the sum of a number of terms each of which is the product of w factors of the type $(\alpha\beta)$, and the result on the right-hand side is a numerical multiple of $F(a_0 a_1 \ldots a_n)$.

Hence we have expressed $F(a_0, a_1, \ldots a_n)$ in the symbolical form peculiar to invariants.

The proof as given applies to invariants of one binary form; it is the same, word for word, for any number of binary forms, for the left-hand side is still of equal degree in ξ and η, and on the right-hand side we have the determinant $(\xi\eta)$ occurring to a power equal to this degree. Hence operating as above the required symbolical expression is obtained.

29. The proof for covariants is of the same nature as that for invariants, although a little more care is required in the manipulation of the symbols; after what has been said on invariants we may confine our attention to covariants of a single form.

Suppose that $\qquad F(a_0, a_1, \ldots a_n, x_1, x_2)$

is a homogeneous covariant of order m of

$$(a_0, a_1, \ldots a_n \! \lbrace\!\! x_1, x_2)^n = \alpha_x{}^n = \beta_x{}^n = \text{etc.}$$

Then using the same notation as before

$$F(A_0, A_1, \ldots A_n, X_1, X_2) = (\xi_1 \eta_2 - \xi_2 \eta_1)^w F(a_0, a_1, \ldots a_n, x_1, x_2).$$

If the A's are replaced by their symbolical expressions we get

$$\Sigma P \,.\, Q X_1^{m_1} X_2^{m_2}$$

where P involves only factors of the type α_ξ and Q only those of the type α_η.

But on solution we have

$$X_1 = \frac{\eta_2 x_1 - \eta_1 x_2}{\xi_1\eta_2 - \xi_2\eta_1}, \quad X_2 = \frac{\xi_1 x_2 - \xi_2 x_1}{\xi_1\eta_2 - \xi_2\eta_1}.$$

Now for convenience we shall replace x_1 by u_2 and x_2 by $-u_1$, so that

$$X_1 = \frac{u_\eta}{(\xi_1\eta_2 - \xi_2\eta_1)}, \quad X_2 = -\frac{u_\xi}{(\xi_1\eta_2 - \xi_2\eta_1)}.$$

Substituting these values and multiplying up by $(\xi\eta)^m$ we obtain the identity

$$\Sigma\,(-1)^{m_2}\, P \,.\, Q u_\eta^{m_1} u_\xi^{m_2} = (\xi\eta)^{w+m}\, F.$$

We may write the left-hand side $\Sigma P' \,.\, Q'$ where P' only contains factors with suffix ξ and moreover exactly $(w+m)$ factors, and Q' contains $(w+m)$ factors with suffix η.

Accordingly after operating with Ω^{w+m} each term will involve $(w+m)$ determinantal factors, and the right-hand side will be a numerical multiple of F. Now of the $(w+m)$ factors, there are w of the type $(\alpha\beta)$ and m of the type (αu), for u must occur to degree m in the final as well as in the original expression.

But (αu) is $-\alpha_x$; hence replacing the u's by the x's throughout we have the symbolical expression for F.

30. Since we have proved that all invariants and covariants of one or more binary forms can be completely represented by products of factors of the types $(\alpha\beta)$ and α_x, and further that every expression which can be so represented is an invariant or covariant—provided it possesses an actual significance,—it follows at once that all properties of invariants and covariants are implicitly contained in the symbolical representation and can be deduced therefrom.

31. Let us examine somewhat more closely the constitution of invariants of a single binary form

$$f \equiv (a_0, a_1, \ldots a_n \between x_1, x_2)^n = \alpha_x{}^n = \beta_x{}^n = \gamma_x{}^n \text{ etc.}$$

Suppose that an invariant I is an aggregate of products of factors $(\alpha\beta)$ such that in every term there are w factors, then inasmuch as each symbol occurs n times in I we must have

$$ni = 2w,$$

where i is the number of different symbols.

Now the weight of a_r is r by definition and its symbolical equivalent is $\alpha_1{}^{n-r}\alpha_2{}^r$; hence the weight of any product of the a's is the sum of the weights of the factors and is therefore equal to the total degree to which the letters $\alpha_2, \beta_2, \gamma_2, \ldots$ occur in the symbolical equivalent.

In the case of an invariant such as I each term in the symbolical expression when multiplied out is the product of w symbols with suffix 1 by w symbols with suffix 2; hence the weight is w. Further, the multiplying power of the determinant for I is also w.

Consider next a covariant of degree i and order m. It is an aggregate of terms, each of which is the product of the same number (say p) of factors of the type $(\alpha\beta)$, by the same number (say q) of factors of the type α_x.

We deduce at once the relations

$$q = m, \quad 2p + q = ni,$$

for each member of the latter equation represents the total degree of the covariant in the symbols $\alpha, \beta, \gamma, \ldots$.

Thus $$m = q = ni - 2p.$$

32. The leading coefficient of the covariant is called a seminvariant—it is found at once from the covariant by putting

$$x_1 = 1, \quad x_2 = 0$$

and therefore represented symbolically it is an aggregate of products of p factors of the type $(\alpha\beta)$ by q factors of the type α_1.

The weight of this seminvariant is accordingly p, and hence we infer that if w be the weight of a seminvariant and i its degree the order of the corresponding covariant is $ni - 2w$.

Thus for example in connection with the cubic

$$a_0x_1^3 + 3a_1x_1^2x_2 + 3a_2x_1x_2^2 + a_3x_2^3 = \alpha_x^3 = \beta_x^3$$

we have the covariant $(\alpha\beta)^2\alpha_x\beta_x$.

The seminvariant is

$$(\alpha\beta)^2\alpha_1\beta_1 = \alpha_1\beta_1(\alpha_1^2\beta_1^2 - 2\alpha_1\alpha_2\beta_1\beta_2 + \beta_2^2\alpha_2^2) = 2(a_0a_2 - a_1^2),$$

i.e. its weight is 2, as we should have inferred from the number of factors $(\alpha\beta)$ in the covariant.

The order of the covariant is 2 and here we have

$$i = 2, \quad n = 3, \quad m = 2, \quad w = 2,$$

so that $$m = ni - 2w.$$

In like manner if the leading coefficient of a joint covariant of two quantics of orders n_1 and n_2 be of degrees i_1, i_2 in the respective coefficients and of total weight w in these coefficients conjointly, then the order of the covariant is

$$n_1i_1 + n_2i_2 - 2w.$$

The reader will readily establish this theorem and extend it to the case of any number of quantics by using the symbolical notation. On putting $m = 0$ we get a relation connecting the degrees and weight of an invariant.

33. Deduction of a covariant from its leading coefficient. As we have seen, each term in the symbolical expression for the seminvariant must be the product of w factors of the type $(\alpha\beta)$ by m factors of the type α_1.

Now suppose that in the seminvariant we replace α_1 by α_x, β_1 by β_x, etc., and leave unaltered α_2, β_2, etc., then $(\alpha\beta)$ becomes

$$(\alpha_1x_1 + \alpha_2x_2)\beta_2 - (\beta_1x_1 + \beta_2x_2)\alpha_2 = (\alpha\beta)x_1;$$

hence the seminvariant S is clearly changed into x_1^w multiplied by the corresponding covariant—*e.g.* in the cubic, $(\alpha\beta)^2\alpha_1\beta_1$ becomes $x_1^2 \times (\alpha\beta)^2\alpha_x\beta_x$. We have thus a simple means of passing from the leading coefficient to the covariant. A similar result for invariants may be obtained by taking the particular case $m=0$; here the leading coefficient is of course the invariant itself.

Let there be an identical rational algebraic relation among a number of seminvariants $S_1, S_2, \ldots S_r$ and let $C_1, C_2, \ldots C_r$ be the corresponding covariants.

If the relation be

$$\Sigma\, S_1^{\mu_1} S_2^{\mu_2} \ldots S_r^{\mu_r} = 0$$

and w_p be the weight of S_p, then the sum

$$\mu_1 w_1 + \mu_2 w_2 + \ldots + \mu_r w_r$$

must be the same for every term—hence if we put the left-hand side of the relation into symbols and then change α_1 into α_x, β_1 into β_x as above, we have

$$x_1^{\mu_1 w_1 + \mu_2 w_2 + \ldots + \mu_r w_r}\, \Sigma\, C_1^{\mu_1} C_2^{\mu_2} \ldots C_r^{\mu_r} = 0,$$

or

$$\Sigma\, C_1^{\mu_1} C_2^{\mu_2} \ldots C_r^{\mu_r} = 0,$$

i.e. the covariants are connected by the same relation as the seminvariants.

34. Again when we replace α_1 by α_x and leave α_2 unaltered we replace the coefficient a_r by

$$\alpha_x^{n-r} \alpha_2^{r} = \frac{(n-r)!}{n!} \frac{\partial^r f}{\partial x_2^r}.$$

Hence except for a multiplier, which is a power of x, a covariant is the same function of

$$f,\ \frac{1}{n}\frac{\partial f}{\partial x_2},\ \frac{1}{n(n-1)}\frac{\partial^2 f}{\partial x_2^2}, \ldots \frac{(n-r)!}{n!}\frac{\partial^r f}{\partial x_2^r}, \ldots \frac{1}{n!}\frac{\partial^n f}{\partial x_2^n}$$

as the corresponding seminvariant is of

$$a_0,\ a_1,\ a_2, \ldots a_r, \ldots a_n.$$

Ex. (i). If in a seminvariant of weight w we replace α_2 by α_x, β_2 by β_x, etc., and leave $\alpha_1, \beta_1 \ldots$ unaltered, then the result is the seminvariant multiplied by x_2^w.

What is the corresponding transformation of the actual coefficients?

Ex. (ii). Find the result of replacing α_1 by α_ξ, α_2 by α_η, β_1 by β_ξ, β_2 by β_η, etc., in a seminvariant, and give the corresponding transformation of the coefficients.

Ex. (iii). Extend all the above results to the case of two or more binary forms.

Ex. (iv). Prove that if in an invariant of a single binary form a_r be replaced by $\frac{(n-r)!}{n!}\frac{\partial^r f}{\partial x_2{}^r}$ and so on, the result is the invariant multiplied by $x_2{}^w$. State the result of replacing a_r by $\frac{(n-r)!}{n!}\frac{\partial^2 f}{\partial x_1{}^r}$ and extend the argument to any number of binary forms.

35. Alternative proof of the Fundamental Theorem—The Aronhold Operator.

We shall now give another proof of the theorem of § 28 in which the original argument of Clebsch will be followed.

Let ϕ be a covariant of a form

$$f = (a_0, a_1, \dots a_n \between x_1, x_2)^n$$

which is homogeneous in both the coefficients and the variables, and of degree i in the former.

If $$F = (A_0, A_1, \dots A_n \between X_1, X_2)^n$$

be the transformed quantic, we have

$$\phi(A_0, A_1, \dots A_n) = \mu\phi(a_0, a_1, \dots a_n)$$

where μ depends only on the transformation.

Now if $$(b_0, b_1, \dots b_n \between x_1, x_2)^n$$

be a second form which transforms into

$$(B_0, B_1, \dots B_n \between X_1, X_2)^n,$$

then $$(a_0 + \lambda b_0,\ a_1 + \lambda b_1,\ \dots a_n + \lambda b_n \between x_1, x_2)^n$$

transforms into

$$(A_0 + \lambda B_0,\ A_1 + \lambda B_1,\ \dots A_n + \lambda B_n \between X_1, X_2)^n.$$

Therefore

$$\phi(A_0 + \lambda B_0,\ A_1 + \lambda B_1,\ \dots A_n + \lambda B_n)$$
$$= \mu\phi(a_0 + \lambda b_0,\ a_1 + \lambda b_1,\ \dots a_n + \lambda b_n),$$

hence expanding by Taylor's Theorem and equating coefficients of λ we have

$$\left(B_0\frac{\partial}{\partial A_0} + B_1\frac{\partial}{\partial A_1} + \dots + B_n\frac{\partial}{\partial A_n}\right)\phi$$
$$= \mu\left(b_0\frac{\partial}{\partial a_0} + b_1\frac{\partial}{\partial a_1} + \dots + b_n\frac{\partial}{\partial a_n}\right)\phi,$$

therefore the expression on the right is a joint covariant of the two forms; in other words the property of invariance is not affected by an operator like

$$\left(b\frac{\partial}{\partial a}\right) \equiv \left(b_0\frac{\partial}{\partial a_0} + b_1\frac{\partial}{\partial a_1} + \dots + b_n\frac{\partial}{\partial a_n}\right).$$

Hence, proceeding exactly as in § 14, we can construct a covariant of i different quantics which is linear in the coefficients of each, and which becomes a numerical multiple of ϕ when we replace each of the i quantics by f.

The operator $\left(b\frac{\partial}{\partial a}\right)$ is called the Aronhold operator; its importance lies in the fact that it enables us to construct simultaneous invariants or covariants of several binary forms of the same order when any invariants or covariants are known for a simple form.

Thus, for example, since $a_0a_2-a_1^2$ is an invariant of the quadratic

$$a_0x_1^2+2a_1x_1x_2+a_2x_2^2$$

the expression $a_0b_2+a_2b_0-2a_1b_1$ is a simultaneous invariant of the two quadratics

$$(a_0, a_1, a_2 \between x_1, x_2)^2 \text{ and } (b_0, b_1, b_2 \between x_1, x_2)^2.$$

The construction of other illustrations will present no difficulty.

36. It has been proved in § 14 that any covariant of degree i can be symbolically represented as a function ϕ of degree n in the coefficients of each of n different linear forms

$$\alpha_x, \quad \beta_x, \quad \gamma_x, \quad \dots;$$

since ϕ is a covariant of the original quantic it is unaltered by any linear transformation, except for a factor which depends only on the transformation, hence also it is a covariant of the i linear forms.

By further use of the Aronhold operator we can now find a covariant of ni different linear forms

$$\alpha_x^{(1)}, \quad \alpha_x^{(2)}, \quad \dots \quad \alpha_x^{(n)}$$
$$\beta_x^{(1)}, \quad \beta_x^{(2)}, \quad \dots \quad \beta_x^{(n)}$$
$$\text{etc.}$$

linear in the coefficients of each form, and such that it becomes a numerical multiple of the original covariant when each of the symbols

$$\alpha^{(1)}, \quad \alpha^{(2)}, \quad \dots \quad \alpha^{(n)}$$

is replaced by α, each of the symbols

$$\beta^{(1)}, \quad \beta^{(2)}, \quad \dots \quad \beta^{(n)}$$

by β, and so on.

We need therefore only consider linear covariants of linear forms in the sequel; we shall prove that every covariant of a system of linear forms is a rational integral function of invariants of the type $(\alpha\beta)$ and covariants of the type α_x. Once this is established the general theorem follows immediately.

37. System of Linear Forms. First consider a single linear form

$$a_1x_1+a_2x_2$$

and let ϕ be any invariant or covariant.

If we use the linear transformation given by

$$a_1x_1+a_2x_2=X_1$$
$$bx_2=X_2,$$

where b is any constant, then the new linear form is X_1 and we have the equation

$$\Phi=\mu\phi$$

where μ depends only on the transformation, *i.e.* only on a_1, a_2, b.

Now let $$\phi=\psi_0a_1^r+\psi_1a_1^{r-1}a_2+\ldots+\psi_ra_2^r$$

where the ψ's do not depend on a_1, a_2 but contain only x_1, x_2.

Then $$\Phi=\Psi_0A_1^r+\Psi_1A_1^{r-1}A_2+\ldots+\Psi_rA_2^r$$

where Ψ_r is the same function of X_1, X_2 as ψ_r is of x_1, x_2; and further $A_1=1$, $A_2=0$ since the transformed form is X_1.

Hence $$\Psi_0=\mu\phi=\mu(\psi_0a_1^r+\psi_1a_1^{r-1}a_2+\ldots+\psi_ra_2^r) \quad \ldots\ldots\ldots\ldots(I).$$

Now Ψ_0 depends only on X_1, X_2, therefore it is of the form

$$C_0X_1^m+C_1X_1^{m-1}X_2+\ldots+C_mX_2^m,$$

the C's being numerical, hence equating coefficients of x_1^m in the equation (I) we find

$$C_0a_1^m=\mu X$$

where X does not depend on b.

Consequently μ does not depend on b and therefore on making $b=0$ in (I) we find

$$C_0X_1^m=\mu\phi,$$

for X_2 is now zero.

Hence μ is constant and ϕ is a numerical multiple of X_1^m, *i.e.* of

$$(a_1x_1+a_2x_2)^m.$$

Thus a single linear form has no invariants and the only covariants are powers of the form itself.

We shall now assume that a covariant of any number, less than n, of linear forms which is linear in the coefficients of each form can be expressed in terms of invariants of the type $(\alpha\beta)$ and covariants of the type α_x.

Let ϕ be a covariant of the same nature of n linear forms

$$\alpha_x,\quad \beta_x,\quad \gamma_x,\quad \delta_x,\quad \ldots,$$

let ϕ_1 be the result of putting $\beta=a$ in ϕ, ϕ_2 the result of putting $\gamma=a$ in ϕ_1, and so on, so that ϕ_{n-1} is the covariant of the single form a_x obtained by making

$$a=\beta=\gamma=\ldots$$

in ϕ.

Now since ϕ is linear in β we have

$$\phi_1=\left(a_1\frac{\partial}{\partial\beta_1}+a_2\frac{\partial}{\partial\beta_2}\right)\phi$$

and in like manner

$$\phi_2=\left(a_1\frac{\partial}{\partial\gamma_1}+a_2\frac{\partial}{\partial\gamma_2}\right)\phi_1$$

etc.

Consider $$\left(a_1\frac{\partial}{\partial\beta_1}+a_2\frac{\partial}{\partial\beta_2}\right)\phi=\phi_1$$

as a differential equation for ϕ, it being given that ϕ_1 does not contain β. As a particular solution we have

$$\phi=\frac{1}{S}\left(\beta_1\frac{\partial}{\partial a_1}+\beta_2\frac{\partial}{\partial a_2}\right)\phi_1$$

where S is the degree of ϕ_1 in a, *i.e.* $S=2$ in our case, therefore

$$\phi=\frac{1}{2}\left(\beta\frac{\partial}{\partial a}\right)\phi_1+\psi_1$$

where $$\left(\beta\frac{\partial}{\partial a}\right)\equiv\beta_1\frac{\partial}{\partial a_1}+\beta_2\frac{\partial}{\partial a_2},$$

$$\left(a_1\frac{\partial}{\partial\beta_1}+a_2\frac{\partial}{\partial\beta_2}\right)\psi_1=0$$

and further ψ_1 is linear in β.

Accordingly $\psi_1=P_1\beta_1+P_2\beta_2$, where P_1, P_2 do not contain β and

$$a_1P_1+a_2P_2=0,$$

$$\therefore\ \frac{P_1}{a_2}=-\frac{P_2}{a_1}=\chi_1,$$

where χ_1 does not contain β and is integral in a, γ, δ, etc.

Hence $$\psi_1=P_1\beta_1+P_2\beta_2=(a\beta)\chi_1$$

and $$\phi=\frac{1}{2}\left(\beta\frac{\partial}{\partial a}\right)\phi_1+(a\beta)\chi_1.$$

But ϕ is a covariant, and $\left(\beta\frac{\partial}{\partial a}\right)\phi_1$ is a covariant; therefore $(a\beta)\chi_1$ is a covariant; again, $(a\beta)$ is an invariant, therefore χ_1 is a covariant. Moreover since $\chi_1(a\beta)$ is linear in a, β, γ, δ, ..., χ_1 is linear in γ, δ, ...; therefore by hypothesis it can be expressed in terms of $(\gamma\delta)$, γ_x, δ_x, etc.

Thus $\phi=\frac{1}{2}\left(\beta\frac{\partial}{\partial a}\right)\phi_1$ together with an expression depending only on $(a\beta)$, a_x, and factors of these types.

Then in like manner

$$\phi_1 = \frac{1}{3}\left(\gamma \frac{\partial}{\partial a}\right)\phi_2 + (a\gamma)\chi_1,$$

therefore
$$\phi = \frac{1}{2.3}\left(\beta \frac{\partial}{\partial a}\right)\left(\gamma \frac{\partial}{\partial a}\right)\phi_2$$

together with terms of the required form.

Proceeding in this way we finally have

$$\phi = \frac{1}{n!}\left(\beta \frac{\partial}{\partial a}\right)\left(\gamma \frac{\partial}{\partial a}\right)\left(\delta \frac{\partial}{\partial a}\right)\dots\phi_n$$

together with terms of the required form.

But ϕ_n being a covariant of a_x is a numerical multiple of

$$a_x^n$$

and therefore
$$\left(\beta \frac{\partial}{\partial a}\right)\left(\gamma \frac{\partial}{\partial a}\right)\left(\delta \frac{\partial}{\partial a}\right)\dots\phi_n$$

is a numerical multiple of $a_x\beta_x\gamma_x\dots$.

Thus we have expressed the covariant ϕ completely in terms of the two types of factors $(a\beta)$ and a_x, that is, if the theorem is true for less than n forms it is true for n forms; but it has been proved true for one form, hence it is true universally.

Q. E. D.

CHAPTER III.

POLARS AND TRANSVECTANTS.

38. Two sets of variables

$$x_1, x_2, \dots x_n,$$
$$y_1, y_2, \dots y_n,$$

there being the same number of variables in each set, are said to be *cogredient*, if, when one set is transformed by any linear transformation, the other set is transformed by the same transformation; thus if $x_1, x_2, \dots x_n$ become $X_1, X_2, \dots X_n$ where

$$\left.\begin{aligned} x_1 &= l_{1,1}X_1 + l_{1,2}X_2 + \dots + l_{1,n}X_n \\ x_2 &= l_{2,1}X_1 + l_{2,2}X_2 + \dots + l_{2,n}X_n \\ &\dots\dots\dots\dots\dots\dots\dots\dots \\ x_n &= l_{n,1}X_1 + l_{n,2}X_2 + \dots + l_{n,n}X_n \end{aligned}\right\} \dots\dots\dots\dots\text{(I)},$$

then $y_1, y_2, \dots y_n$ will become $Y_1, Y_2, \dots Y_n$ where

$$\left.\begin{aligned} y_1 &= l_{1,1}Y_1 + l_{1,2}Y_2 + \dots + l_{1,n}Y_n \\ y_2 &= l_{2,1}Y_1 + l_{2,2}Y_2 + \dots + l_{2,n}Y_n \\ &\dots\dots\dots\dots\dots\dots\dots\dots \\ y_n &= l_{n,1}Y_1 + l_{n,2}Y_2 + \dots + l_{n,n}Y_n \end{aligned}\right\} \dots\dots\dots\dots\text{(II)}.$$

Two sets of variables

$$x_1, x_2, \dots x_n$$
$$y_1, y_2, \dots y_n$$

are said to be *contragredient* if, whenever the first set is transformed by the equations (I), then the second set of transformed variables $Y_1, Y_2, \dots Y_n$ is given by the equations

$$\left.\begin{aligned} Y_1 &= l_{1,1}y_1 + l_{2,1}y_2 + \dots + l_{n,1}y_n \\ Y_2 &= l_{1,2}y_1 + l_{2,2}y_2 + \dots + l_{n,2}y_n \\ &\dots\dots\dots\dots\dots\dots\dots\dots \\ Y_n &= l_{1,n}y_1 + l_{2,n}y_2 + \dots + l_{n,n}y_n \end{aligned}\right\} \dots\dots\dots\dots\text{(III)}.$$

It is easy to see that if the set of variables $x_1, x_2, \ldots x_n$ is contragredient to $y_1, y_2, \ldots y_n$; then $y_1, y_2, \ldots y_n$ is contragredient to $x_1, x_2, \ldots x_n$. For example the symbols α and x in a symbolical product are contragredient.

39. From these definitions we deduce the following theorem:

If $x_1, x_2, \ldots x_n$; $y_1, y_2, \ldots y_n$ are two contragredient sets of variables, then

$$x_1y_1 + x_2y_2 + \ldots + x_ny_n$$

is unaltered by any linear transformation.

Conversely, if

$$x_1y_1 + x_2y_2 + \ldots + x_ny_n$$

is unaltered by any linear transformation, the two sets of variables $x_1, x_2, \ldots x_n$; $y_1, y_2, \ldots y_n$ are contragredient.

Let the two sets of variables be contragredient, then if (I) and (III) be the equations of transformation,

$$\begin{aligned} & x_1y_1 + x_2y_2 + \ldots + x_ny_n \\ &= y_1(l_{1,1}X_1 + l_{1,2}X_2 + \ldots + l_{1,n}X_n) \\ &+ y_2(l_{2,1}X_1 + l_{2,2}X_2 + \ldots + l_{2,n}X_n) \\ &+ \ldots\ldots\ldots\ldots\ldots\ldots\ldots\ldots \\ &+ y_n(l_{n,1}X_1 + l_{n,2}X_2 + \ldots + l_{n,n}X_n) \\ &= X_1(l_{1,1}y_1 + l_{2,1}y_2 + \ldots + l_{n,1}y_n) \\ &+ X_2(l_{1,2}y_1 + l_{2,2}y_2 + \ldots + l_{n,2}y_n) \\ &+ \ldots\ldots\ldots\ldots\ldots\ldots\ldots\ldots \\ &+ X_n(l_{1,n}y_1 + l_{2,n}y_2 + \ldots + l_{n,n}y_n) \\ &= X_1Y_1 + X_2Y_2 + \ldots + X_nY_n. \end{aligned}$$

Q. E. D.

Conversely let

$$x_1y_1 + x_2y_2 + \ldots + x_ny_n$$

be invariantive, then if $x_1, x_2, \ldots x_n$ are transformed by equations (I)

$$\begin{aligned} & X_1Y_1 + X_2Y_2 + \ldots + X_nY_n \\ &= x_1y_1 + x_2y_2 + \ldots + x_ny_n \\ &= y_1(l_{1,1}X_1 + l_{1,2}X_2 + \ldots + l_{1,n}X_n) \\ &+ y_2(l_{2,1}X_1 + l_{2,2}X_2 + \ldots + l_{2,n}X_n) \\ &+ \ldots\ldots\ldots\ldots\ldots\ldots\ldots\ldots \\ &+ y_n(l_{n,1}X_1 + l_{n,2}X_2 + \ldots + l_{n,n}X_n). \end{aligned}$$

But this is an identity true for all values of $X_1, X_2, \dots X_n$; hence the coefficients of $X_1, X_2, \dots X_n$ on the two sides of the identity are equal; *i.e.*

$$Y_1 = l_{1,1}\,y_1 + l_{2,1}\,y_2 + \dots + l_{n,1}\,y_n$$
$$Y_2 = l_{1,2}\,y_1 + l_{2,2}\,y_2 + \dots + l_{n,2}\,y_n$$
$$\dots\dots\dots\dots\dots\dots\dots\dots\dots\dots$$
$$Y_n = l_{1,n}\,y_1 + l_{2,n}\,y_2 + \dots + l_{n,n}\,y_n.$$

This set of equations is the same as (III), and hence the two sets of variables are contragredient. Q. E. D.

It may happen that a set of variables $x_1, x_2, \dots x_n$ is subject to a restricted group of linear transformations, and that

$$x_1 y_1 + x_2 y_2 + \dots + x_n y_n$$

is an invariant for all transformations which are allowed. The above proof still holds that $y_1, y_2, \dots y_n$ is contragredient to $x_1, x_2, \dots x_n$. But $y_1, y_2, \dots y_n$ is subject to a restricted group of transformations.

Ex. (i). If the binary quantic

$$(a_0, a_1, a_2, \dots a_n \between x_1, x_2)^n$$

become after any linear transformation

$$(A_0, A_1, A_2, \dots A_n \between X_1, X_2)^n,$$

then $$(a_0, a_1, \dots a_n \between x_1, x_2)^n = (A_0, A_1, \dots A_n \between X_1, X_2)^n.$$

Hence

$$A_0, \binom{n}{1} A_1, \binom{n}{2} A_2, \dots, A_n,$$
$$x_1^n, x_1^{n-1}x_2, \dots, x_2^n,$$

are two contragredient sets of variables, subject to a restricted group of linear transformations. The linear equations connecting the original with the transformed coefficients of a binary form may be deduced from this fact (cf. § 16)*.

* By means of the group here indicated binary invariants and covariants are brought under Lie's general theory of invariants of continuous groups. If we pass from the variables $\zeta_1, \zeta_2, \dots \zeta_m$ to the variables $\zeta_1', \zeta_2', \dots \zeta_m'$ by means of the transformations,

$$\zeta_p' = f_p(\zeta_1, \zeta_2, \dots \zeta_m;\ a_1, a_2, \dots a_r),$$

and these transformations form a group when the parameters a vary, then a function

$$F(\zeta_1, \zeta_2, \dots \zeta_m)$$

Ex. (ii). Two sets of variables both contragredient to a third set are cogredient with one another.

Ex. (iii). If the determinant of transformation of a set of variables be μ, the determinant of transformation of the contragredient set is $\frac{1}{\mu}$.

40. *If $F(a_0, a_1, a_2, \ldots x_1, x_2)$ be any covariant of a binary quantic $(a_0, a_1, a_2, \ldots a_n \between x_1, x_2)^n$ and if y_1, y_2 be a pair of variables cogredient with x_1, x_2, then*

$$\left[y_1 \frac{\partial}{\partial x_1} + y_2 \frac{\partial}{\partial x_2}\right] F$$

is unchanged by any linear transformation of the variables, except that it is multiplied by the same power of the determinant of transformation as that by which F is multiplied after transformation.

For if

$$F(A_0, A_1, A_2, \ldots X_1, X_2) = \mu F(a_0, a_1, a_2, \ldots x_1, x_2)$$

where $A_0, A_1, A_2, \ldots, X_1, X_2$ are the transformed coefficients and variables, and if z_1, z_2 be any pair of variables cogredient with x_1, x_2 which become Z_1, Z_2 after transformation, then

$$F(A_0, A_1, \ldots Z_1, Z_2) = \mu F(a_0, a_1, \ldots z_1, z_2),$$

μ being a power of the determinant of transformation.

But $x_1 + \lambda y_1$, $x_2 + \lambda y_2$, λ being any constant, are a pair of variables cogredient with x_1, x_2, hence

$$F(A_0, A_1, \ldots, X_1 + \lambda Y_1, X_2 + \lambda Y_2) = \mu F(a_0, a_1, \ldots, x_1 + \lambda y_1, x_2 + \lambda y_2).$$

Now F being a rational integral algebraic function of all its variables, we may expand each side of the above equation in powers of λ; since λ is arbitrary, the coefficients of the different

---

is said to be an invariant of the group if

$$F(\zeta_1', \zeta_2', \ldots \zeta_m') = F(\zeta_1, \zeta_2, \ldots \zeta_m).$$

In our case the variables are $(n+3)$ in number, viz. $a_0, a_1, \ldots a_n, x_1, x_2$, and the transforming equations are all linear. The parameters are the coefficients in the original transformation of x_1, x_2, and if they be so chosen that $\xi_1\eta_2 - \xi_2\eta_1 = 1$, then an invariant or covariant of the binary form is an invariant of the above group in $(n+3)$ variables. The reader will find it interesting to write down the actual transformations for the a's and thence to verify that they form a group. Cf. Lie, *Vorlesungen über continuierliche Gruppen*, p. 718 etc.

powers of λ must be the same on the two sides of the equation. Hence, using Taylor's theorem:

$$F(A_0, A_1, \ldots, X_1, X_2) = \mu F(a_0, a_1, \ldots, x_1, x_2)$$

$$\left(Y_1 \frac{\partial}{\partial X_1} + Y_2 \frac{\partial}{\partial X_2}\right) F(A_0, A_1, \ldots, X_1, X_2)$$
$$= \mu \left(y_1 \frac{\partial}{\partial x_1} + y_2 \frac{\partial}{\partial x_2}\right) F(a_0, a_1, \ldots, x_1, x_2)$$

....

from which we see that $\left(y_1 \frac{\partial}{\partial x_1} + y_2 \frac{\partial}{\partial x_2}\right) F$ is invariantive, if F is a covariant.

41. The definition of covariants may now be extended thus: Any function of the coefficients of a single quantic, or of a simultaneous system of quantics, and of sets of variables,—all sets being cogredient with the variables of the quantics,—which is such that, when any linear transformation is made and the original coefficients and variables replaced by the transformed coefficients and variables, it is unaltered except for a factor depending on the coefficients of transformation.

Let $x_1, x_2; y_1, y_2$ be two cogredient sets of variables, then

$$(x_1 y_2 - x_2 y_1) = (xy)$$

is unaltered, except for a factor—which is the modulus itself—by any transformation.

Hence, if it be necessary to consider covariants with sets of cogredient variables, we may replace all the sets of variables but one by the coefficients of linear forms added to the system. For we may regard (xy) as a linear form—since it becomes (XY) by transformation—and hence replace the variables $y_2, -y_1$ by its coefficients.

Hence covariants having more than one set of variables may be represented symbolically as a sum of products of symbolical factors of the types $(\alpha\beta)$, α_x, α_y, (xy), ...

42. Convolution*. The word convolution is used as a name for the process of obtaining from a given symbolical product (representing a covariant) another symbolical product

* German *Faltung*.

(representing another covariant) by removing two of its factors of the form $\alpha_x\beta_x$ and replacing them by a single factor of the form $(\alpha\beta)$. Any symbolical product P' obtained from the product P by means of this process either once or several times repeated, is said to be obtained from P by convolution. Thus the covariant $(ab)^2(ac)\,a_x b_x^2 c_x^3$ of the quartic is obtained by convolution from the covariant $(ab)^2 a_x^2 b_x^2 c_x^4$; the factor $a_x c_x$ being replaced by (ac). It may also be obtained by convolution from $(ab)(ac)\,a_x^2 b_x^3 c_x^3$; also from $(ab)\,a_x^3 b_x^3 c_x^4$, or from $a_x^4 b_x^4 c_x^4$.

This process, it should be noticed, is purely a symbolical process, and has no analogue in the non-symbolical treatment of modern algebra.

43. Polars. The operator

$$\left(y\frac{\partial}{\partial x}\right) \equiv y_1\frac{\partial}{\partial x_1} + y_2\frac{\partial}{\partial x_2}$$

has already been introduced.

In § 15 we defined the form

$$\frac{(n-r)!}{n!}\left(y\frac{\partial}{\partial x}\right)^r F \quad\ldots\ldots\ldots\ldots\ldots\ldots\text{(IV)}$$

to be the rth polar of F; F being a binary form of order n.

Again in § 40 of this chapter it was proved that if F is a covariant and the variables y_1, y_2 are cogredient with x_1, x_2, then the form (IV) is also a covariant.

For the purpose of calculating polars, we may use a theorem identical with that of Leibnitz for ordinary differentiation.

Thus if the rth polar of $\quad a_x^m b_x^n$

be required, it may be obtained by operating on this expression with

$$\frac{(m+n-r)!}{(m+n)!}(D_1+D_2)^r$$

where $D_1 = y_1\dfrac{\partial}{\partial x_1} + y_2\dfrac{\partial}{\partial x_2}$, but operates only on a_x^m; and D_2 is the same expression but operates only on b_x^n.

Ex. (i). Find the second polar of $(ab)^2 a_x^2 b_x^2$.

Ex. (ii). Find the rth polar of $a_x^m b_x^n c_x^p$.

44. In view of the fact that covariants will generally be given in terms of symbolical letters which refer to the original quantic or quantics, and that in this case the factors α_x are not all the same, we shall consider the general case in which these factors are all different. Results, proved for this case, may be obtained for any other by simply equating two or more letters.

Consider now the form

$$P \,.\, \alpha_{1_x} \alpha_{2_x} \ldots \alpha_{n_x} \equiv F_x^{\,n},$$

where P is a product of symbolical factors not containing x_1, x_2.

The first polar is

$$\begin{aligned} F_x^{\,n-1} F_y = \frac{1}{n} [\alpha_{1_y} \alpha_{2_x} \alpha_{3_x} \ldots \alpha_{n_x} & \\ + \alpha_{1_x} \alpha_{2_y} \alpha_{3_x} \ldots \alpha_{n_x} & \\ + \ldots\ldots\ldots\ldots\ldots\ldots & \\ + \alpha_{1_x} \alpha_{2_x} \ldots \alpha_{n-1_x} \alpha_{n_y}] & \\ = \frac{1}{n} \Sigma \left[\frac{F_x^{\,n} \,.\, \alpha_{r_y}}{\alpha_{r_x}} \right] & \end{aligned}$$

the term in the bracket being simply an abbreviation for

$$P\alpha_{1_x} \alpha_{2_x} \ldots \alpha_{r-1_x} \alpha_{r_y} \alpha_{r-1_x} \ldots \alpha_{n_x}.$$

We notice in this expression that:

(i) The difference between two terms

$$\left[\frac{F_x^{\,n} \,.\, \alpha_{r_y}}{\alpha_{r_x}} \right] - \left[\frac{F_x^{\,n} \,.\, \alpha_{s_y}}{\alpha_{s_x}} \right] = \left[\frac{F_x^{\,n} (\alpha_{r_y} \alpha_{s_x} - \alpha_{r_x} \alpha_{s_y})}{\alpha_{r_x} \alpha_{s_x}} \right]$$

$$= (xy) \left[\frac{(\alpha_r \alpha_s) F_x^{\,n}}{\alpha_{r_x} \alpha_{s_x}} \right] = (xy) X,$$

where X is an expression obtained from F by convolution.

(ii) The difference between the whole polar and one of its terms

$$= \frac{1}{n} \Sigma \left[\frac{F_x^{\,n} \,.\, \alpha_{r_y}}{\alpha_r} \right] - \left[\frac{F_x^{\,n} \,.\, \alpha_{s_y}}{\alpha_{s_x}} \right]$$

$$= \frac{1}{n} \sum_r \left[\frac{F_x^{\,n} \,.\, \alpha_{r_y}}{\alpha_{r_x}} - \frac{F_x^{\,n} \,.\, \alpha_{s_y}}{\alpha_{s_x}} \right] = \frac{1}{n} \sum_r \left[\frac{(xy)(\alpha_r \alpha_s) F_x^{\,n}}{\alpha_{r_x} \alpha_{s_x}} \right] = (xy) X,$$

where X is a sum of terms each obtained from F by convolution, each such term being possibly multiplied by a constant.

Similarly the rth polar of F_x^n is

$$F_x^{n-r} F_y^r = \frac{r!\,(n-r)!}{n!} \Sigma \left[F_x^n \frac{\alpha_{1_y} \alpha_{2_y} \dots \alpha_{r_y}}{\alpha_{1_x} \alpha_{2_x} \dots \alpha_{r_x}} \right] \dots\dots \text{(V)},$$

where the summation extends to all possible sets of r factors taken from $\alpha_{1_x} \alpha_{2_x} \dots \alpha_{n_x}$. To see the truth of this suppose it true for the rth polar, then the $\overline{r+1}$th polar may be obtained from the rth by operating with $\frac{1}{n-r}\left(y \frac{\partial}{\partial x}\right)$. On the left this gives $F_x^{n-r-1} F_y^{r+1}$. Consider one term on the right, that expressed above; by polarizing we obtain

$$\frac{r!\,(n-r-1)!}{n!} \sum_{i=r+1}^{i=n} \left[F_x^n \frac{\alpha_{1_y} \alpha_{2_y} \dots \alpha_{r_y} \alpha_{i_y}}{\alpha_{1_x} \alpha_{2_x} \dots \alpha_{r_x} \alpha_{i_x}} \right].$$

Each term of (V) gives rise to a similar expression. Now any particular term of the $\overline{r+1}$th polar, arises $\overline{r+1}$ times, once from each of $r+1$ terms of the rth polar, thus

$$\left[F_x^n \cdot \frac{\alpha_{1_y} \alpha_{2_y} \dots \alpha_{r_y} \alpha_{r+1_y}}{\alpha_{1_x} \alpha_{2_x} \dots \alpha_{r_x} \alpha_{r+1_x}} \right]$$

arises from each of those terms of the rth polar, obtained by omitting one of the factors in the numerator and the corresponding factor of the denominator of this expression. Hence

$$F_x^{n-r-1} F_y^{r+1} = \frac{(r+1)!\,(n-r-1)!}{n!} \Sigma \left[F_x^n \frac{\alpha_{1_y} \alpha_{2_y} \dots \alpha_{r+1_y}}{\alpha_{1_x} \alpha_{2_x} \dots \alpha_{r+1_x}} \right].$$

Now it has been seen that this is the correct form for the first polar; hence it is the correct form for the second, and so on; the formula is therefore true in general.

45. The number of different terms

$$\left[F_x^n \frac{\alpha_{1_y} \alpha_{2_y} \dots \alpha_{r_y}}{\alpha_{1_x} \alpha_{2_x} \dots \alpha_{r_x}} \right]$$

is $\binom{n}{r}$. The coefficient of each term as it appears in the polar

$$F_x^{n-r} F_y^r$$

is

$$\frac{1}{\binom{n}{r}};$$

hence the sum of the coefficients of all the terms of a polar is unity.

This remains true if some of the letters α become equal, for no term of the polar can vanish.

46. Two terms of the rth polar are said to be *adjacent*, when they differ only in that one has a factor of the form $\alpha_{h_x}\alpha_{k_y}$ while in the other this factor is replaced by $\alpha_{h_y}\alpha_{k_x}$.

We shall now prove that:

(i) *The difference between any two terms of the* rth *polar of* F_x^n *is equal to* $(xy)X$, *where* X *is a sum of terms each of which is a term of the* $\overline{r-1}$th *polar of an expression obtained from* F_x^n *by convolution.*

(ii) *The difference between the* rth *polar of* F_x^n *and any one of its terms is equal to* $(xy)X$, *where* X *is a sum of terms each of which is a term (multiplied by some constant) of the* $\overline{r-1}$th *polar of an expression obtained from* F_x^n *by convolution.*

The difference between any two adjacent terms

$$\alpha_{h_x}\alpha_{k_y}F_1 - \alpha_{h_y}\alpha_{k_x}F_1 = (xy)(\alpha_h\alpha_k)F_1 = (xy)X$$

where X is a term of the $\overline{r-1}$th polar of $(\alpha_h\alpha_k)\left[\dfrac{F_x^n}{\alpha_{h_x}\alpha_{k_x}}\right]$, *i.e.* a term of the $\overline{r-1}$th polar of an expression obtained from F_x^n by convolution.

Now any term of the rth polar of F_x^n may be obtained from any other by means of a finite number of interchanges of letters such as α_h, α_k. Hence between any two terms T_1, T_2 of the rth polar a series of terms $T_{1,1}$, $T_{1,2}$, ... $T_{1,i}$ may be placed such that each term of the series

$$T_1, T_{1,1}, T_{1,2}, \ldots T_{1,i}, T_2$$

is adjacent to that on either side of it. Hence the difference between two terms

$$T_1 - T_2 = (T_1 - T_{1,1}) + (T_{1,1} - T_{1,2}) + (T_{1,2} - T_{1,3}) + \ldots + (T_{1,i} - T_2)$$
$$= (xy)X$$

where X is a sum of terms each of which is a term of the $\overline{r-1}$th polar of an expression obtained from F_x^n by convolution.

Again, the difference between the complete polar and any single term T

$$= \sum_{i=1}^{i=\binom{n}{r}} \left[\frac{T_i}{\binom{n}{r}}\right] - T$$

T_i representing the general term in the polar

$$= \sum_{i=1}^{i=\binom{n}{r}} \left[\frac{T_i - T}{\binom{n}{r}}\right]$$

and hence the theorem (ii) follows at once by (i).

It should be observed that if two or more factors are now made identically equal, the above proof is not affected; we may for the sake of argument suppose them all different, and make them equal in the final result. The only effect of the equality of factors is that some of the terms obtained by convolution from F_x^n will vanish. The propositions are true, then, as stated, for any symbolical product; and consequently for any covariant form.

47. Let T be any term of the rth polar of F_x^n, then by proposition (ii) of the last paragraph

$$T - F_x^{n-r} F_y^r = (xy) \Sigma \lambda_{r-1} \phi_{r-1}$$

where ϕ_{r-1} is a term of the $\overline{r-1}$th polar of a form obtained from F_x^n by convolution and λ_{r-1} is numerical. Let the $\overline{r-1}$th polar, of which ϕ_{r-1} is a term, be ψ_{r-1}. Then applying (ii) again we have

$$\phi_{r-1} - \psi_{r-1} = (xy) \Sigma \lambda_{r-2} \phi_{r-2}$$

where ϕ_{r-2} is a term of the $\overline{r-2}$th polar of a form obtained by convolution from F_x^n.

Proceeding thus we see that

$$T = F_x^{n-r} F_y^r + (xy) \Sigma \lambda_{r-1} \psi_{r-1} + (xy)^2 \Sigma \lambda_{r-2} \psi_{r-2} + \dots + (xy)^r \psi_0$$

where ψ_k is the kth polar of a form obtained from F_x^n by convolution; and λ_k is numerical.

The terms of this series, the existence of which has just been demonstrated, will be accurately determined later. The series is known as Gordan's series.

48. Transvectants*. If a_x^m, b_x^n be any two binary quantics, the form

$$(ab)^r a_x^{m-r} b_x^{n-r}$$

is called their rth transvectant, or their transvectant of *index* r.

The symbol

$$(f, \phi)^r$$

is used to denote the rth transvectant of two forms f, ϕ.

Thus

$$(a_x^m, b_x^n)^r = (ab)^r a_x^{m-r} b_x^{n-r}.$$

The definition of a transvectant just given is symbolical: the process of forming a transvectant is however not a purely symbolical one,—like that of convolution. In order that we may be able to obtain transvectants of any two forms, we make use of the differential operator, introduced in Chapter I.:

$$\Omega = \frac{\partial^2}{\partial x_1 \partial y_2} - \frac{\partial^2}{\partial x_2 \partial y_1}.$$

Thus

$$\Omega^r a_x^m b_y^n = \frac{m!}{(m-r)!} \frac{n!}{(n-r)!} (ab)^r a_x^{m-r} b_y^{n-r}.$$

Hence if $f(x)$, $\phi(x)$ be any two forms of orders m, n respectively, then the rth transvectant of f and ϕ may be obtained by operating with $\frac{(m-r)!}{m!} \cdot \frac{(n-r)!}{n!} \Omega^r$ on $f(x) . \phi(y)$ and after operation replacing y by x.

Thus

$$(f(x), \phi(x))^r = \frac{(m-r)!}{m!} \cdot \frac{(n-r)!}{n!} [\Omega^r f(x) . \phi(y)]_{y=x}.$$

Ex. For the cubic the first and second transvectants of the Hessian with the cubic itself are

$$((ab)^2 a_x b_x, c_x^3)^1 = \tfrac{1}{2} (ab)^2 (ac) b_x c_x^2 + \tfrac{1}{2} (ab)^2 (bc) a_x c_x^2,$$

$$((ab)^2 a_x b_x, c_x^3)^2 = (ab)^2 (ac) (bc) c_x;$$

as may be seen by using the differential operator.

* German *Überschiebung.*

It is useful in calculating transvectants to notice that, just as in the case of polars, the sum of the coefficients of the various terms of a transvectant is unity.

49. For the purpose of calculating transvectants the following method is extremely useful.

Consider the rth transvectant of two forms a_x^m, b_x^n; it is $(ab)^r a_x^{m-r} b_x^{n-r}$.

It may be obtained by the following rule:

Polarize a_x^m r times with respect to y, we obtain $a_x^{m-r} a_y^r$; then replace y_1 by b_2, y_2 by $-b_1$ and multiply by b_x^{n-r}, the result is $(ab)^r a_x^{m-r} b_x^{n-r}$ which is the rth transvectant of a_x^m and b_x^n.

We proceed to illustrate the method:

(i) Consider the second transvectant of the Hessian of a quartic with the quartic itself,

$$((ab)^2 a_x^2 b_x^2, c_x^4)^2.$$

The second polar of $(ab)^2 a_x^2 b_x^2$ is

$$\tfrac{1}{6}(ab)^2 a_y^2 b_x^2 + \tfrac{2}{3}(ab)^2 a_x b_x a_y b_y + \tfrac{1}{6}(ab)^2 a_x^2 b_y^2.$$

Hence the transvectant required

$$= \tfrac{1}{6}(ab)^2 (ac)^2 b_x^2 c_x^2 + \tfrac{2}{3}(ab)^2 (ac)(bc) a_x b_x c_x^2 + \tfrac{1}{6}(ab)^2 (bc)^2 a_x^2 c_x^2$$
$$= \tfrac{1}{3}(ab)^2 (ac)^2 b_x^2 c_x^2 + \tfrac{2}{3}(ab)^2 (ac)(bc) a_x b_x c_x^2$$

since a and b are equivalent symbols.

(ii) The third transvectant of these two forms is

$$(ab)^2 (ac)^2 (bc) b_x c_x.$$

(iii) To obtain the second transvectant of the Hessian of the quartic with itself: *i.e.*

$$((ab)^2 a_x^2 b_x^2, (cd)^2 c_x^2 d_x^2)^2.$$

Let us write $(cd)^2 c_x^2 d_x^2 \equiv h_x^4$, where the symbolical letter h refers to the coefficients of the Hessian considered as a separate binary form. Then as in (i)

$$((ab)^2 a_x^2 b_x^2, h_x^4)^2$$
$$= \tfrac{1}{3}(ab)^2 (ah)^2 b_x^2 h_x^2 + \tfrac{2}{3}(ab)^2 (ah)(bh) a_x b_x h_x^2.$$

To obtain the first term we polarize h_x^4 twice, replace y_1 by $-a_2$

and y_2 by $+a_1$ and then multiply the result by $\frac{1}{2}(ab)^2 b_x^2$. Thus the second polar of h_x^4

$$=\tfrac{1}{6}(cd)^2 c_y^2 d_x^2 + \tfrac{2}{3}(cd)^2 c_y d_y c_x d_x + \tfrac{1}{3}(cd)^2 c_x^2 d_y^2.$$

Hence

$$\tfrac{1}{3}(ab)^2 (ah)^2 b_x^2 h_x^2$$
$$= \tfrac{1}{18}(ab)^2 (ac)^2 (cd)^2 b_x^2 d_x^2 + \tfrac{2}{9}(ab)^2 (ac)(ad)(cd)^2 c_x d_x b_x^2$$
$$+ \tfrac{1}{18}(ab)^2 (cd)^2 (ad)^2 b_x^2 c_x^2.$$

To obtain the second term we polarize h_x^4 once with respect to y and once with respect to z, and then replace y by a and z by b.

The polar is

$$\tfrac{1}{6}(cd)^2 c_y c_z d_x^2 + \tfrac{1}{3}(cd)^2 c_y d_z c_x d_x + \tfrac{1}{3}(cd)^2 c_z d_y c_x d_x + \tfrac{1}{6}(cd)^2 c_x^2 d_y d_z.$$

Hence

$$\tfrac{2}{3}(ab)^2 (ah)(bh) a_x b_x h_x^2 = \tfrac{1}{9}(ab)^2 (cd)^2 (ac)(bc) a_x b_x d_x^2$$
$$+ \tfrac{2}{9}(ab)^2 (cd)^2 (ac)(bd) a_x b_x c_x d_x + \tfrac{2}{9}(ab)^2 (cd)^2 (ad)(bc) a_x b_x c_x d_x$$
$$+ \tfrac{1}{9}(ab)^2 (cd)^2 (ad)(bd) a_x b_x c_x^2.$$

Hence remembering that all four letters are equivalent, we obtain

$$((ab)^2 a_x^2 b_x^2,\ (cd)^2 c_x^2 d_x^2)^2$$
$$= \tfrac{1}{9}(ab)^2 (ac)^2 (cd)^2 b_x^2 d_x^2 + \tfrac{4}{9}(ab)^2 (cd)^2 (ac)(ad) c_x d_x b_x^2$$
$$+ \tfrac{4}{9}(ab)^2 (cd)^2 (ac)(bd) a_x b_x c_x d_x.$$

(iv) Calculate

$$((ab)^2 (bc) a_x c_x^2,\ (de)^2 d_x e_x)^2.$$

(v) Transvectants of the following form are of frequent occurrence:

$$(a_x^n b_x^m,\ c_x^p)^r$$

$$= \sum_{\lambda+\mu=r} \frac{\binom{n}{\lambda}\binom{m}{\mu}}{\binom{n+m}{r}} (ac)^\lambda (bc)^\mu a_x^{n-\lambda} b_x^{m-\mu} c_x^{p-r}.$$

The result may be obtained at once by polarization.

(vi) From this may be deduced the value of

$$(a_x^n b_x^m,\ c_x^p d_x^q)^r \equiv T.$$

Let $$c_x^p d_x^q \equiv h_x^{p+q}.$$

Then from (v),

$$T = \sum_{\sigma+\tau=r} \frac{\binom{n}{\sigma}\binom{m}{\tau}}{\binom{n+m}{r}} (ah)^\sigma (bh)^\tau a_x^{n-\sigma} b_x^{m-\tau} h_x^{p+q-r}.$$

But polarizing h_x^{p+q} σ times with respect to y and τ times with respect to z we have

$$h_y^\sigma h_z^\tau h_x^{p+q-r}$$

$$= \sum_{\substack{\lambda+\mu=\sigma \\ \nu+\varpi=\tau}} \frac{\dfrac{p!}{\lambda!\,\nu!\,(p-\lambda-\nu)!} \; \dfrac{q!}{\mu!\,\varpi!\,(q-\mu-\varpi)!}}{\dfrac{(p+q)!}{\sigma!\,\tau!\,(p+q-\sigma-\tau)!}} c_y^\lambda d_y^\mu c_z^\nu d_z^\varpi c_x^{p-\lambda-\nu} d_x^{q-\mu-\varpi}.$$

Hence replacing the y's by a's and the z's by b's

$$T = \sum_{\lambda+\mu+\nu+\varpi=r} \frac{r!}{\lambda!\,\mu!\,\nu!\,\varpi!} \frac{\dfrac{n!}{(n-\lambda-\mu)!} \; \dfrac{m!}{(m-\nu-\varpi)!} \; \dfrac{p!}{(p-\lambda-\nu)!} \; \dfrac{q!}{(q-\mu-\varpi)!}}{\dfrac{(m+n)!}{(m+n-r)!} \cdot \dfrac{(p+q)!}{(p+q-r)!}}$$

$$(ac)^\lambda (ad)^\mu (bc)^\nu (bd)^\varpi a_x^{n-\lambda-\mu} b_x^{m-\nu-\varpi} c_x^{p-\lambda-\nu} d_x^{q-\mu-\varpi}.$$

(vii) If $f = a_{1_x} a_{2_x} \ldots a_{m_x}$, $\phi = \beta_{1_x} \beta_{2_x} \ldots \beta_{n_x}$,

$$(f, \phi)^r = \frac{1}{r!\binom{m}{r}\binom{n}{r}} \Sigma \left[\frac{(a_1\beta_1)(a_2\beta_2)\ldots(a_r\beta_r)}{a_{1_x} a_{2_x} \ldots a_{r_x} \beta_{1_x} \beta_{2_x} \ldots \beta_{r_x}} f \cdot \phi \right],$$

where the Σ extends to all possible arrangements of the letters $a_1, a_2, \ldots a_n$; $\beta_1, \beta_2, \ldots \beta_n$. § 44.

50. Two important theorems relating to the difference between terms of a transvectant, must now be proved. They are exactly analogous to those already obtained for polars in § 46.

(i) *The difference between any two terms of a transvectant is equal to a sum of terms each of which is a term of a lower transvectant of forms obtained by convolution from the original forms.*

(ii) *The difference between the whole transvectant and any one of its terms is equal to a sum of terms each of which is a term of a lower transvectant of forms obtained by convolution from the original forms.*

Here, as in the case of polars, we introduce the idea of *adjacent* terms. Two terms of a transvectant are said to be *adjacent* when they differ merely in the arrangement of the letters in a pair of symbolical factors. Two terms can be *adjacent* in any one of the following ways:

(i) $P(\alpha_i\beta_j)(\alpha_h\beta_k)$ and $P(\alpha_i\beta_k)(\alpha_h\beta_j)$,

(ii) $P(\alpha_i\beta_j)\,\alpha_{h_x}$ and $P(\alpha_h\beta_j)\,\alpha_{i_x}$,

(iii) $P(\alpha_i\beta_j)\,\beta_{k_x}$ and $P(\alpha_i\beta_k)\,\beta_{j_x}$, where the letters $\alpha_1, \alpha_2, \ldots$ belong to the first of the two forms in the transvectant, while $\beta_1, \beta_2, \ldots$ belong to the second form.

The difference between two adjacent terms is in the three cases seen to be

(i) $P(\alpha_i\alpha_h)(\beta_j\beta_k)$,

(ii) $P(\alpha_i\alpha_h)\beta_{j_x}$,

(iii) $P(\beta_k\beta_j)\alpha_{i_x}$.

To fix ideas we shall suppose that the transvectant we are considering is the rth transvectant of

$$f = A \,.\, \alpha_{1_x}\alpha_{2_x}\ldots\alpha_{m_x},$$

and

$$\phi = B \,.\, \beta_{1_x}\beta_{2_x}\ldots\beta_{n_x},$$

where A and B are products of symbolical factors of the type $(\gamma\delta)$, and all the factors of the type γ_x are different.

Then

$$(f, \phi)^r = \frac{1}{r!\binom{m}{r}\binom{n}{r}}\Sigma\left[\frac{(\alpha_1\beta_1)(\alpha_2\beta_2)\ldots(\alpha_r\beta_r)}{\alpha_{1_x}\alpha_{2_x}\ldots\alpha_{r_x}\beta_{1_x}\beta_{2_x}\ldots\beta_{r_x}}f\,.\,\phi\right],$$

where the Σ extends to all possible arrangements of the letters $\alpha_1, \alpha_2 \ldots \alpha_m$; $\beta_1, \beta_2 \ldots \beta_n$,—§ 49 (vii).

The truth of this statement may also be seen by operating with Ω^r on $f(x)\,\phi(y)$; and remembering that if we write

$$\alpha_1 = \alpha_2 = \ldots = \alpha_m = \alpha, \quad \beta_1 = \beta_2 = \ldots = \beta_n = \beta,$$

then

$$(f, \phi)^r = (\alpha\beta)^r \alpha_x^{m-r}\beta_x^{n-r}.$$

The difference between two adjacent terms of the above transvectant is a term in which at least one factor of the type $(\alpha\beta)$ is replaced by a factor of the type $(\alpha\alpha')$, or else of the type $(\beta\beta')$. There are then not more than $(r-1)$ factors of the type $(\alpha\beta)$.

Hence the difference between two adjacent terms is a term of a transvectant of index less than r, of forms obtained by convolution from f and ϕ.

Thus, for example,

$$(ab)^2(ac)^2 b_x^2 c_x^2, \quad (ab)^2(ac)(bc)\,a_x b_x c_x^2$$

are adjacent terms of the transvectant

$$((ab)^2 a_x^2 b_x^2, c_x^4)^2,$$

and their difference

$$(ab)^3(ac)\,b_x c_x^3$$

is a term of

$$((ab)^3 a_x b_x, c_x^4).$$

Now between any two terms T_1, T_2 we may place a series of terms

$$T_{1,1}, \ T_{1,2}, \ldots T_{1,i}$$

such that any term in the series

$$T_1, \ T_{1,1}, \ T_{1,2}, \ldots T_{1,i}, \ T_2$$

is adjacent to that on either side of it.

For we may obtain T_2 from T_1 by a finite number of interchanges of pairs of letters, a pair being composed either of two α's or else of two β's; since each letter occurs only once—in our argument,—the terms which differ by the interchange of a single pair are adjacent. Hence the difference between any two terms T_1, T_2

$$T_1 - T_2 = (T_1 - T_{1,1}) + (T_{1,1} - T_{1,2}) + \ldots + (T_{1,i} - T_2),$$

i.e. a sum of terms each of which is a term of a transvectant of lower index of forms obtained from the original forms by convolution. This is the first theorem.

Again if T be any term of the transvectant, then

$$(f, \phi)^r - T = \frac{1}{r!\binom{m}{r}\binom{n}{r}} \Sigma T' - T$$

$$= \frac{1}{r!\binom{m}{r}\binom{n}{r}} \Sigma (T' - T)$$

$$\left(\text{since the number of terms } T' \text{ is } r!\binom{m}{r}\binom{n}{r}\right)$$

and this is equal to a linear function of terms of transvectants of lower index of forms obtained by convolution from f and ϕ.

51. This theorem may be extended by applying it again to each of the terms of transvectants of lower index on the right-hand side. This may be done repeatedly. Each time the process is applied the *terms* of transvectants on the right are replaced by the transvectants themselves, and linear functions of terms of transvectants of lower index. After not more than r applications of the process, we have on the right a linear function of transvectants of forms obtained from f, ϕ by convolution, whose index is less than r, and of terms of transvectants of zero index.

But a transvectant of zero index of two forms is simply the product of the forms; a term of such a transvectant is merely the same product, and therefore the transvectant itself.

We obtain then the following important theorem: *The difference between any transvectant and one of its terms is a linear function of transvectants of lower index of forms obtained from the original forms by convolution.*

Ex. (i).
$$(ab)^2 (bc)^2 a_x^2 c_x^2$$
$$= ((ab)^2 a_x^2 b_x^2, c_x^4)^2 - \tfrac{5}{6} (ab)^3 (bc) a_x c_x^3 - \tfrac{1}{6} (ab)^3 (ac) b_x c_x^3$$
$$= ((ab)^2 a_x^2 b_x^2, c_x^4)^2 - ((ab)^3 a_x b_x, c_x^4) + \tfrac{2}{3} (ab)^4 . c_x^4.$$

Ex. (ii). Prove that
$$((ab) a_x^3 b_x^3, (cd) c_x^3 d_x^3)^6$$
$$= \tfrac{1}{20} (ab)(cd) \{(bc)^3 (ad)^3 + 9 (bc)^2 (ad)^2 (ac)(bd)$$
$$+ 9 (bc)(ad)(ac)^2 (bd)^2 + (ac)^3 (bd)^3\}$$
$$= \tfrac{1}{4} \{(bc)^4 (ad)^4 - (ac)^4 (bd)^4\} - \tfrac{1}{5} (ab)^2 (cd)^2 \{(bc)^2 (ad)^2 - (ac)^2 (bd)^2\}$$
$$= \tfrac{1}{4} \{(bc)^4 (ad)^4 - (ac)^4 (bd)^4\} - \tfrac{1}{5} (ab)^3 (cd)^3 \{(bc)(ad) + (ac)(bd)\}$$
$$= \tfrac{1}{4} \{(bc)^4 (ad)^4 - (ac)^4 (bd)^4\} - \tfrac{2}{5} ((ab)^3 a_x b_x, (cd)^3 c_x d_x)^2.$$

Ex. (iii). If f be a cubic and H its Hessian, prove that $(H, f)^2 = 0$.

Ex. (iv). If f be a quartic and H its Hessian, prove that $(H, f)^3 = 0$.

Ex. (v). If b_x^4 be the Hessian of a_x^4, and $(ac)^4 = 0 = (ad)^4$, then
$$((ab) a_x^3 b_x^3, (cd) c_x^3 d_x^3)^6 = 0.$$

Ex. (vi). The second transvectant of a quantic f of order n and its Hessian H is
$$(H, f)^2 = \frac{n-3}{2(2n-5)} if,$$
where
$$i = (f, f)^4.$$

For
$$(H, f)^2 = \frac{1}{\binom{2n-4}{2}} \left\{ 2\binom{n-2}{2} (ac)^2 b_x + (n-2)^2 (ac)(bc) a_x \right\} (ab)^2 a_x^{n-4} b_x^{n-3} c_x^{n-2}$$
$$= (ab)^2 (ac)^2 a_x^{m-4} b_x^{n-2} c_x^{n-2} - \frac{n-2}{2(2n-5)} if$$
on using (III) § 22.

Finally the result is obtained by the help of the last identity of § 22.

CHAPTER IV.

GORDAN'S SERIES.

52. **Gordan's series.** In § 47 it was proved that any term T of the mth polar of a symbolical product F can be expressed in the form $\Sigma (xy)^i F_i$; where F_i is the $(m-i)$th polar of a sum of symbolical products—possibly multiplied by positive or negative constants—each of which is obtained from F by convolution. The product $a_x{}^n b_y{}^m$ is a term of the mth polar of $a_x{}^n b_x{}^m$, hence

$$a_x{}^n b_y{}^m = \sum_{i=0}^{i=m} (xy)^i F^{(i)}{}_{y^{m-i}} \ldots\ldots\ldots\ldots\ldots\ldots\ldots\text{(I)};$$

where the suffix indicates that $F^{(i)}{}_{y^{m-i}}$ is the $(m-i)$th polar of $F^{(i)}$.

This is an identity; it must therefore be true when $y = x$*; hence

$$a_x{}^n b_x{}^m = [F^{(0)}{}_{y^m}]_{y=x},$$

for when $y = x$, $(xy) = 0$. Now if in a function of x polarized with respect to y, y is replaced by x, the result is the same as the original expression. Thus the nth polar of $a_x{}^{m+n}$ is $a_x{}^m a_y{}^n$, which becomes $a_x{}^{m+n}$ when y is replaced by x. Hence

$$a_x{}^n b_x{}^m = [F^{(0)}{}_{y^m}]_{y=x} = F^0 \ \ldots\ldots\ldots\ldots\ldots \ \text{(II)}.$$

Thus the first term of the series (I) is obtained, the other terms may be obtained in a similar manner. Operate on (I) with Ω^j. Consider first the effect of this operation on the term

$$(xy)^i F^{(i)}{}_{y^{m-i}}.$$

* This of course means $\dfrac{y_1}{y_2} = \dfrac{x_1}{x_2}$.

By comparing the degrees in x on the two sides of (I) it may be seen that the degree of $F^{(i)}{}_{y^{m-i}}$ in x is $n-i$. Hence we may write

$$F^{(i)} = \alpha_x^{m+n-2i},$$

and

$$F^{(i)}{}_{y^{m-i}} = \alpha_x^{n-i}\alpha_y^{m-i}.$$

Now

$$\Omega\,(xy)^i\alpha_x^{n-i}\alpha_y^{m-i}$$
$$= i\,(m+n-i+1)\,(xy)^{i-1}\alpha_x^{n-i}\alpha_y^{m-i},$$

a result obtained by ordinary differentiation as in §§ 26, 27.

Hence by repeated operation of Ω we obtain

$$\Omega^j\,(xy)^i\alpha_x^{n-i}\alpha_y^{m-i} = \frac{i!}{(i-j)!}\cdot\frac{(m+n-i+1)!}{(m+n-i-j+1)!}(xy)^{i-j}\alpha_x^{n-i}\alpha_y^{m-i},$$

when $i \geqslant j$; but when $i<j$

$$\Omega^j\,(xy)^i\alpha_x^{n-i}\alpha_y^{m-i} = 0.$$

As regards the right-hand side of the equation (I) we know, § 27, Ex. (i), that

$$\Omega^j a_x^n b_y^m = \frac{n!}{(n-j)!}\cdot\frac{m!}{(m-j)!}\,(ab)^j a_x^{n-j} b_y^{m-j}.$$

The result of this operation is, then,

$$\frac{n!}{(n-j)!}\cdot\frac{m!}{(m-j)!}(ab)^j a_x^{n-j}b_y^{m-j}$$
$$= \sum_{i=j}^{i=m}\frac{i!\,(m+n-i+1)!}{(i-j)!\,(m+n-i-j+1)!}(xy)^{i-j}F^{(i)}{}_{y^{m-i}}.$$

In this put $y=x$, and we obtain

$$\frac{n!}{(n-j)!}\cdot\frac{m!}{(m-j)!}\,(ab)^j a_x^{n-j}b_x^{m-j} = \frac{j!\,(m+n-j+1)!}{(m+n-2j+1)!}F^{(j)},$$

and hence

$$F^{(j)} = \frac{\binom{n}{j}\binom{m}{j}}{\binom{m+n-j+1}{j}}(ab)^j a_x^{n-j}b_x^{m-j}$$

$$= \frac{\binom{n}{j}\binom{m}{j}}{\binom{m+n-j+1}{j}}(a_x^n,\ b_x^m)^j.$$

Substituting this value of $F^{(j)}$ in (I) we have Gordan's series:—

$$a_x^n b_y^m = \sum_{i=0}^{i=m} \frac{\binom{n}{i}\binom{m}{i}}{\binom{m+n-i+1}{i}} (xy)^i (a_x^n, b_x^m)^i_{y^{m-i}} \ldots\ldots (\text{III}).$$

53. This series may be put into other, rather more general, forms.

Multiply by $(ab)^r$, write $n+r$ for n, and $m+r$ for m, then

$$(ab)^r a_x^{n-r} b_y^{m-r} = \sum_i \frac{\binom{n-r}{i}\binom{m-r}{i}}{\binom{m+n-2r-i+1}{i}} (xy)^i (a_x^n, b_x^m)^{i+r}_{y^{m-i-r}} \ldots (\text{IV}).$$

Operate with $\left(x\frac{\partial}{\partial y}\right)^k$, and we obtain

$$(ab)^r a_x^{n-r} b_x^k b_y^{m-r-k}$$

$$= \sum_i \frac{\binom{n-r}{i}\binom{m-r-k}{i}}{\binom{m+n-2r-i+1}{i}} (xy)^i (a_x^n, b_x^m)^{i+r}_{y^{m-i-r-k}} \ldots (\text{V}).$$

Operate with $\left(y\frac{\partial}{\partial x}\right)^k$ on (IV), then

$$(ab)^r a_x^{n-r-k} a_y^k b_y^{m-r}$$

$$= \sum_i \frac{\binom{n-r-k}{i}\binom{m-r}{i}}{\binom{m+n-2r-i+1}{i}} (xy)^i (a_x^n, b_x^m)^{i+r}_{y^{m-i-r+k}} \ldots (\text{VI}).$$

In (V) replace y_1 by c_2, y_2 by $-c_1$, and multiply by $c_x^{p-m+r+k}$, we obtain—see § 49—

$$(ab)^r (bc)^{m-r-k} a_x^{n-r} b_x^k c_x^{p-m+r+k}$$

$$= \Sigma (-1)^i \frac{\binom{n-r}{i}\binom{m-r-k}{i}}{\binom{m+n-2r-i+1}{i}} ((a_x^n, b_x^m)^{i+r}, c_x^p)^{m-i-r-k} \ldots (\text{VII}).$$

The equation (VI) gives a similar result when the weight of the covariants under consideration is greater than m the order of b_x^m: viz.—

$$(ab)^r (bc)^{m-r} (ac)^k a_x^{n-r-k} c_x^{p-m+r-k}$$

$$= \Sigma (-1)^i \frac{\binom{n-r-k}{i}\binom{m-r}{i}}{\binom{m+n-2r-i+1}{i}} ((a_x^n, b_x^m)^{i+r}, c_x^p)^{m-i-r+k} \quad \text{... (VIII).}$$

54. Now if in (VII) a and c, n and p, r and $m-r-k$ are interchanged, the left-hand side of the relation is unaltered, except for a factor $(-1)^{m-k}$. Hence

$$\Sigma \frac{\binom{n-r}{i}\binom{m-r-k}{i}}{\binom{m+n-2r-i+1}{i}} ((b_x^m, a_x^n)^{i+r}, c_x^p)^{m-i-r-k}$$

$$= \Sigma \frac{\binom{p+r+k-m}{i}\binom{r}{i}}{\binom{p+2r+2k-m-i+1}{i}} ((b_x^m, c_x^p)^{i+m-r-k}, a_x^n)^{r-i}.$$

On the supposition that

$$f = b_x^m, \quad \phi = a_x^n, \quad \psi = c_x^p, \quad \alpha_1 = 0, \quad \alpha_2 = m-r-k, \quad \alpha_3 = r,$$

this relation is the same as

$$\Sigma \frac{\binom{n-\alpha_1-\alpha_3}{i}\binom{\alpha_2}{i}}{\binom{m+n-2\alpha_3-i+1}{i}} ((f, \phi)^{\alpha_3+i}, \psi)^{\alpha_1+\alpha_2-i}$$

$$= (-1)^{\alpha_1} \Sigma \frac{\binom{p-\alpha_1-\alpha_2}{i}\binom{\alpha_3}{i}}{\binom{m+p-2\alpha_2-i+1}{i}} ((f, \psi)^{\alpha_2+i}, \phi)^{\alpha_1+\alpha_3-i} \quad \text{... (IX).}$$

Again, if in (VIII) we interchange a and c, n and p, r and $m-r$ respectively, the left-hand side of the relation is unaltered except for a factor $(-1)^{m-r}$.

Hence we obtain a relation of the same form as (IX), but in which

$$\alpha_1 = k, \quad \alpha_2 = m-r, \quad \alpha_3 = r.$$

Thus the identity (IX) is true in two cases. In the first $\alpha_1 = 0$: α_3 cannot exceed either m or n, for we use in (VII) the factor $(ab)^r$: $\alpha_2 + \alpha_3 = m - r$ and therefore cannot exceed m: and finally α_2 cannot exceed p.

With these restrictions α_2, α_3 may take any positive integral values.

In the second case

$$\alpha_2 + \alpha_3 = m.$$

$\alpha_3 + \alpha_1$ cannot exceed n, and $\alpha_1 + \alpha_2$ cannot exceed p; subject to these restrictions α_1, α_2, α_3 may take any positive integral values.

The series (IX) is one of great importance for calculating transvectants. It is usually quoted in the form

$$\begin{pmatrix} f & \phi & \psi \\ m & n & p \\ \alpha_1 & \alpha_2 & \alpha_3 \end{pmatrix},$$

where

$$\alpha_2 + \alpha_3 \not> m, \quad \alpha_3 + \alpha_1 \not> n, \quad \alpha_1 + \alpha_2 \not> p,$$

and either

(i) $$\alpha_1 = 0$$

or

(ii) $$\alpha_2 + \alpha_3 = m.$$

55. In order to illustrate the use of the series, we shall now calculate some transvectants which are covariants of the sextic f*. Let us write

$$H = (f, f)^2, \quad i = (f, f)^4, \quad A = (f, f)^6, \quad t = (f, H).$$

Then to calculate the transvectant $(H, f)^3$ we use the series

$$\begin{pmatrix} f & f & f \\ 6 & 6 & 6 \\ 0 & 3 & 2 \end{pmatrix},$$

$$\Sigma \frac{\binom{4}{i}\binom{3}{i}}{\binom{9-i}{i}}((f, f)^{2+i}, f)^{3-i} = \Sigma \frac{\binom{3}{i}\binom{2}{i}}{\binom{7-i}{i}}((f, f)^{3+i}, f)^{2-i},$$

* Other examples will be found in Chapter V.

that is

$$(H, f)^3 + \frac{6}{7}(i, f) = (i, f),$$

or

$$(H, f)^3 = \frac{1}{7}(i, f).$$

The object usually in view when calculating transvectants, is to express them in terms of other transvectants of lower index.

There is no difficulty in choosing the series which will give the desired result, in general we select it so that the transvectant to be calculated appears as the first term on one side of the identity, while all terms on the other side are of lower index.

The transvectant $(H, f)^4$ is given by

$$\begin{pmatrix} f & f & f \\ 6 & 6 & 6 \\ 0 & 4 & 2 \end{pmatrix}$$

$$\Sigma \frac{\binom{4}{i}\binom{4}{i}}{\binom{9-i}{i}}((f, f)^{2+i}, f)^{4-i} = \Sigma \frac{\binom{2}{i}\binom{2}{i}}{\binom{5-i}{i}}((f, f)^{4+i}, f)^{2-i},$$

hence

$$(H, f)^4 + \frac{12}{7}(i, f)^2 + \frac{1}{5}A \,.f = (i, f)^2 + \frac{1}{3}A\,.f,$$

or

$$(H, f)^4 = \frac{-5}{7}(i, f)^2 + \frac{2}{15}A\,.f.$$

To calculate $(H, f)^5$ we cannot write $\alpha_1 = 0$, for then the condition $\alpha_2 + \alpha_3 \not> m$ would be violated.

We must then make $\alpha_2 + \alpha_3 = m = 6$ and use the series

$$\begin{pmatrix} f & f & f \\ 6 & 6 & 6 \\ 1 & 4 & 2 \end{pmatrix}$$

$$\Sigma \frac{\binom{3}{i}\binom{4}{i}}{\binom{9-i}{i}}((f, f)^{2+i}, f)^{5-i} = -\Sigma \frac{\binom{1}{i}\binom{2}{i}}{\binom{5-i}{i}}((f, f)^{4+i}, f)^{3-i},$$

hence

$$(H, f)^5 = \frac{-8}{7}(i, f)^3 *.$$

* This form vanishes identically—see Ex. (ii).

To find $(t, f)^2$ we use

$$\begin{pmatrix} f & H & f \\ 6 & 8 & 6 \\ 0 & 2 & 1 \end{pmatrix}$$

whence using the result, § 51, Ex. (vi)

$$(f, H)^2 = \frac{3}{14} if,$$

we obtain

$$(t, f)^2 = \frac{12}{165} f \cdot (f, i).$$

Ex. (i). Prove that for the sextic

$$(H, f)^6 = \frac{5}{7} (i, f)^4.$$

Ex. (ii). By means of the series

$$\begin{pmatrix} f & f & f \\ 6 & 6 & 6 \\ 1 & 3 & 3 \end{pmatrix}$$

prove that $(i, f)^3 = 0$.

Ex. (iii). Prove that

$$((f, i)^2_, f)^3 = -\frac{3}{10} ((f, i)^4, f),$$

$$((f, i)^4, f)^2 = 2 (i, i)^2 + \frac{1}{3} A \cdot i,$$

$$((f, i)^2, f)^4 = -\frac{1}{5} (i, i)^2 + \frac{2}{15} A \cdot i,$$

$$(H, i)^4 = \frac{2}{7} (i, i)^2 + \frac{2}{15} A \cdot i.$$

56. The series we have been illustrating does not give all the relations between the covariants of f, ϕ, ψ which are of weight $\alpha_1 + \alpha_2 + \alpha_3$.

For example, we cannot by means of this series reduce the form $((f, f)^2, f)^2$ which is reducible whatever be the order of f, § 51, Ex. (vi).

57. The series (III) may be inverted with the following result

$$(a_x{}^m, b_x{}^n)^0{}_{y^i} = \sum_i (-1)^i \frac{\binom{m}{i}\binom{n}{i}}{\binom{m+n}{i}} (ab)^i a_x{}^{m-i} b_y{}^{n-i} (xy)^i \text{......(X)}.$$

(Gordan, *Invarianten-theorie*, § 7, pp. 89, 90.)

To prove this, we first establish the existence of an expansion of the form indicated; and secondly we find the coefficients. To prove the possibility of such an expansion, we observe that

$$\begin{aligned}(a_x{}^m, b_x{}^n)^0{}_{y^n} &= \Sigma\lambda_i a_x{}^{m-i} b_x{}^i a_y{}^i b_y{}^{n-i}\\ &= \Sigma\lambda_i a_x{}^{m-i} b_y{}^{n-i}\{a_x b_y - (ab)(xy)\}^i\\ &= \Sigma\mu_j (ab)^j (xy)^j a_x{}^{m-j} b_y{}^{n-j}.\end{aligned}$$

To determine the coefficients μ, we operate on the relation just proved with Ω; the left-hand side of the relation vanishes, and hence by equating the coefficients on the right to zero we obtain

$$\mu_i i(m+n-i+1) = -\mu_{i-1}(m-i+1)(n-i+1).$$

Replacing y by x, we see that $\mu_0 = 1$, hence in general

$$\mu_i = (-)^i \frac{\binom{m}{i}\binom{n}{i}}{\binom{m+n}{i}}.$$

Other series may be deduced from (X), by the same methods as were used in § 53.

Thus

$$(a_x{}^m, b_x{}^n)^k{}_{y^{n-k}} = \Sigma(-)^i \frac{\binom{m-k}{i}\binom{n-k}{i}}{\binom{m+n-2k}{i}} (ab)^{i+k} a_x{}^{m-i-k} b_y{}^{n-i-k} (xy)^i,$$

$$((a_x{}^m, b_x{}^n)^k, c_x{}^p)^r$$

$$= \Sigma \frac{\binom{m-k}{i}\binom{r}{i}}{\binom{m+n-2k}{i}} (ab)^{i+k} (bc)^{r-i} a_x{}^{m-i-k} b_x{}^{n-r-k} c_x{}^{p-r+i} \ldots\text{(XI)},$$

when $r \not> n-k$.

$$((a_x{}^m, b_x{}^n)^k, c_x{}^p)^r = \Sigma \frac{\binom{m+n-2k-r}{i}\binom{n-k}{i}}{\binom{m+n-2k}{i}}$$

$$\times (ab)^{i+k} (bc)^{n-k-i} (ac)^{k+r-n} a_x{}^{m+n-2k-r-i} c_x{}^{p-r+i} \quad \ldots\ldots\text{(XII)}.$$

58. It has already been pointed out, § 41, that the system of concomitants belonging to a binary form with two or more cogredient sets of variables, is the same as that obtained when certain binary forms with only one set of variables are taken for simultaneous ground-forms. Gordan's series (III) shews us which the ground-forms must be; thus for the form

$$a_x^2 b_y$$

we consider the system

$$a_x^2 b_x,\ (ab)\, a_x,\ (xy).$$

Each member of this system is unaltered by linear transformation, and $a_x^2 b_y$ can be expressed in terms of the forms given.

59. Ex. (i). Any symbolical product having two cogredient sets of variables x, y, can be expressed in the form

$$\Sigma\, (xy)^i\, P_{y^{m-i}}$$

where P is a symbolical product containing only one set of variables x.

If we put

$$x_1 = 1,\ \ x_2 = 0,\ \ y_1 = 0,\ \ y_2 = 1,$$

then

$$P_{y^{m-i}}$$

becomes the coefficient of $x_1^{n-m+i} x_2^{m-i}$ in P.

Hence we may express any rational integral function of the coefficients of a binary form linearly in terms of the coefficients of its covariants. (Elliott, *Proc. London Math. Soc.* vol. XXXII. p. 213.)

Ex. (ii). Express the product $a_0 a_1 a_2$ of the coefficients of the cubic

$$(a_0,\ a_1,\ a_2,\ a_3 \between x_1,\ x_2)^3$$

linearly in terms of coefficients of its covariants.

60. *If a symbolical product representing a covariant of a single binary form contain a factor* $(ab)^{2\lambda-1}$, *it may be expressed in terms of products each containing a factor* $(ab)^{2\lambda}$.

Let the order of the binary form be n; and let P be the covariant in question. Then since P contains the factor $(ab)^{2\lambda-1}$ it is evidently a term of a transvectant

$$((ab)^{2\lambda-1} a_x^{n+1-2\lambda} b_x^{n+1-2\lambda},\ \phi)^\rho,$$

where ϕ is some other covariant.

Hence by § 51

$$P = ((ab)^{2\lambda-1} a_x^{n+1-2\lambda} b_x^{n+1-2\lambda},\ \phi)^\rho$$
$$+ \Sigma C\, (\overline{(ab)^{2\lambda-1} a_x^{n+1-2\lambda} b_x^{n+1-2\lambda}},\ \overline{\phi})^{\rho'} \ldots\ldots\ldots \text{(XIII)},$$

where C is a constant, and $\overline{\phi}$ denotes any function obtained from ϕ by convolution.

Now

$$(ab)^{2\lambda-1} a_x^{n+1-2\lambda} b_x^{n+1-2\lambda} = -(ba)^{2\lambda-1} a_x^{n+1-2\lambda} b_x^{n+1-2\lambda} = 0,$$

for b and a are equivalent symbols. Hence all those transvectants in (VIII) in which no convolution has taken place in the first form are zero. But every form obtained by convolution from $(ab)^{2\lambda-1} a_x^{n+1-2\lambda} b_x^{n+1-2\lambda}$ has a factor $(ab)^{2\lambda}$. Hence every transvectant in (VIII) either vanishes or is the transvectant of a form having a factor $(ab)^{2\lambda}$, with another form: and hence P can be expressed as a sum of terms each of which has a factor $(ab)^{2\lambda}$.

A convenient way of expressing this is to write

$$P \equiv 0, \bmod (ab)^{2\lambda}.$$

61. Owing to the large number of symbolical products which represent covariants of given degree and order it is important to obtain methods of classifying them. For this purpose the greatest index of any determinant factor in the symbolical product is chosen. We shall use the word *grade* to denote this index. If the symbolical product is a covariant of a single binary form, then, by the theorem just proved, covariants of odd grade may be expressed in terms of covariants of higher even grade. On this account the German equivalent *Stüfe* is used for half the index. We prefer to define grade as the index itself, since the classification is useful when the symbolical product is not merely a covariant of a single form.

62. Covariants of degree 3. We proceed to obtain criteria, by means of which it may be at once determined, whether or not a given covariant of degree 3 can be expressed in terms of covariants of higher grade. These were first obtained by Jordan* in 1876. They were independently discovered by Stroh†; and his method of investigation is given here.

We consider first the covariants of weight w which are linear in the coefficients of each of three binary quantics

$$f_1 = a_x^{n_1},\ f_2 = b_x^{n_2},\ f_3 = c_x^{n_3}.$$

Let us write u_1 for $(bc)\,a_x$, u_2 for $(ca)\,b_x$, u_3 for $(ab)\,c_x$; then

$$u_1 + u_2 + u_3 = 0.$$

* *Liouville*, 2 Sér. III. 1876.

† *Math. Ann.* Bd. XXXI. p. 444 *et seq.*

Hence

$$u_1^{g-i}u_2^i = (-1)^{g-i}\,u_2^i\,(u_2+u_3)^{g-i}$$

$$= (-1)^{g-i}\sum_{\lambda=0}^{g-i}\binom{g-i}{\lambda}u_2^{g-\lambda}u_3^{\lambda}$$

$$= (-1)^{g-i}\sum_{\lambda=0}^{g-i}\frac{\binom{g}{\lambda}\binom{g-\lambda}{i}}{\binom{g}{i}}u_2^{g-\lambda}u_3^{\lambda}.$$

Multiply this expression for $u_1^{g-i}u_2^i$ by

$$(-1)^{k_2-i}\binom{g}{i}\binom{-k_1-1}{k_2-i}^{*},$$

where k_1, k_2 are any positive integers, and take the sum of each side of the result from $i=0$ to $i=k_2$: hence

$$\sum_{i=0}^{k_2}(-1)^{k_2-i}\binom{g}{i}\binom{-k_1-1}{k_2-i}u_1^{g-i}u_2^i$$

$$=(-1)^{k_2+g}\sum_{i=0}^{k_2}\sum_{\lambda=0}^{g-i}\binom{g}{\lambda}\binom{g-\lambda}{i}\binom{-k_1-1}{k_2-i}u_2^{g-\lambda}u_3^{\lambda}$$

$$=(-1)^{k_2+g}\sum_{\lambda=0}^{g}\binom{g}{\lambda}u_2^{g-\lambda}u_3^{\lambda}\sum_{i}\binom{g-\lambda}{i}\binom{-k_1-1}{k_2-i}$$

$$=(-1)^{k_2+g}\sum_{\lambda=0}^{g}\binom{g}{\lambda}\binom{g-\lambda-k_1-1}{k_2}u_2^{g-\lambda}u_3^{\lambda}\,\dagger \quad\text{......(XIV).}$$

Now when

$$g-k_1>\lambda>g-k_1-k_2-1,$$

$$\binom{g-\lambda-k_1-1}{k_2}=0;$$

therefore the last sum written down may be divided into two series, viz. $\lambda=0$ up to $g-k_1-k_2-1$, and $\lambda=g-k_1$ up to g.

Let $$g-k_1-k_2-1\equiv k_3.$$

Then the first of these series is (writing i for λ)

$$\sum_{i=0}^{k_3}\binom{g}{i}\binom{k_2+k_3-i}{k_2}u_2^{g-i}u_3^{i}.$$

In the second series, write $g-i$ for λ, then it becomes

$$\sum_{i=0}^{k_1}\binom{g}{i}\binom{i-k_1-1}{k_2}u_3^{g-i}u_2^{i}\quad\text{..............(XV).}$$

* $\binom{l}{m}$ is the coefficient of x^m in the expansion of $(1+x)^l$.

† For $\sum_i\binom{g-\lambda}{i}\binom{-k_1-1}{k_2-1}$ is the coefficient of x^{k_2} in $(1+x)^{g-\lambda}\,.\,(1+x)^{-k_1-1}$.

Now if l and m be positive integers

$$\binom{-l}{m} = (-)^m \frac{(l+m-1)!}{m!\,(l-1)!} = (-)^m \binom{l+m-1}{m}$$
$$= (-)^m \binom{l+m-1}{l-1}$$
$$= (-)^{m+l-1} \binom{-m-1}{l-1} \text{ ....(XVI).}$$

Hence the series (XV) becomes

$$\sum_{i=0}^{k_1} (-)^{k_2+k_1-i} \binom{g}{i} \binom{-k_2-1}{k_1-i} u_3^{g-i} u_2^{i}$$
$$= (-)^{k_2+k_1} \sum_{i=0}^{k_1} \binom{g}{i} \binom{-k_2-1}{k_1-i} u_3^{g-i} (u_1+u_3)^i$$
$$= (-)^{k_2+k_1} \sum_{\lambda=0}^{k_1} u_3^{g-\lambda} u_1^{\lambda} \sum_i \binom{g}{i} \binom{i}{\lambda} \binom{-k_2-1}{k_1-i}.$$

But the coefficient of $u_3^{g-\lambda} u_1^{\lambda}$ is here equal to

$$\sum_i \binom{g}{\lambda} \binom{g-\lambda}{i-\lambda} \binom{-k_2-1}{k_1-i}$$
$$= \binom{g}{\lambda} \binom{g-\lambda-k_2-1}{k_1-\lambda}$$
$$= \binom{g}{\lambda} \binom{k_1+k_3-\lambda}{k_3}.$$

Hence the series (XV) becomes

$$(-)^{k_1+k_2} \sum_{i=0}^{k_1} \binom{g}{i} \binom{k_1+k_3-i}{k_3} u_3^{g-\lambda} u_1^{\lambda}.$$

By means of the relations (XVI), the series on the left-hand side of (XIV) becomes

$$\sum_{i=0}^{k_2} \binom{g}{i} \binom{k_1+k_2-i}{k_1} u_1^{g-i} u_2^{i}.$$

Hence the equation (XIV) may now be written

$$(-)^{k_2} \sum_{i=0}^{k_1} \binom{g}{i} \binom{k_1+k_3-i}{k_3} u_3^{g-i} u_1^{i}$$
$$+ (-)^{k_3} \sum_{i=0}^{k_2} \binom{g}{i} \binom{k_2+k_1-i}{k_1} u_1^{g-i} u_2^{i}$$
$$+ (-)^{k_1} \sum_{i=0}^{k_3} \binom{g}{i} \binom{k_3+k_2-i}{k_2} u_2^{g-i} u_3^{i} = 0 \text{ ......(XVII).}$$

This equation is true for all positive integral values of k_1, k_2, k_3 for which

$$k_1 + k_2 + k_3 = g - 1.$$

63. Let us suppose that w, the weight of the covariants under consideration, is not greater than the order of any of the three quantics.

Then any one of the covariants may be expressed in terms of the set

$$(ab)^{w-i}(bc)^i a_x^{n_1+i-w} b_x^{n_2-w} c_x^{n_3-i},$$

where $i = 0, 1, \ldots w$. For since $n_2 \nless w$, whenever a factor (ca) appears in a covariant of this weight, it may be removed by means of the identity

$$(ca)\, b_x = -(bc)\, a_x - (ab)\, c_x.$$

Similarly all the covariants may be expressed linearly in terms of the set

$$(bc)^{w-i}(ca)^i a_x^{n_1-i} b_x^{n_2+i-w} c_x^{n_3-w},$$

or of the set

$$(ca)^{w-i}(ab)^i a_x^{n_1-w} b_x^{n_2-i} c_x^{n_3+i-w}.$$

We shall suppose the members of each of these sets arranged in order according to increasing values of i. *Then the forms of any one set are linearly independent.* For suppose a relation to exist between the forms of the first of the above sets. Let the terms in this relation be arranged in the order indicated. The relation still remains true if we take for f_1, f_2, f_3 special quantics instead of general ones. We shall suppose them to be merely powers of linear forms; the result of this is that the letters a, b, c are no longer purely symbolical. Hence if

$$\lambda_i (ab)^{w-i}(bc)^i a_x^{n_1+i-w} b_x^{n_2-w} c_x^{n_3-i}$$

be the first term, we may divide the identity by $(bc)^i$. Every term of the quotient except the first contains (bc) as a factor; but the quotient must be zero, hence the first term must vanish, when we make $b = c$,

i.e. $$(ab)^{w-i} a_x^{n_1+i-w} b_x^{n_2+n_3-w-i} \equiv 0$$

which is clearly untrue. Hence no linear relation can exist between the forms of one set. And therefore there are exactly $w + 1$ linearly independent covariants, which are of the first degree in the coefficients of each of three quantics, and of weight $w\ (\ngtr n_1, n_2 \text{ or } n_3)$.

64. Writing in (XVII), for u_1, u_2, u_3 their values in terms of a, b, c, we obtain a linear relation between the first $k_1 + 1$ members of the first set, the first $k_2 + 1$ members of the second set, and the first $k_3 + 1$ members of the third set, where

$$k_1 + k_2 + k_3 = g - 1 = w - 1.$$

Thus we obtain a relation between $w + 2$ forms.

Hence if we take the first m_1 forms of the first set, the first m_2 of the second, and the first m_3 of the third, where

$$m_1 + m_2 + m_3 = w + 1,$$

we have a set in terms of which all other of these covariants can be expressed. *This set may be chosen so that it contains no covariant of grade less than* $\frac{2w}{3}$.

For if $w = 3m - 1$, we may take

$$m_1 = m_2 = m_3 = m$$

and we see that all covariants can be expressed in terms of those whose grade is $2m$ at least.

If $w = 3m - 2$, we may take

$$m_1 = m_2 = m, \quad m_3 = m - 1$$

and we may express all covariants in terms of those whose grade is $2m - 1$ at least.

If $w = 3m$ we take

$$m_1 = m + 1, \quad m_2 = m_3 = m$$

and we have to include one covariant of grade $2m$, the rest being of grade $2m + 1$ at least.

In the second case there is one relation between the $3m$ covariants whose grade is $\not< 2m - 1$; viz. by (XVII),

$$\begin{aligned} &(ab)^{2m-1}(bc)^{m-1} a_x^{n_1 - 2m + 1} b_x^{n_2 - 3m + 2} c_x^{n_3 - m + 1} \\ &+ (bc)^{2m-1}(ca)^{m-1} a_x^{n_1 - m + 1} b_x^{n_2 - 2m + 1} c_x^{n_3 - 3m + 2} \\ &+ (ca)^{2m-1}(ab)^{m-1} a_x^{n_1 - 3m + 2} b_x^{n_2 - m + 1} c_x^{n_3 - 2m + 1} \\ &= \Sigma C_{2m} \quad \text{.......................................(XVIII)}, \end{aligned}$$

where C_{2m} denotes a covariant whose grade is $\not< 2m$.

In the third case there are two relations between the covariants of grade $\not< 2m$; these will be found from (XVII) to express that the difference between any two covariants of grade $2m$ and weight $3m$,

$$= \Sigma C_{2m+1}.$$

It would be sufficient for the general theory to prove the theorem:—If $u_1 + u_2 + u_3 = 0$ then there are $w + 1$ linearly independent products of u_1, u_2, u_3 of order w such that each contains an exponent of $\frac{2w}{3}$ at least: this is not easy.

65. Next let w be greater than the order of one or more of the quantics.

If $w > n_1$, the sum of the indices of (ab) and (ac) cannot be greater than n_1, and hence (bc) must have an index equal to $w - n_1$ at least.

We define the quantity ϵ_i to be $w - n_i$ if $w > n_i$, and to be zero in the contrary case. Then each of our covariants of weight w must have a symbolical factor

$$(bc)^{\epsilon_1}(ca)^{\epsilon_2}(ab)^{\epsilon_3}.$$

The remaining factor will be a symbolical product representing a covariant of weight

$$w - \epsilon_1 - \epsilon_2 - \epsilon_3 = \varpi,$$

of the quantics

$$a_x^{n_1-\epsilon_2-\epsilon_3}, \quad b_x^{n_2-\epsilon_3-\epsilon_1}, \quad c_x^{n_3-\epsilon_1-\epsilon_2}.$$

But the weight ϖ is not greater than the order of any of the three quantics: hence we may apply the results of the last paragraph. That is, all the covariants of weight w may be expressed in terms of $\varpi + 1$ of them. If ϵ_1, ϵ_2, ϵ_3 are unequal our choice of covariants in terms of which the rest are to be expressed may not be the same as before. But the student will have no difficulty in writing down the result as regards grade.

Ex. Prove that any covariant of degree three can be expressed in terms of covariants of grade $\frac{\varpi + w}{3}$ at least.

66. We have now two different series which may be used for obtaining relations between covariants involving three symbols. These are series (IX) due to Gordan, and Stroh's series (XVII)

$$(-1)^{k_2}\sum_{i=0}^{k_1}\binom{w}{i}\binom{k_1+k_3-i}{k_3}(ab)^{w-i}(bc)^i a_x^{n_1-w+i} b_x^{n_2-w} c_x^{n_3-i} + \ldots = 0,$$

where $$k_1 + k_2 + k_3 = w - 1$$

and w is not greater than any one of the numbers

$$n_1, n_2, n_3.$$

If this condition as regards w be not satisfied we write

$$w - n_1 = \epsilon_1, \quad w - n_2 = \epsilon_2, \quad w - n_3 = \epsilon_3$$

where it is understood that $\epsilon=0$ if $n>w$. Then the reduced weight ϖ is

$$w-\epsilon_1-\epsilon_2-\epsilon_3.$$

In this case Stroh's series is obtained from (XVII) by writing ϖ for w, $n_1-\epsilon_2-\epsilon_3$ for n_1, $n_2-\epsilon_3-\epsilon_1$ for n_2, $n_3-\epsilon_1-\epsilon_2$ for n_3, and multiplying the result by

$$(bc)^{\epsilon_1}(ca)^{\epsilon_2}(ab)^{\epsilon_3}.$$

Here k_1, k_2, k_3 satisfy the relation

$$k_1+k_2+k_3=\varpi-1.$$

The advantage of Stroh's series is that it gives all possible relations between the covariants under discussion. It is, however, generally more convenient to have relations between transvectants than to have them between symbolical products. Thus although series (IX) does not give all possible relations, yet it is frequently the more convenient one to use.

By means of series (VII) and (VIII) we may translate Stroh's series into a relation between transvectants. In fact this is what Stroh himself does. This relation has the disadvantage that the coefficients in it are themselves series.

It is convenient to have a short method of referring to Stroh's series; we therefore introduce the scheme used by Stroh

$$\begin{pmatrix} f_1 & f_2 & f_3 \\ k_1 & k_2 & k_3 \end{pmatrix}_w,$$

which is distinguishable from Gordan's scheme by the weight being indicated outside the bracket.

67. The quantics of low order furnish very few examples of covariants containing three symbols concerning which Gordan's series gives incomplete information. We have mentioned $(f, H)^2$ as one such case. The covariant $((f, i)^2, f)^2$ of the sextic f—where $i=(f, f)^4$—is another. The reader will have no difficulty in proving that

$$((f, i)^2, f)^2=\frac{1}{6}i^2+\frac{1}{15}(f, i)^4.f.$$

Ex. Prove that the covariant

$$(ab)^6(ac)^4 b_x{}^4 c_x{}^6$$

of the binary form $$a_x{}^{10}=b_x{}^{10}=c_x{}^{10}$$

can be expressed in terms of covariants of higher grade; and that the covariant

$$(ab)^6 (ac)^4 (bc)^3 b_x^4 c_x^3$$

vanishes identically.

68. Let the three quantics f_1, f_2, f_3 be made identical, so that

$$f_1 = f_2 = f_3 = a_x^n = b_x^n = c_x^n = f.$$

Then if $w \not> n$, all covariants of weight w and degree n can be expressed linearly in terms of those whose grade is not less than

$$\frac{2w}{3}.$$

If $w = 3m - 2$, it is possible to go a step further; we may express all these covariants in terms of such as are of grade $2m$ at least; since covariants of a single quantic of odd grade may be expressed in terms of covariants of higher even grade.

Hence *if* $w = 3m - \delta \not> n$, $\delta < 3$, *all covariants of degree three and weight* w *can be expressed linearly in terms of those whose grade is* $2m$ *at least.*

Similarly if $w > n$, we have

$$\epsilon_1 = \epsilon_2 = \epsilon_3 = \epsilon = w - n,$$

$$\varpi = w - 3\epsilon.$$

Then we may express all covariants of degree 3 and weight w in terms of those whose grade is not less than $\frac{2\varpi}{3} + \epsilon = \frac{2w}{3} - \epsilon.$ If ϵ is odd the lowest grade given by this is odd unless $\varpi = 3m - 2$, in which case we see on multiplying (XVIII) by

$$(ab)^\epsilon (bc)^\epsilon (ca)^\epsilon$$

that we may express covariants of grade $2m - 1$, in terms of covariants of higher grade. Hence if ϵ is odd, all these covariants may be expressed in terms of those whose grade $\not< \frac{2w}{3} - \epsilon + 1.$

If ϵ is even, then as before we take $w = 3m - \delta$, $\delta < 3$, and the minimum grade becomes $2m - \epsilon$.

69. There is one further point in this matter to be noticed. It has been proved that all covariants which are linear in the coefficients of each of three given quantics and are of given weight w can be expressed in terms of those whose grade is higher than a certain number. Further it has been shewn that among the covariants actually retained, no linear relations can exist. We passed to covariants of a single quantic, and deduced that all covariants of degree 3 and weight w can be expressed in terms of those whose grade is higher than a given number. Is it possible that amongst the covariants retained here, there may exist other relations which do not appear in the general case? To answer this question we express the symbolical products as transvectants by means of § 53. The product

$$(ab)^{w-i}(bc)^i a_x^{n_1+i-w} b_x^{n_2-w} c_x^{n_3-i}$$
$$= ((f_1, f_2)^{w-i}, f_3)^i + \Sigma C_{w-i+1},$$

where as before C_m indicates a covariant whose grade is not less than m.

The covariants retained will be—when $w \not> n$—

$$((f_1, f_2)^\alpha, f_3)^\beta, \quad ((f_2, f_3)^\alpha, f_1)^\beta, \quad ((f_3, f_1)^\alpha, f_2)^\beta$$

where $\alpha \not< \frac{2w}{3}$. If $w = 3m - 2$, there will be one relation: and if $w = 3m$ there will be two relations between these covariants.

Let us suppose all the quantics to become identical. Then those covariants for which α is odd vanish. Let us suppose that amongst the remaining covariants for which $\alpha \not< \frac{2w}{3}$ a relation exists, say

$$\Sigma\lambda_i ((f, f)^{\alpha_i}, f)^{\beta_i} = 0.$$

Then using Aronhold's operators which may be written for short $\left(f_1 \frac{\partial}{\partial f}\right)$, $\left(f_2 \frac{\partial}{\partial f}\right)$, $\left(f_3 \frac{\partial}{\partial f}\right)$ we obtain

$$\Sigma\lambda_i [((f_1, f_2)^{\alpha_i}, f_3)^{\beta_i} + ((f_2, f_3)^{\alpha_i}, f_1)^{\beta_i} + ((f_3, f_1)^{\alpha_i}, f_2)^{\beta_i}] = 0$$

—since α_i is supposed even.

Hence corresponding to a relation between the covariants for a single quantic, we may deduce a relation between the corresponding covariants of three different quantics. The results obtained then, for a single quantic, are as complete as those from which they were deduced; and no linear relation can exist between the retained covariants.

70. *The covariant*

$$(ab)^\lambda (bc)^\mu (ca)^\nu a_x^{n-\nu-\lambda} b_x^{n-\lambda-\mu} c_x^{n-\mu-\nu}$$

of the binary form $a_x^n = b_x^n = c_x^n$, *can be expressed in terms of covariants whose grade is greater than* λ, *provided that* $\lambda \not> \frac{n}{2}$ *and* $\mu + \nu > \frac{\lambda}{2}$; *unless*

$$\lambda = \mu = \nu = \frac{n}{2}.$$

The verification of the above important theorem is left to the reader; it is really only a restatement of the theorem of § 68.

It should be noticed also that a similar theorem is true when the letters a, b, c do not refer to the same quantic; it is

$$(ab)^\lambda (bc)^\mu (ca)^\nu a_x^{n_1-\nu-\lambda} b_x^{n_2-\lambda-\mu} c_x^{n_3-\mu-\nu} = \Sigma C_{\lambda+1},$$

provided that

$$\mu + \nu > \frac{\lambda}{2}.$$

71. Ex. Prove that the following covariants of $f = a_x^n$ vanish identically:

$$((f, f)^{2\lambda}, f)^{n-1} \quad \text{when } n = 4\lambda,\ 4\lambda - 1, \text{ or } 4\lambda - 2,$$
$$((f, f)^{2\lambda}, f)^{n} \quad \text{when } n = 4\lambda + 1,$$
$$((f, f)^{2\lambda+2}, f)^{n-3} \quad \text{when } n = 4\lambda + 2.$$

And shew that no other covariants of degree three vanish except those included in the form

$$((f, f)^{2a+1}, f)^{\beta}.$$

Stroh.

72. Covariants of degree four. It is the object of the present paragraph to determine the conditions that a covariant of degree four and grade λ $\left(\text{where } \lambda \not> \frac{n}{2}\right)$ may be expressible in the form

$$\Sigma C_{\lambda+1} + (ab)^{\frac{n}{2}} (bc)^{\frac{n}{2}} (ca)^{\frac{n}{2}} . \Psi:$$

the expression $C_{\lambda+1}$ denotes—as before—a covariant of grade not less than $\lambda + 1$, and Ψ being a covariant of degree one can only be the quantic itself. Of course the second term cannot appear if n is odd.

It has already been proved that any covariant

$$C = (ab)^\lambda (bc)^\mu (ca)^\nu a_x^{n-\nu-\lambda} b_x^{n-\lambda-\mu} c_x^{n-\mu-\nu},$$

where $\mu + \nu > \frac{\lambda}{2}$, is of grade greater than λ, unless

$$\lambda = \mu = \nu = \frac{n}{2}.$$

Any covariant obtained by convolution from C is of the same form as C, but the indices of some of the determinant factors are increased; hence any covariant obtained by convolution from C either is of grade greater than λ, or else is the invariant

$$(ab)^{\frac{n}{2}} (bc)^{\frac{n}{2}} (ca)^{\frac{n}{2}}.$$

Any transvectant $(C, F)^\rho$, where F is any binary form, is also of grade greater than λ or else, if $\rho = 0$ and $\lambda = \mu = \nu = \frac{n}{2}$, has a factor

$$(ab)^{\frac{n}{2}} (bc)^{\frac{n}{2}} (ca)^{\frac{n}{2}}.$$

Further, any term of this transvectant differs from the whole transvectant, by a linear function of transvectants

$$(\bar{C}, \bar{F})^{\rho'};$$

where $\bar{C}$ and $\bar{F}$ are obtained by convolution from C and F. Hence any term of a transvectant $(C, F)^\rho$ is equal to

$$\Sigma C_{\lambda+1} + (ab)^{\frac{n}{2}} (bc)^{\frac{n}{2}} (ca)^{\frac{n}{2}} . \Psi,$$

where as before $C_{\lambda+1}$ is used to denote a covariant of grade $\lambda + 1$ at least.

Again, any covariant having a symbolical factor

$$(ab)^\lambda (bc)^\mu (ca)^\nu,$$

where $\mu + \nu > \frac{\lambda}{2}$, is a term of a transvectant $(C, F)^\rho$, and hence may be expressed in the form

$$\Sigma C_{\lambda+1} + (ab)^{\frac{n}{2}} (bc)^{\frac{n}{2}} (ca)^{\frac{n}{2}} . \Psi.$$

Consider now the covariant of degree four where $\nu \not> \lambda$

$$K = (ab)^\lambda (bc)^\mu (cd)^\nu a_x^{n-\lambda} b_x^{n-\lambda-\mu} c_x^{n-\mu-\nu} d_x^{n-\nu}.$$

If $\mu > \frac{\lambda}{2}$, then

$$K = \Sigma C_{\lambda+1} + (ab)^{\frac{n}{2}} (bc)^{\frac{n}{2}} (ca)^{\frac{n}{2}} . \Psi.$$

Otherwise by means of the relation

$$(cd)\, a_x = (ad)\, c_x - (ac)\, d_x$$

we obtain, since $\frac{n}{2} \not< \lambda \not< \nu$,

$$K = \Sigma (-)^i \binom{\nu}{i} L_i,$$

where

$$L_i = (ab)^\lambda (bc)^\mu (ac)^{\nu-i} (ad)^i a_x^{n-\lambda-\nu} b_x^{n-\lambda-\mu} c_x^{n-\mu-\nu+i} d_x^{n-i}.$$

If either

$$\mu + \nu - i > \frac{\lambda}{2},$$

or

$$i > \frac{\lambda}{2},$$

then

$$L_i = \Sigma C_{\lambda+1} + (ab)^{\frac{n}{2}} (bc)^{\frac{n}{2}} (ca)^{\frac{n}{2}} . \Psi.$$

But if

$$\mu + \nu > \lambda,$$

one of the above inequalities must be satisfied; hence in this case

$$K = \Sigma C_{\lambda+1} + (ab)^{\frac{n}{2}} (bc)^{\frac{n}{2}} (ca)^{\frac{n}{2}} . \Psi.$$

Next consider the most general symbolical product K of degree four; it is sufficient to write down its determinant factors, which we take to be

$$(ab)^\lambda (ac)^{\mu_1} (bc)^{\mu_2} (ad)^{\nu_1} (bd)^{\nu_2} (cd)^{\nu_3}.$$

It is supposed here that no index is greater than λ, which is itself not greater than $\frac{n}{2}$.

By means of the identities

$$(cd)\, a_x = (ad)\, c_x - (ac)\, d_x,$$

$$(cd)\, b_x = (bd)\, c_x - (bc)\, d_x,$$

we can express K in terms of covariants L in which either (i) the index of (cd) is zero, or (ii) the indices of both a_x and b_x are zero.

The indices for the covariants L will be denoted by accented letters. In the second case

$$\lambda + \mu_1' + \nu_1' = n, \quad \lambda + \mu_2' + \nu_2' = n,$$

hence

$$\mu_1' + \mu_2' + \nu_1' + \nu_2' = 2n - 2\lambda \nless 2\lambda,$$

consequently either

$$\mu_1' + \mu_2' > \frac{\lambda}{2} \quad \text{or} \quad \nu_1' + \nu_2' > \frac{\lambda}{2}.$$

Therefore

$$K = \Sigma L = \Sigma C_{\lambda+1} + (ab)^{\frac{n}{2}} (bc)^{\frac{n}{2}} (ca)^{\frac{n}{2}} \Psi.$$

In the first case

$$\mu_1' + \mu_2' + \nu_1' + \nu_2' = \mu_1 + \mu_2 + \nu_1 + \nu_2 + \nu_3,$$

and if this sum is greater than λ, the same result is true.

Hence—*When* $\mu_1 + \mu_2 + \nu_1 + \nu_2 + \nu_3 > \lambda$,

$$\lambda \ngtr \frac{n}{2},$$

the covariant of degree four whose determinant factors are

$$(ab)^\lambda (ac)^{\mu_1} (bc)^{\mu_2} (ad)^{\nu_1} (bd)^{\nu_2} (cd)^{\nu_3},$$

may be expressed in the form

$$\Sigma C_{\lambda+1} + (ab)^{\frac{n}{2}} (bc)^{\frac{n}{2}} (ca)^{\frac{n}{2}} \Psi.$$

73. *Any covariant which contains the symbolical factor*

$$(ab)^\lambda (bc)^\mu (cd)^\nu,$$

where $\mu + \nu > \lambda$, *and* $\lambda \ngtr \frac{n}{2}$, *may be expressed in the form*

$$\Sigma C_{\lambda+1} + (ab)^{\frac{n}{2}} (bc)^{\frac{n}{2}} (ca)^{\frac{n}{2}} \Psi.$$

For if Γ be such a covariant, and

$$K = (ab)^\lambda (bc)^\mu (cd)^\nu a_x^{n-\lambda} b_x^{n-\lambda-\mu} c_x^{n-\mu-\nu} d_x^{n-\nu};$$

then

$$\Gamma = (K, \Phi)^\rho + \Sigma (\bar{K}, \bar{\Phi})^{\rho'}.$$

Each term of this sum has just been proved to be of the required form.

74. *Any covariant of the covariant*

$$k_x^{2n-2r} \equiv (ab)^r a_x^{n-r} b_x^{n-r}, \left(r \not> \frac{n}{2}\right)$$

of the binary quantic $f = a_x^n = b_x^n = \dots$, *may be expressed in the form*

$$\Sigma C_{r+1} + (ab)^{\frac{n}{2}} (bc)^{\frac{n}{2}} (ca)^{\frac{n}{2}} \Psi,$$

where C_{r+1} *represents a covariant of grade not less than* $r+1$.

To prove this we observe that any form obtained by convolution from a product

$$(ab)^r a_x^{n-r} b_x^{n-r} (cd)^r c_x^{n-r} d_x^{n-r} \dots$$

is either of grade greater than r or else has a factor of the form

$$(ab)^r (bc)^\rho (cd)^r.$$

In the latter case by § 73 this covariant may be expressed in the form

$$\Sigma C_{r+1} + (ab)^{\frac{n}{2}} (bc)^{\frac{n}{2}} (ca)^{\frac{n}{2}} . \Psi.$$

Now any covariant of k_x^{2n-2r} may be expressed in terms of transvectants of the form

$$((k, k)^\sigma, \Phi)^\tau.$$

But the transvectant $(k, k)^\sigma$ is a linear function of covariants of f obtained by convolution from

$$(ab)^r a_x^{n-r} b_x^{n-r} . (cd)^r c_x^{n-r} d_x^{n-r}.$$

Hence the theorem is true for covariants of the second degree; and therefore for all covariants of k_x^{2n-2r}.

75. It is well to notice that nowhere in the last three paragraphs has it been assumed that two different symbolical letters refer to the same quantic. The theorems are thus true when some of the letters refer to different quantics. For example the theorem of § 74 is true for covariants of

$$k_x^{2n-2r} = (ab)^r a_x^{n-r} b_x^{n-r}$$

when the quantics a_x^n, b_x^n are different.

76. The discussion of covariants of degree four may be carried on a step further.

Thus if λ is even, and

$$\mu + \nu = \lambda \not> \frac{n}{2},$$

then the covariant

$$K=(ab)^\lambda (bc)^\mu (cd)^\nu a_x^{n-\lambda} b_x^{n-\lambda-\mu} c_x^{n-\mu-\nu} d_x^{n-\nu}$$

differs from

$$\frac{\binom{\nu}{\frac{\lambda}{2}}}{\binom{\lambda}{\frac{\lambda}{2}}} (ab)^\lambda a_x^{n-\lambda} b_x^{n-\lambda} . (cd)^\lambda c_x^{n-\lambda} d_x^{n-\lambda},$$

by an expression of the form

$$\Sigma C_{\lambda+1}.$$

To prove this we notice that the index of b_x in K is $n-\lambda-\mu$, and this is not less than ν; for by the inequalities above we see that

$$\lambda+\mu+\nu \ngtr n;$$

hence we may use the identity

$$(cd)\, b_x = (bd)\, c_x - (bc)\, d_x$$

to obtain

$$K = \Sigma(-1)^i \binom{\nu}{i} L_i$$

where

$$L_i = (ab)^\lambda (bc)^{\mu+\nu-i} (bd)^i a_x^{n-\lambda} b_x^{n-\lambda-\mu-\nu} c_x^{n-\mu-\nu+i} d_x^{n-i}.$$

Now if

$$i > \frac{\lambda}{2}$$

or

$$\mu+\nu-i > \frac{\lambda}{2},$$

then

$$L_i = \Sigma C_{\lambda+1}.$$

(The weight of K is $2\lambda \ngtr n$; hence no term with the factor

$$(ab)^{\frac{n}{2}} (bc)^{\frac{n}{2}} (ca)^{\frac{n}{2}}$$

can appear.)

Then the only term we need consider is

$$L_{\frac{\lambda}{2}} = (ab)^\lambda (bc)^{\frac{\lambda}{2}} (bd)^{\frac{\lambda}{2}} a_x^{n-\lambda} b_x^{n-2\lambda} c_x^{n-\frac{\lambda}{2}} d_x^{n-\frac{\lambda}{2}}.$$

Again, let K' be that particular covariant of the form K for which $\mu=0$, $\nu=\lambda$. Then

$$K' = \Sigma(-1)^i \binom{\lambda}{i} L_i$$

$$= (-1)^{\frac{\lambda}{2}} \binom{\lambda}{\frac{\lambda}{2}} L_{\frac{\lambda}{2}} + \Sigma C_{\lambda+1}.$$

Hence

$$\binom{\lambda}{\frac{\lambda}{2}} K = \binom{\nu}{\frac{\lambda}{2}} K' + \Sigma C_{\lambda+1}.$$

In just the same way it may be proved that if λ is odd, and

$$\text{(i)} \quad \mu+\nu=\lambda \ngtr \frac{n}{2},$$

then

$$K = \Sigma C_{\lambda+1};$$

or

$$\text{(ii)} \quad \mu+\nu+1=\lambda \ngtr \frac{n}{2},$$

then

$$K = \frac{\binom{\nu}{\frac{\lambda-1}{2}}}{\binom{\lambda}{\frac{\lambda-1}{2}}} (ab)^{\lambda} a_x^{n-\lambda} b_x^{n-\lambda} . (cd)^{\lambda-1} c_x^{n-\lambda+1} d_x^{n-\lambda+1} + \Sigma C_{\lambda+1}.$$

The results of this paragraph may be stated as follows:

(i) *If λ be even and $\mu+\nu=\lambda$, then the covariant*

$$(ab)^{\lambda} (bc)^{\mu} (cd)^{\nu} a_x^{n-\lambda} b_x^{n-\lambda-\mu} c_x^{n-\mu-\nu} d_x^{n-\nu}$$

can be expressed as a sum of a reducible covariant and covariants of grade greater than λ.

(ii) *If λ be odd and $\mu+\nu=\lambda$, then the above covariant can be expressed as a sum of covariants of grade greater than λ.*

If λ be odd and $\mu+\nu=\lambda-1$, the covariant can be expressed as a sum of a reducible covariant and covariants of grade greater than λ.

Ex. (i). Any covariant which contains the factor

$$(ab)^{\lambda} (ac)^{\mu_1} (bc)^{\mu_2} (ad)^{\nu_1} (bd)^{\nu_2} (cd)^{\nu_3},$$

where

$$\mu_1+\mu_2+\nu_1+\nu_2+\nu_3 > \lambda \text{ and } \lambda < \frac{n}{2},$$

can be expressed linearly in terms of covariants whose grade is greater than λ.

Ex. (ii). Prove that if a covariant C of degree 5 has the factor written down in the last question, for which

$$\mu_1+\mu_2+\nu_1+\nu_2+\nu_3=\lambda, \text{ and } \lambda \ngtr \frac{n}{2},$$

then $C=\Sigma C_{\lambda+1}+$ reducible terms.

77. Jacobians. The first transvectant of two binary forms f_x^m, ϕ_x^n is called their Jacobian. It is, in fact, equal to

$$\frac{1}{mn} \frac{\partial (f_x^m, \phi_x^n)}{\partial (x_1, x_2)}.$$

The following properties of Jacobians are important.

(i) *If f, ϕ, ψ be three binary forms, each of order greater than unity, the Jacobian of the Jacobian of f and ϕ with ψ is reducible.*

Let $$f = a_x^m, \quad \phi = b_x^n, \quad \psi = c_x^p$$

$$(f, \phi) = (ab)\, a_x^{m-1} b_x^{n-1}.$$

Polarize once with respect to y, the result is

$$(f, \phi)_y = \frac{m-1}{m+n-2}(ab)\, a_x^{m-2} b_x^{n-1} a_y + \frac{n-1}{m+n-2}(ab)\, a_x^{m-1} b_x^{n-2} b_y.$$

Hence

$$(m+n-2)((f, \phi), \psi)$$

$$= (m-1)(ab)(ac)\, a_x^{m-2} b_x^{n-1} c_x^{p-1} + (n-1)(ab)(bc)\, a_x^{m-1} b_x^{n-2} c_x^{p-1}.$$

But $$2(ab)(ac)\, b_x c_x = (ab)^2 c_x^2 + (ac)^2 b_x^2 - (bc)^2 a_x^2$$

and $$2(ab)(bc)\, a_x c_x = -(ab)^2 c_x^2 - (bc)^2 a_x^2 + (ac)^2 b_x^2\ *.$$

Therefore

$$((f, \phi), \psi)$$

$$= \tfrac{1}{2} a_x^{m-2} b_x^{n-2} c_x^{p-2} \left\{ \frac{m-n}{m+n-2}(ab)^2 c_x^2 + (ac)^2 b_x^2 - (bc)^2 a_x^2 \right\}$$

$$= \frac{m-n}{2(m+n-2)}(f, \phi)^2 . \psi + \tfrac{1}{2}(f, \psi)^2 . \phi - \tfrac{1}{2}(\phi, \psi)^2 . f \ldots \text{(XIX)}.$$

(ii) *The product of two Jacobians may be expressed as a sum of products of covariants, there being at least three covariants in each product; provided the forms of which the Jacobians are taken are all of order greater than unity.*

To prove this, we first establish a useful identity between symbolical forms.

Consider the determinant

$$\begin{vmatrix} a_1^2 & a_1 a_2 & a_2^2 \\ b_1^2 & b_1 b_2 & b_2^2 \\ c_1^2 & c_1 c_2 & c_2^2 \end{vmatrix};$$

it vanishes if $\frac{a_1}{b_1} = \frac{a_2}{b_2}$, hence (ab) is a factor; similarly (bc), (ca) are factors. There can only be besides a numerical factor, which may

* See Chapter I. § 22.

be determined by considering the coefficient of $a_1^2 b_1 b_2 c_2^2$. The determinant is therefore equal to

$$-(ab)(bc)(ca).$$

Hence $$2.(ab)(bc)(ca).(de)(ef)(fd)$$

$$= \begin{vmatrix} a_1^2 & a_1 a_2 & a_2^2 \\ b_1^2 & b_1 b_2 & b_2^2 \\ c_1^2 & c_1 c_2 & c_2^2 \end{vmatrix} . \begin{vmatrix} d_2^2 & -2d_1 d_2 & d_1^2 \\ e_2^2 & -2e_1 e_2 & e_1^2 \\ f_2^2 & -2f_1 f_2 & f_1^2 \end{vmatrix}$$

$$= \begin{vmatrix} (ad)^2 & (ae)^2 & (af)^2 \\ (bd)^2 & (be)^2 & (bf)^2 \\ (cd)^2 & (ce)^2 & (cf)^2 \end{vmatrix} \quad \ldots\ldots\ldots\ldots\ldots(\text{XX}).$$

In this identity let us put

$$c_1 = -x_2,\ c_2 = x_1;\quad f_1 = -x_2,\ f_2 = x_1.$$

Then

$$2(ab)\,a_x b_x.(de)\,d_x e_x = \begin{vmatrix} (ad)^2 & (ae)^2 & a_x^2 \\ (bd)^2 & (be)^2 & b_x^2 \\ d_x^2 & e_x^2 & 0 \end{vmatrix}.$$

Consider now two Jacobians (f, ϕ), (ψ, χ); where

$$f = a_x^m, \quad \phi = b_x^n,$$

$$\psi = d_x^p, \quad \chi = e_x^q.$$

Then

$$(f, \phi).(\psi, \chi) = (ab)\,a_x^{m-1} b_x^{n-1}.(de)\,d_x^{p-1} e_x^{q-1}$$

$$= \tfrac{1}{2} a_x^{m-2} b_x^{n-2} d_x^{p-2} e_x^{q-2} \begin{vmatrix} (ad)^2 & (ae)^2 & a_x^2 \\ (bd)^2 & (be)^2 & b_x^2 \\ d_x^2 & e_x^2 & 0 \end{vmatrix}$$

$$= -\tfrac{1}{2}(f, \psi)^2 \phi\chi + \tfrac{1}{2}(f, \chi)^2 \phi\psi$$

$$+ \tfrac{1}{2}(\phi, \psi)^2 f\chi - \tfrac{1}{2}(\phi, \chi)^2 f\psi \quad \ldots\ldots\ldots\ldots\ldots(\text{XXI}).$$

For the sake of generality the forms f, ϕ, ψ, χ have been supposed to be all different. The theorem is still true if two are equal to one another. In particular, it is true for the square of a Jacobian.

Another symbolical identity of interest is obtained thus: Form by means of the ordinary law of multiplication the product of the following vanishing determinants:

$$\begin{vmatrix} a_1^2 & a_1a_2 & a_2^2 & 0 \\ b_1^2 & b_1b_2 & b_2^2 & 0 \\ c_1^2 & c_1c_2 & c_2^2 & 0 \\ d_1^2 & d_1d_2 & d_2^2 & 0 \end{vmatrix} \cdot \begin{vmatrix} e_2^2 & -2e_1e_2 & e_1^2 & 0 \\ f_2^2 & -2f_1f_2 & f_1^2 & 0 \\ g_2^2 & -2g_1g_2 & g_1^2 & 0 \\ h_2^2 & -2h_1h_2 & h_1^2 & 0 \end{vmatrix}$$

$$= \begin{vmatrix} (ae)^2 & (af)^2 & (ag)^2 & (ah)^2 \\ (be)^2 & (bf)^2 & (bg)^2 & (bh)^2 \\ (ce)^2 & (cf)^2 & (cg)^2 & (ch)^2 \\ (de)^2 & (df)^2 & (dg)^2 & (dh)^2 \end{vmatrix} = 0 \quad \text{............(XXII)}.$$

Hence if
$$f=a_x^m, \quad \phi=b_x^n, \quad \psi=c_x^p, \quad \chi=d_x^q$$
$$F=e_x^M, \quad \Phi=f_x^N, \quad \Psi=g_x^P, \quad X=h_x^Q$$

$$\begin{vmatrix} (f, F)^2 & (f, \Phi)^2 & (f, \Psi)^2 & (f, X)^2 \\ (\phi, F)^2 & (\phi, \Phi)^2 & (\phi, \Psi)^2 & (\phi, X)^2 \\ (\psi, F)^2 & (\psi, \Phi)^2 & (\psi, \Psi)^2 & (\psi, X)^2 \\ (\chi, F)^2 & (\chi, \Phi)^2 & (\chi, \Psi)^2 & (\chi, X)^2 \end{vmatrix} = 0.$$

Here, as before, the forms are not necessarily all different.

78. The expression for the product of two Jacobians may also be obtained as follows.

We have

$$(f, \phi) \times (\psi, \chi) = (ab)\, a_x^{m-1} b_x^{n-1} . (de)\, d_x^{p-1} e_x^{q-1}$$
$$= (ab)(ae)\, a_x^{m-2} b_x^{n-1} e_x^{q-1} . \psi - (ab)(ad)\, a_x^{m-2} b_x^{n-1} d_x^{p-1} . \chi.$$

and, by means of the identities of the type

$$(ab)(ae)\, b_x e_x = (ab)^2 e_x^2 + (ae)^2 b_x^2 - (be)^2 a_x^2,$$

the right-hand side becomes

$$\tfrac{1}{2}(f, \chi)^2 \phi\psi + \tfrac{1}{2}(\phi, \psi)^2 f\chi - \tfrac{1}{2}(f, \psi)^2 \phi\chi - \tfrac{1}{2}(\phi, \chi)^2 f\psi$$

as before.

Thus if J be the Jacobian of f and ϕ

$$-2J^2 = (f, f)^2 \phi^2 + (\phi, \phi)^2 f^2 - 2(f, \phi)^2 f\phi.$$

79. Copied forms. If in the symbolical expression of a covariant Π of a binary form F_x^p, the symbolical letters are taken to refer to another binary form ϕ_x^m where $m \geqslant p$, and Π is multiplied by

$$a_x^{m-p} b_x^{m-p} \ldots$$

—where $a, b, \ldots$ are the letters occurring in Π—then the resulting form Π' is called a *copied form.* The original form Π is called the *model form.*

Ex. The Hessian of any binary form $\phi = a_x^m = b_x^m$ is

$$(ab)^2 a_x^{m-2} b_x^{m-2}.$$

It is formed on the model of the Hessian of the quadratic.

The use which we are going to make of the idea of copied forms is for the case when the orders p and m of the fundamental quantics of the two systems are the same. The form ϕ will be taken to be a covariant of degree two and order m of a binary form f_x^n: then

$$\phi = (f, f)^{2\sigma} = (ab)^{2\sigma} a_x^{n-2\sigma} b_x^{n-2\sigma} = \phi_x^{2n-4\sigma}, \quad 2n - 4\sigma = m.$$

On the model of the complete system for the general binary form F_x^m, we may construct a complete system for ϕ. If when this is done the symbolical letters $\phi_1, \phi_2, \ldots$ which refer to ϕ are replaced by the symbolical letters $a, b, \ldots$ which refer to f, each covariant of ϕ will become a covariant of f, which consists of a sum of symbolical products instead of a single term. These separate products may be arranged in a series of adjacent terms, the difference between any two of which contains a factor of the form $(ab)^{2\sigma+1}$. Thus if we reject covariants of grade greater than 2σ, any one of the terms given by a covariant of ϕ may be taken to represent this covariant. Before proceeding to a rigid proof of this statement, let us consider an example.

The covariant

$$\phi = (ab)^4 a_x^2 b_x^2$$

of the sextic

$$a_x^6 = b_x^6 = \ldots$$

is a quartic. Let us consider the Hessian of ϕ,

$$(\phi_1 \phi_2)^2 \phi^2_{1_x} \phi^2_{2_x} = (\phi^4_{1_x}, \phi^4_{2_x})^2 = ((ab)^4 a_x^2 b_x^2, (cd)^4 c_x^2 d_x^2)^2.$$

It consists of a linear function of ten covariants of the following types:

$$(ab)^4 (cd)^4 (ac)^2 b_x^2 d_x^2,$$

$$(ab)^4 (cd)^4 (ac)(ad) b_x^2 c_x d_x,$$

$$(ab)^4 (cd)^4 (ac)(bd) a_x b_x c_x d_x.$$

Any one of these may be taken for the copied form, for each differs from the whole transvectant by transvectants of forms obtained from

$$(ab)^4 a_x^2 b_x^2, \quad (cd)^4 c_x^2 d_x^2$$

by convolution. Hence when forms of grade higher than 4 are neglected, the whole transvectant and each of its terms are equivalent.

80. Consider any symbolical product P, the symbols of which refer to

$$\phi = (ab)^{2\sigma} a_x^{n-2\sigma} b_x^{n-2\sigma},$$

in this product let us replace any particular letter $\phi^{(1)}$ by a new variable y; (we replace ϕ_{12} by y_1 and ϕ_{11} by $-y_2$). Let $P_{y^{2n-4\sigma}}$ be the resulting expression, then

$$\begin{aligned} P &= (P_{y^{2n-4\sigma}}, (\phi_{1y})^{2n-4\sigma})^{2n-4\sigma} \\ &= (P_{y^{2n-4\sigma}}, (ab)^{2\sigma} a_y^{n-2\sigma} b_y^{n-2\sigma})^{2n-4\sigma}. \end{aligned}$$

Any term of this transvectant differs from the whole transvectant by terms involving a factor $(ab)^{2\sigma+1}$: for since y is absent when the transvectant is expanded, the only kind of adjacent terms are of the form

$$(\alpha a)(\beta b) M, \quad (\alpha b)(\beta a) M \qquad \text{(see § 50).}$$

The above argument is not affected if we suppose that symbolical letters are present in P which do not refer to ϕ. Hence we may replace each of the letters which refer to ϕ in turn by letters which refer to f; and in doing this we may at each stage choose any one term to represent the whole expression P, provided that those terms which involve $(ab)^{2\sigma+1}$ are rejected. Thus taking the quartic invariant $(ab)^2 (bc)^2 (ca)^2$ for model, we obtain the copied invariant of

$$\phi = (ab)^4 a_x^2 b_x^2,$$

viz.

$$(\phi_1 \phi_2)^2 (\phi_2 \phi_3)^2 (\phi_3 \phi_1)^2.$$

Any one of the sextic invariants

$$(ab)^4 (cd)^4 (ef)^4 (ac)^2 (df)^2 (eb)^2,$$

$$(ab)^4 (cd)^4 (ef)^4 (ad)^2 (ce)^2 (fb)^2,$$

$$(ab)^4 (cd)^4 (ef)^4 (ac)(ad)(ce)(de)(fb)^2$$

differs from the invariant of ϕ, by terms involving the factor $(ab)^6$.

Conversely in any covariant of a binary form $a_x^n = b_x^n = \ldots$, which has a symbolical factor $(ab)^{2\sigma}$, we may replace the letters a, b wherever they occur in the symbolical product by a single letter ϕ, and remove the factor $(ab)^{2\sigma}$ altogether: on the understanding that forms of grade greater than 2σ are being rejected, and that ϕ refers to the form $(ab)^{2\sigma} a_x^{n-2\sigma} b_x^{n-2\sigma}$. The reader will have no difficulty in verifying this statement.

81. Generalized Transvectants. Consider a product

$$P = a_x^m b_y^n c_z^p$$

of any binary forms,

$$f_1(x) = a_x^m, \quad f_2(y) = b_y^n, \quad f_3(z) = c_z^p.$$

The result of operating on P with

$$\frac{(m-\lambda-\mu)!}{m!} \cdot \frac{(n-\lambda-\nu)!}{n!} \cdot \frac{(p-\mu-\nu)!}{p!} \Omega^\lambda{}_{xy} \Omega^\mu{}_{xz} \Omega^\nu{}_{zy}$$

where

$$\Omega_{xy} = \frac{\partial^2}{\partial x_1 \partial y_2} - \frac{\partial^2}{\partial x_2 \partial y_1},$$

after operation y and z being replaced by x, is

$$(ab)^\lambda (ac)^\mu (cb)^\nu a_x^{m-\lambda-\mu} b_x^{n-\lambda-\nu} c_x^{p-\mu-\nu}.$$

This we define to be a generalized transvectant.

Instead of taking forms each having a single symbolical letter, we may construct generalized transvectants of forms each of which has two or more symbolical letters. Thus we may replace a_x^m by

$$d_x^q e_x^{m-q}.$$

The generalized transvectant may then be expanded as a linear function of certain symbolical products. Just as in the case of ordinary transvectants, any term of a generalized transvectant differs from the whole transvectant by lower transvectants of forms obtained from the original forms by convolution. We leave the verification of this statement to the reader.

Further it is evident that any symbolical product may be regarded as a generalized transvectant. A copied form is then merely the same generalized transvectant with a new form taken for ground form.

The theorem of § 80—that any single term of a copied form, when a covariant of the second degree is the new ground form, may be taken to represent the whole form, provided that forms of higher grade than that of the fundamental ground form are to be neglected—is merely a particular case of that just enunciated.

82. Hyperdeterminants. When the forms of a generalized transvectant are all the same, it will be noticed that the transvectant is entirely given by the differential operator. Thus a covariant of the binary form $f=a_x^n$, is completely defined by the operator

$$\Omega^\lambda_{xy}\,\Omega^\mu_{xz}\,\Omega^\nu_{zy}.$$

Cayley used the notation

$$\overline{12}^\lambda\,\overline{13}^\mu\,\overline{32}^\nu$$

to define such a covariant. These symbolical forms are called hyperdeterminants. Cayley introduced his calculus of hyperdeterminants some years before the symbolical notation was invented by Aronhold.

The hyperdeterminant notation was introduced for a single binary form merely for convenience. It is evident that it may be used perfectly well for covariants of two or more different forms.

It is interesting to notice that the letters of a symbolical product may be regarded as differential operators. Thus if

$$\alpha_1=\frac{\partial}{\partial\xi_1},\qquad \beta_1=\frac{\partial}{\partial\eta_1},\qquad \gamma_1=\frac{\partial}{\partial\zeta_1},$$

$$\alpha_2=\frac{\partial}{\partial\xi_2},\qquad \beta_2=\frac{\partial}{\partial\eta_2},\qquad \gamma_2=\frac{\partial}{\partial\zeta_2},$$

then $$(\alpha\beta)^\lambda(\alpha\gamma)^\mu(\gamma\beta)^\nu$$

operating on the product of

$$f_1(\xi)=a_\xi^m,\quad f_2(\eta)=b_\eta^n,\quad f_3(\zeta)=c_\zeta^p$$

produces the covariant

$$(ab)^\lambda(ac)^\mu(cb)^\nu\,a_x^{m-\lambda-\mu}\,b_x^{n-\lambda-\nu}\,c_x^{p-\mu-\nu}$$

multiplied by

$$\frac{m!}{(m-\lambda-\mu)!}\,\frac{n!}{(n-\lambda-\nu)!}\,\frac{p!}{(p-\mu-\nu)!},$$

provided that after operation ξ, η, ζ are each replaced by x.

Further the operator

$$(\alpha\beta)^\lambda(\alpha\gamma)^\mu(\gamma\beta)^\nu\,\alpha_x^{m-\lambda-\mu}\,\beta_x^{n-\lambda-\nu}\,\gamma_x^{p-\mu-\nu}$$

acting on the same product produces the same covariant multiplied by

$$m!\,n!\,p!.$$

In this case ξ, η, ζ all disappear after operation, so there is no question of replacing them by x.

CHAPTER V.

ELEMENTARY COMPLETE SYSTEMS.

83. Complete Systems of irreducible covariants. We shall devote this chapter to a detailed discussion of the invariants and covariants of single binary forms of the first four orders; in particular, we shall obtain what are known as the complete systems of covariants for such forms. It has been observed, in fact, that the symbolical notation enables us to construct an infinite number of covariants of any form f, but, as was first proved by Gordan, all these are rational integral functions of a finite number of covariants of f; this finite number is said to constitute the complete system of irreducible concomitants, or, more briefly, the complete system of concomitants of the form. The general proof of Gordan's Theorem will be given in the next chapter. For the present we shall content ourselves with explaining easier methods of obtaining the complete systems in the simpler cases, and proving of course that such systems are actually complete.

Inasmuch as every covariant can be expressed as an aggregate of symbolical products, we need only retain such as consist of one product in seeking for the complete system.

84. Linear Form. The discussion of a single linear form

$$f = a_x = b_x = \text{etc.}$$

presents no difficulty.

For a symbolical product either contains a factor of the type (ab) or it does not. If it does so it is zero because

$$(ab) = 0$$

and if it does not it is simply a power of f.

Hence every covariant of a linear form is a power of the form itself, or in other words, the form constitutes the complete system.

Cor. The same argument applies to any number of linear forms, for every symbolical product is a rational integral function of invariants of the type (ab) and covariants of the type a_x. Hence the complete system for n linear forms consists of the n forms themselves and the $\frac{1}{2}n(n-1)$ non-vanishing invariants of the type (ab).

This result has already been established in § 37 where it forms the lemma preliminary to the proof of the fundamental theorem.

85. Quadratic Form. Suppose the form is

$$f = a_x^2 = b_x^2 = c_x^2 = \text{etc.}$$

Then if a symbolical product contain no factor of the type (ab) it is a power of f; if on the contrary it contains such a factor (ab) it can be transformed so as to contain $(ab)^2$ (§ 60), which is an invariant. Thus every invariant and covariant except

$$a_x^2, \quad (ab)^2$$

can be expressed in terms of covariants of lower degree, hence by continued reduction we infer that every such form is a rational integral function of

$$f, (f, f)^2 = \Delta,$$

the latter being the discriminant of the quadratic. In other words, the form and its discriminant constitute the complete system.

Ex. Prove that

$$(ab)(ac)(bd)(ce)\, d_x e_x = \tfrac{1}{4}\Delta^2 f.$$

(By interchanging a and b put the factor $(ab)^2$ in evidence.)

86. Before proceeding to the discussion of the cubic and the quartic we shall explain some general principles relating to the formation of the irreducible covariants of any given degree of a binary form.

Suppose that the form in question is

$$f = a_x^n = b_x^n = c_x^n = \text{etc.}$$

then the only irreducible covariant of degree one is f.

Next, the only covariants of degree two are those of the type

$$(ab)^r a_x^{n-r} b_x^{n-r}; \quad r = 0, 1, 2, \dots n.$$

If r be odd this covariant vanishes, and if r be zero it is reducible since it is equal to f^2; the remaining forms corresponding to even values of r constitute the complete set of irreducible covariants of the second degree.

Now assuming a knowledge of all the irreducible covariants of degree less than m we shall shew how to find the irreducible covariants of degree m.

Suppose that the given irreducible forms are

$$f, \phi_1, \phi_2, \dots \phi_r,$$

then any covariant of degree less than m is a rational integral function of f and the ϕ's. Now a covariant of degree m is an aggregate of symbolical products containing m letters; let C_m be one of the products and k one of the symbols involved, then C_m is a term in a transvectant

$$(C_{m-1}, k_x^n)^\rho,$$

where C_{m-1} is a product containing only $m-1$ letters, that is, it is a covariant of degree $m-1$.

Thus $$C_m = (C_{m-1}, f)^\rho + \Sigma\, (\overline{C_{m-1}}, f)^{\rho'}, \; \rho' < \rho$$

and $\overline{C_{m-1}}$ is derived from C_{m-1} by convolution.

But C_{m-1}, $\overline{C_{m-1}}$ being covariants of degree $m-1$ are rational integral functions of the forms

$$f, \phi_1, \phi_2, \dots \phi_r,$$

i.e. they are aggregates of terms of the types

$$U_{m-1} = f^a \phi_1^{a_1} \dots \phi_r^{a_r}$$

of degree $m-1$.

Consequently C_m is a sum of transvectants of the form

$$(U_{m-1}, f)^\rho,$$

where of course $\rho \not> n$. Since this is true for every separate term in a covariant of degree m it is true for the whole; or, in other words, every covariant of degree m is expressible in terms of transvectants

$$(U_{m-1}, f)^\rho,$$

where U_{m-1} is a product of the form

$$f^{a}\,\phi_1^{a_1}\,\phi_2^{a_2}\ldots\phi_r^{a_r}$$

and is of degree $m-1$.

Hence to find all the *irreducible* covariants of degree m we have to write down all transvectants of the form

$$(U_{m-1}, f)^{\rho}$$

and reduce as many of them as possible. The remaining ones are the irreducible forms of degree m, for any covariant of degree m can be expressed in terms of them and covariants of lower degree.

For example, in the case of a binary quintic the irreducible forms of degrees one and two are

$$f,\ H=(ab)^2 a_x^3 b_x^3,\quad i=(ab)^4 a_x b_x.$$

The only products of powers of these of degree two are f^2, H and i, so that all the irreducible covariants of degree three are included in

$$(f^2, f)^{\rho},\ (H, f)^{\rho},\ (i, f)^{\rho},$$

where $\rho \not> 5$ for the first two transvectants and $\rho \not> 2$ for the third, since i is a quadratic.

87. Let us now return to the transvectants

$$(U_{m-1}, f)^{\rho}$$

which we have to reduce as far as possible. That many of them are reducible follows from the following principles.

Suppose that $$U_{m-1} = VW$$

where V and W are likewise products of powers of f, ϕ_1, ϕ_2, ... ϕ_r, but of smaller degree than U_{m-1}, and further suppose the order of W is not less than ρ.

Then if T_1 be any term belonging to V and T_2 any term of the transvectant $(W, f)^{\rho}$, which is a possible transvectant because the order of W is at least equal to ρ, T_1T_2 will be a term in the transvectant

$$(U_{m-1}, f)^{\rho}.$$

Hence

$$(U_{m-1}, f)^{\rho} = T_1T_2 + \Sigma\,(\overline{U_{m-1}}, f)^{\rho'},\ \rho' < \rho$$

and T_1, T_2 being both covariants of degree less than m are expressible in terms of f and the ϕ's.

Now in discussing the reducibility of the transvectants

$$(U_{m-1}, f)^\rho$$

let us consider them in the order of their indices, *e.g.* we examine all those of index one before we proceed to any of index two, and so on.

Then since $\overline{U_{m-1}}$ is a covariant of degree $m-1$ it is an aggregate of products of the type U_{m-1}, and since $\rho' < \rho$ it follows from the equation

$$(U_{m-1}, f)^\rho = T_1 T_2 + \Sigma\,(\overline{U_{m-1}}, f)^{\rho'}$$

that the transvectant on the left is completely expressible in terms of covariants of less degree and transvectants previously considered, or to put the matter briefly, it is reducible for it certainly cannot give rise to a new irreducible form.

Hence the irreducible covariants of degree m can only arise from such transvectants

$$(U_{m-1}, f)^\rho$$

for which U_{m-1} has not a factor of order greater than ρ. Thus in the case of the quintic no transvectant of the type

$$(f^2, f)^\rho$$

is irreducible because $\rho \not> 5$ and the term f^2 contains a factor f whose order is 5.

In the general case if U_{m-1} possess a factor W whose order is not less than n, then the product U_{m-1} can give rise to irreducible covariants for no value of ρ; it may therefore be neglected entirely in the search for irreducible covariants.

As an application of this remark we note that if the order of one of the ϕ's, say ϕ_p, be at least equal to n, then we need not consider the product of this form with any others.

Finally if ϕ_q be an invariant we may leave it out of account in forming the transvectants, because it would occur as a factor in each transvectant in which it appeared and so the transvectant would be reducible.

88. Irreducible system for the binary cubic. After these preliminary explanations the deduction of the complete system of a cubic presents little difficulty.

Let the form be

$$f = a_x^3 = b_x^3 = c_x^3 = \text{etc.}$$

then the only irreducible form of degree two is

$$H = (ab)^2 a_x b_x = h_x^2 = h'^2_x.$$

To find the irreducible forms of degree three we note that the product f^2 is negligeable, so the only possible irreducible forms are

$$(H, f), \quad (H, f)^2.$$

Now

$$(H, f) = (ab)^2 (ac) b_x c_x^2 = -t,$$

where t is an irreducible covariant, and

$$\begin{aligned}(H, f)^2 &= \{(ab)^2 a_x b_x, c_x^3\}^2 = (ab)^2 (ac)(bc) c_x \\ &= -(bc)(ca)(ab)\{(ab) c_x\} \\ &= -\tfrac{1}{3}(bc)(ca)(ab)\{(ab) c_x + (bc) a_x + (ab) c_x\},\end{aligned}$$

as we see by interchanging a, c and b, c and adding the results.

Hence $(H, f)^2 = 0$ and the only new irreducible form is t.

The products of degree three are

$$f^3, \ Hf, \ t,$$

of which the first two may be neglected; hence the irreducible forms of degree four are included in

$$(t, f), \quad (t, f)^2, \quad (t, f)^3.$$

Of these (t, f) is the Jacobian of a Jacobian and hence can be expressed in terms of forms of lower degree, § 77.

In fact, to give the actual expression, we have

$$\{(H, f), \psi\} = \tfrac{1}{2}(f, \psi)^2 H - \tfrac{1}{2}(H, \psi)^2 f,$$

whatever ψ may be, since $(H, f)^2 = 0$.

Therefore $$-(t, f) = \tfrac{1}{2}(f, f)^2 H - \tfrac{1}{2}(H, f)^2 f,$$

or $$(t, f) = -\tfrac{1}{2} H^2.$$

Next

$$\begin{aligned}(t, f)^2 &= -\{(ha) h_x a_x^2, b_x^3\}^2 \\ &= -(ha)(hb)(ab) a_x b_x + \lambda \{(ha)^2 a_x, b_x^3\},\end{aligned}$$

since $$(ha)(hb)(ab) a_x b_x$$

is one term in the transvectant.

Now this term vanishes, and since

$$(ha)^2 a_x = (H, f)^2 = 0$$

we have

$$(t, f)^2 = 0.$$

Finally

$$(t, f)^3 = -\{(ab)^2 (ac) b_x c_x^2, d_x^3\}^3 = -(ab)^2 (ac)(bd)(cd)^2 = -\Delta,$$

so that $(f, t)^3 = \Delta$ an invariant which proves to be irreducible.

The products formed from f, H, t which are of degree four are

$$f^4,\ f^2H,\ ft,\ H^2,$$

and all except H^2 may be rejected because t and f are both of order three.

Further $(H^2, f)^\rho$ is reducible unless $\rho > 2$ because H^2 contains the factor H whose order is two.

Hence the only possible irreducible form of degree five is $(H^2, f)^3$.

But

$$\begin{aligned}(H^2, f)^3 &= (h_x^2 h'^2_x, a_x^3)^3 = (ha)^2 (h'a) h'_x \\ &= -\{(ha)^2 a_x, h'^2_x\} = 0 \text{ since } (ha)^2 a_x = 0,\end{aligned}$$

hence there are no irreducible forms of degree five, and in fact it is easy to see that there are no more irreducible forms. For if there were, the one coming next in ascending degree would be of the form

$$(f^\alpha H^\beta t^\gamma, f)^\rho.$$

The only products that can lead to irreducible forms are f, t, H and H^2, because when $\beta > 2$, H^β involves the factor H^2 whose order is greater than three; but the transvectants arising from each of these products have already been considered, hence there are no more irreducible forms; in other words, every invariant or covariant of the cubic is a rational integral function of f, H, t and Δ.

89. Irreducible system for the quartic. If the form be

$$f = a_x^4 = b_x^4 = c_x^4,$$

then the irreducible forms of degree two are

$$H = (ab)^2 a_x^2 b_x^2 \text{ and } i = (ab)^4,$$

H being of order four and i an invariant.

The only product of degree two that we need consider is H, and hence the irreducible forms of degree three are included in $(H,f)^1$, $(H,f)^2$, $(H,f)^3$, $(H,f)^4$.

Now

$$(H,f)^1 = (ab)^2 (ac)\, a_x b_x^2 c_x^3 = -t$$

where t is an irreducible covariant.

$$(H,f)^2 = \{(ab)^2 a_x^2 b_x^2,\ c_x^4\}^2$$

$$= \tfrac{2}{12}(ab)^2 (ac)^2 b_x^2 c_x^2 + \tfrac{8}{12}(ab)^2 (ac)(bc)\, a_x b_x c_x^2 + \tfrac{2}{12}(ab)^2 (bc)^2 a_x^2 c_x^2$$

$$= \tfrac{1}{3}(ab)^2 (ac)^2 b_x^2 c_x^2 + \tfrac{2}{3}(ab)^2 (ac)(bc)\, a_x b_x c_x^2$$

$$= \tfrac{1}{3}(ab)^2 (ac)^2 b_x^2 c_x^2 + \tfrac{1}{3}(ab)^2 c_x^2 \{(ac)^2 b_x^2 + (bc)^2 a_x^2 - (ab)^2 c_x^2\}$$

$$= (ab)^2 (ac)^2 b_x^2 c_x^2 - \tfrac{1}{3}(ab)^4 \,.\, c_x^2,$$

since the symbols are all equivalent.

Then since, § 22,

$$(ab)^4 c_x^4 + (bc)^4 a_x^4 + (ca)^4 b_x^4$$
$$= 2(ab)^2 (ac)^2 b_x^2 c_x^2 + 2(ba)^2 (bc)^2 a_x^2 c_x^2 + 2(ca)^2 (cb)^2 a_x^2 b_x^2,$$

we have $$(ab)^2 (ac)^2 b_x^2 c_x^2 = \tfrac{1}{2}(ab)^4 \,.\, c_x^4.$$

Therefore $$(H,f)^2 = \tfrac{1}{2} if - \tfrac{1}{3} if = \tfrac{1}{6} if$$

and is reducible.

Next $(H,f)^3 = \{(ab)^2 a_x^2 b_x^2,\ c_x^4\}^3 = (ab)^2 (bc)^2 (ac)\, b_x$

for the two terms in the transvectant are equivalent. By interchanging a and c it follows that $(H,f)^3 = 0$.

Finally

$$(H,f)^4 = \{(ab)^2 a_x^2 b_x^2,\ c_x^4\}^4 = (ab)^2 (bc)^2 (ca)^2 = j$$

an invariant; hence the only irreducible forms of degree three are t and j.

The only product $f^\alpha H^\beta t^\gamma$ of degree three that we need consider is t, and the only possible irreducible forms of degree four are therefore

$$(t,f)^1,\ (t,f)^2,\ (t,f)^3,\ (t,f)^4.$$

As a matter of fact these are all reducible.

To calculate them we use the series of § 54, with the scheme

$$\begin{pmatrix} f & H & f \\ 4 & 4 & 4 \\ 0 & r & 1 \end{pmatrix},$$

which gives

$$\sum_i \frac{\binom{3}{\iota}\binom{r}{\iota}}{\binom{7-\iota}{\iota}} \{(f, H)^{1+\iota}, f\}^{r-\iota}$$

$$= \sum_\iota \frac{\binom{4-r}{\iota}\binom{1}{\iota}}{\binom{9-2r-\iota}{\iota}} \{(f, f)^{r+\iota}, H\}^{1-\iota}.$$

If $r=1$ we find

$$\{(f, H), f\} + \tfrac{3}{6}\{(f, H)^2, f\}^0 = \{(f, f), H\} + \tfrac{3}{6}\{(f, f)^2, H\}^0,$$

$$(t, f) + \tfrac{1}{12} if^2 = \tfrac{1}{2}H^2,$$

since
$$(f, H)^2 = \tfrac{1}{6}if,$$

hence
$$(t, f) = \tfrac{1}{2}H^2 - \tfrac{1}{12}if^2.$$

If $r=2$,

$$(t, f)^2 + \{(f, H)^2, f\} + \tfrac{3}{10}\{(fH)^3, f\}^0$$

$$= \tfrac{1}{2}(H, H)^2,$$

and since
$$(f, H)^2 = \tfrac{1}{6}if, \quad (fH)^3 = 0,$$

this gives
$$(t, f)^2 = 0.$$

From $r=3$,

$$(t, f)^3 + \tfrac{3}{2}\{(f, H)^2, f\}^2 + \tfrac{9}{10}\{(f, H)^3, f\} + \tfrac{1}{4}(f, H)^4 . f$$

$$= \{(f, f)^3, H\} + \tfrac{1}{2}(f, f)^4 . H,$$

or
$$(t, f)^3 + \tfrac{1}{4}iH + \tfrac{1}{4}jf = \tfrac{1}{2}iH,$$

on putting in the values for the transvectants of f and H.

Thus

$$(t, f)^3 = \tfrac{1}{4}(iH - jf).$$

The series does not apply when $r=4$ because then $r+1>4$, but it is easy to calculate

$$(t, f)^4$$

directly or as in § 94.

For

$$\{(ah)\, a_x^3 h_x^3,\ b_x^4\}^4 = (ah)(ab)^3\, b_x h_x^3 + \lambda\, \{(ah)^2\, a_x^2 h_x^2,\ b_x^4\}^3$$
$$+ \mu\, \{(ah)^3\, a_x h_x,\ b_x^4\}^2 + \nu\, \{(ah)^4,\ b_x^4\},$$

while
$$(ab)^3 (ah)\, b_x^3 h_x^3 = \{(ab)^3\, a_x b_x,\ h_x^4\} = 0,$$
$$(ah)^2\, a_x^2 h_x^2 = \tfrac{1}{6} if;\ \ (f, f)^3 = 0,$$
$$(ah)^3\, a_x h_x = 0,$$
hence
$$(t, f)^4 = 0.$$

Hence there are no irreducible forms of degree four. In fact there are no more irreducible forms, because the next in order of degree would be of the form

$$(f^\alpha H^\beta t^\gamma, f)^\rho.$$

Now since f, H, t are each of order four at least, irreducible forms can only arise from products containing each of the three by itself, and all these, viz.

$$(f, f)^\rho,\ (H, f)^\rho,\ (t, f)^\rho$$

have been already considered. Hence every invariant or covariant of the quartic is a rational integral function of f, H, t, i and j.

90. Quintic. To illustrate still further the method of this chapter we shall apply it to some extent to the binary quintic. The covariants of degree two are

$$(f, f)^2 = H \text{ and } (f, f)^4 = i.$$

The products of powers of f, H and i which are of degree two are f^2, H and i, and to find the covariants of degree three we have to consider transvectants of these three forms with f.

The transvectants arising from f^2 may be neglected, and hence we are left with

$$(H, f)^1,\ (H, f)^2,\ (H, f)^3,\ (H, f)^4,\ (H, f)^5,$$
$$(i, f)^1,\ (i, f)^2$$

as the only possible irreducible covariants of degree three.

Now of these

$$(H, f)^2 = \{(ab)^2\, a_x^3 b_x^3,\ c_x^5\}^2$$

and involves a term

$$(ab)^2 (ac)^2 a_x b_x^3 c_x^3,$$

$$\therefore \ (H, f)^2 = (ab)^2 (ac)^2 a_x b_x^3 c_x^3 + \lambda \{(ab)^4 a_x b_x, c_x^5\}^0$$

$$= \tfrac{1}{3} a_x b_x c_x \{(ab)^2 (ac)^2 b_x^2 c_x^2 + (ba)^2 (bc)^2 c_x^2 a_x^2 + (ca)^2 (cb)^2 a_x^2 b_x^2\} + \lambda if$$

$$= \tfrac{1}{3} a_x b_x c_x \tfrac{1}{2} \{(ab)^4 c_x^4 + (bc)^4 a_x^4 + (ca)^4 b_x^4\} + \lambda if$$

$$= (\lambda + \tfrac{1}{2})\, if,$$

so that $(H, f)^2$ is reducible.

Again $$(H, f)^3 = \{(ab)^2 a_x^3 b_x^3, c_x^5\}^3$$

and contains the term

$$(ab)^2 (ac)^3 b_x^3 c_x^2.$$

This term can be transformed so as to contain $(ac)^4$ and hence must be a multiple of

$$(ab)^4 (ac) b_x c_x^4$$

since the letters are equivalent.

$$\therefore \ (H, f)^3 = \lambda (ab)^4 (ac) b_x c_x^4 + \mu \{(ab)^4 a_x b_x, c_x^5\}$$

$$= (\lambda + \mu) (i, f).$$

Further

$$(H, f)^4 = (ab)^2 (bc)^2 (ac)^2 a_x b_x c_x + \lambda (i, f)^2$$

and

$$(H, f)^5 = (ab)^2 (bc)^3 (ac)^2 a_x$$

$$= \tfrac{1}{3} (bc)^2 (ca)^2 (ab)^2 \{(bc) a_x + (ca) b_x + (ab) c_x\} = 0.$$

Finally $(i, f)^1$ is an irreducible form and

$$(i, f)^2 = (ab)^4 (ac) (bc) c_x^3$$

$$= -\tfrac{1}{3} (bc)(ca)(ab) \{(ab)^3 c_x^3 + (bc)^3 a_x^3 + (ca)^3 b_x^3\}$$

$$= -(bc)^2 (ca)^2 (ab)^2 a_x b_x c_x \quad (\S\, 22).$$

Hence the only irreducible covariants of degree three are

$$(H, f) = t, \ (f, i)$$

and $$(f, i)^2 = -(bc)^2 (ca)^2 (ab)^2 a_x b_x c_x = -j.$$

The reader may now find the irreducible forms of degree four and verify the result by reference to the chapter on the quintic. It will be seen at once that the method leads to much labour, that the reduction processes are not easy to discover, and, when we mention that for the quintic we have to proceed step by step until we get to degree 18 before the irreducible system is obtained, the impracticability of these methods in dealing with forms of order greater than four will be at once admitted.

91. Further Theory of the Cubic. Syzygy among the irreducible forms. There is an identical relation connecting the irreducible concomitants of the binary cubic—the simplest example of what is known as a syzygy among the covariants of a single binary form.

In fact since t is the Jacobian of f and H we have, § 78,

$$-2t^2 = (f, f)^2 H^2 + (H, H)^2 f^2 - 2(H, f)^2 Hf.$$

Now

$$(f, f)^2 = H$$
$$(H, f)^2 = (ah)^2 a_x = 0$$
$$(H, H)^2 = \{(ab)^2 a_x b_x, (cd)^2 c_x d_x\}^2 = (ab)^2 (cd)^2 (ac)(bd) = \Delta,$$

for although there are four terms in the transvectant they are identical in value, and we have

$$-2t^2 = H^3 + \Delta f^2,$$

the relation required.

92. Since every covariant of the cubic is a rational integral function of f, H, t and Δ it follows that all expressions derived by convolution from products of powers of f, H and t can be expressed in terms of f, H, t and Δ.

The form of the expression can be easily inferred by consideration of its degree and order, but the actual determination of the coefficients may be a troublesome process.

As an example consider the Hessian of t, *i.e.* $(t, t)^2$. It is a covariant of degree six and order two since t is of degree three, and the only product of f, H, t, Δ fulfilling these conditions is $H\Delta$. We at once see that $(t, t)^2$ is a numerical multiple of $H\Delta$.

The reader may calculate the actual value directly by using the series of § 54—we give an alternative process.

Let $$J = (t, f),$$

then $$-2J^2 = (t, t)^2 f^2 + (f, f)^2 t^2 - 2(f, t)^2 ft.$$

But $(t, f) = \frac{1}{2} H^2, \quad (f, f)^2 = H, \quad (f, t)^2 = 0$

and $t^2 = -\frac{1}{2} H^3 - \frac{1}{2} \Delta f^2;$

therefore $-\frac{1}{2} H^4 = (t, t)^2 f^2 + H(-\frac{1}{2} H^3 - \frac{1}{2} \Delta f^2);$

that is $(t, t)^2 f^2 = \frac{1}{2} \Delta f^2 H$

or $(t, t)^2 = \frac{1}{2} \Delta H.$

Again, consider the symbolical product

$$(ab)^2 (ac) (bd) (cd) c_x d_x$$

which represents a covariant of degree four and order two.

Since there is no product

$$f^\alpha H^\beta t^\gamma \Delta^\delta$$

of this degree and order the covariant in question must vanish identically.

To verify this we remark that, on interchanging a, b and c, d, the expression

$$(ab)^2 (ac) (bd) (cd) c_x d_x$$

changes sign; hence it vanishes.

Ex. (i). Calculate the following transvectants in terms of f, H, t, Δ, viz.

$$(H, H), \quad (H, H)^2, \quad (t, t), \quad (t, t)^2, \quad (t, t)^3, \quad (H, t), \quad (H, t)^2.$$

(The only one presenting any difficulty is (H, t) and this is the Jacobian of a Jacobian; its value is $-\frac{1}{2} \Delta f$.)

Ex. (ii). Shew that any symbolical product involving the factor $(ah)^2$ vanishes identically.

Ex. (iii). Shew that

$$(H^3, f^2)^5 = 0, \quad (H^3, t^2)^5 = 0, \quad (H^3, f^2) = H^2 f (H, f) = -H^2 ft.$$

93. **Further Theory of the Quartic.** As in the case of the cubic, the square of the covariant t can be expressed rationally in terms of the remaining forms.

In fact we have, § 78,

$$-2t^2 = (f, f)^2 H^2 - 2(H, f)^2 Hf + (H, H)^2 f^2,$$

while $(f, f)^2 = H, \quad (H, f)^2 = \frac{1}{6} if,$

so that it only remains to calculate $(H, H)^2$.

Now using the series of § 54 with the scheme

$$\begin{pmatrix} f, & f, & H \\ 4, & 4, & 4 \\ 0, & 2, & 2 \end{pmatrix}$$

we have

$$\sum_i \frac{\binom{2}{i}\binom{2}{i}}{\binom{5-i}{i}} \{(f, f)^{2+i}, H\}^{2-i} = \sum_i \frac{\binom{2}{i}\binom{2}{i}}{\binom{5-i}{i}} \{(f, H)^{2+i}, f\}^{2-i}$$

or

$$(H, H)^2 + \{(f, f)^3, H\} + \tfrac{1}{3}(f, f)^4 . H$$
$$= \{(f, H)^2, f\}^2 + \{(f, H)^3, f\} + \tfrac{1}{3}(f, H)^4 . f,$$

that is $$(H, H)^2 + \tfrac{1}{3} iH = \tfrac{1}{6} i (f, f)^2 + \tfrac{1}{3} jf,$$

since $$(f, H)^3 = 0 \text{ etc.}$$

Hence $$(H, H)^2 = \tfrac{1}{3} jf - \tfrac{1}{6} iH.$$

Consequently

$$-2t^2 = H^3 - \tfrac{1}{3} if . Hf + (\tfrac{1}{3} jf - \tfrac{1}{6} iH) f^2$$
$$= H^3 - \tfrac{1}{2} iHf^2 + \tfrac{1}{3} jf^3,$$

which is the syzygy required.

94. We shall illustrate the reduction of covariants of the quartic by calculating the values of the transvectants of

$$f, \; H, \; t,$$

taken two together.

The transvectants of f with itself, H and t have already been found.

As regards the transvectant

$$(H, H)^r$$

we remark that it vanishes when r is odd and

$$(H, H)^2 = \tfrac{1}{3} jf - \tfrac{1}{6} iH.$$

There only remains $(H, H)^4$.

This is equal to

$$\{(ab)^2 a_x^2 b_x^2, h_x^4\}^4 = (ab)^2 (ah)^2 (bh)^2 = \{(ah)^2 a_x^2 h_x^2, b_x^4\}^4$$
$$= \{(H, f)^2, f\}^4 = \tfrac{1}{6} i (f, f)^4 = \tfrac{1}{6} i^2.$$

To calculate the transvectants

$$(t, H)^r, \quad r \not> 3,$$

apply the series of § 54 with the scheme

$$\begin{pmatrix} H, & f, & H \\ 4, & 4, & 4 \\ 0, & r, & 1 \end{pmatrix}$$

and we have

$$\Sigma \frac{\binom{3}{i}\binom{r}{i}}{\binom{7-i}{i}} \{(H, f)^{1+i}, H\}^{r-i} = \Sigma \frac{\binom{4-r}{i}\binom{1}{i}}{\binom{9-2r-i}{i}} \{(H, H)^{r+i}, f\}^{1-i}.$$

On taking $r = 1, 2, 3$ successively and putting in the values of the transvectants $(H, H)^{r+i}$ we find

$$(t, H) = \tfrac{1}{6} f (iH - jf)$$

$$(t, H)^2 = 0$$

$$(t, H)^3 = \tfrac{1}{4} jH - \tfrac{1}{24} i^2 f.$$

For $(t, H)^4$, the scheme

$$\begin{pmatrix} H, & f, & H \\ 4, & 4, & 4 \\ 1, & 3, & 1 \end{pmatrix}$$

must be used; or else as in the case of $(t, f)^4$ it is easy to see that

$$(t, H)^4 = 0.$$

To find $(t, t)^2$ we apply the series with the scheme

$$\begin{pmatrix} f, & H, & t \\ 4, & 4, & 6 \\ 0, & 2, & 1 \end{pmatrix}$$

which gives

$$\{(f, H), t\}^2 + \{(f, H)^2, t\} + \tfrac{3}{10} (f, H)^3 t = \{(f, t)^2, H\} + \tfrac{2}{3} (f, t)^3 H$$

or

$$(t, t)^2 + \frac{i}{6} (f, t) = \frac{2}{3} (f, t)^3 . H.$$

On substituting the values for (t, f) and $(t, f)^3$ we find

$$(t, t)^2 = \tfrac{1}{6} (jfH - \tfrac{1}{2} iH^2 - \tfrac{1}{12} i^2 f^2).$$

For $(t, t)^4$ we use the scheme

$$\begin{pmatrix} f, & H, & t \\ 4, & 4, & 6 \\ 1, & 3, & 1 \end{pmatrix}$$

leading to

$$(t, t)^4 + \frac{i}{6}(f, t)^3 = \{(t, f)^3, H\}^2$$

$$= \frac{i}{4}(H, H)^2 - \frac{j}{4}(f, H)^2$$

and hence $$(t, t)^4 = 0.$$

Finally

$$(t, t)^6 = \{(t, f)^3, H\}^4 = \tfrac{1}{4} i (H, H)^4 - \tfrac{1}{4} j (f, H)^4$$

$$= \frac{1}{4}\left(\frac{i^3}{6} - j^2\right).$$

Ex. (i). Deduce the value of $(t, t)^2$ from the relation

$$-2\{(t, f)\}^2 = (t, t)^2 f^2 + (f, f)^2 t^2 - 2(f, t)^2 ft.$$

Ex. (ii). Apply Gordan's series to calculate $(t, t)^2$ for the cubic.

Ex. (iii). Prove that

$$\{(t, f)^3, H\}^2 = \{(t, H)^3, f\}^2.$$

Ex. (iv). Prove that for the quartic

$$\{(H, H)^2, H\} = +\tfrac{1}{3} jt$$

$$\{(H, H)^2, H\}^4 = \tfrac{1}{3} j^2 - \tfrac{1}{36} i^3.$$

Ex. (v). Prove that the Hessian of the Hessian of the Hessian of a quartic f is

$$-\tfrac{1}{216} i^2 jf + \tfrac{1}{9} H (j^2 - \tfrac{1}{24} i^3).$$

Ex. (vi). Calculate the values of $\{(t, t)^2, t\}^r$ for $r = 1, 2, 3, 4, 5, 6$.

Ex. (vii). If a quartic f be the product of a cubic by one of the linear factors of its Hessian, then

$$(f, f)^4 = 0.$$

CHAPTER VI.

GORDAN'S THEOREM.

95. WE have already referred to Gordan's theorem which asserts the existence of a finite complete system of covariants for any binary form, and, in fact, we have illustrated the truth of the theorem in obtaining the complete systems for the quadratic, cubic, and quartic. Our previous method is of little practical utility in dealing with forms of order greater than four, but a comparison between it and the procedure of Gordan may not be without value as a primary indication of the salient features of the latter. In the last chapter covariants were classified according to their degree and we shewed how to obtain those of degree m by transvection from those of less degree. In Gordan's investigation covariants are classified according to their grade—the grade being a definite even number associated with any symbolical product, § 61, and all covariants of grade $2r$ are obtained by transvection from those of inferior grade together with some of grade $2r$.

The advantage of using the grade is that no covariant can be of grade greater than n; accordingly, the number of steps in the process is small, whereas there being no limit, *à priori*, to the degree of an irreducible covariant, and the actual degree reached by irreducible forms of quintics, etc., being very high, the number of steps in the other process is uncertain and at the best large. As will be seen later, on the other hand, the transition from grade $2r-2$ to grade $2r$ is commonly much more difficult than that from one degree to the next higher.

Several preliminary propositions are necessary before we can undertake the actual proof of the existence of the complete system; these we now proceed to explain.

96. The first lemma required belongs to that branch of the theory of numbers known as Diophantine Equations.

For the sake of clearness we shall begin by giving an illustration.

Consider the homogeneous linear equation

$$2x + 5y = 3z;$$

it is easy to see that the number of solutions in positive integers is infinite.

Moreover, if

$$x = p, \quad y = q, \quad z = r;$$

and

$$x = p', \quad y = q', \quad z = r',$$

be two solutions, then

$$x = p + p', \quad y = q + q', \quad z = r + r'$$

is also a solution.

We shall call this latter the sum of the former two solutions, and when any solution can be written as the sum of two smaller solutions (throughout we deal only with solutions in positive integers), it is said to be reducible. Otherwise a solution is irreducible, and the important fact for us is that the number of irreducible solutions is always finite.

Thus for the equation above the only irreducible solutions are

$$x = 3,\ y = 0,\ z = 2;\quad x = 0,\ y = 3,\ z = 5;$$
$$x = 1,\ y = 2,\ z = 4;\quad x = 2,\ y = 1,\ z = 3.$$

In fact if $x > 3$, then $z > 2$, and the solution can be reduced by means of $x = 3$, $y = 0$, $z = 2$; whereas if $y > 3$, then $z > 5$, and the solution can be reduced by means of $x = 0$, $y = 3$, $z = 5$; thus in an irreducible solution neither x nor y can exceed 3, and, as the number of remaining possibilities is finite, the irreducible solutions can be easily found by trial.

By continually reducing a given solution, say $x = p$, $y = q$, $z = r$, we can express it in terms of the irreducible solutions, that is in the form

$$\left.\begin{aligned} p &= 3\lambda + \nu + 2\rho \\ q &= 3\mu + 2\nu + \rho \\ r &= 2\lambda + 5\mu + 4\nu + 3\rho \end{aligned}\right\} \text{.................(A)},$$

where λ, μ, ν, ρ are positive integers:

e.g. take the solution

$$x = 50, \quad y = 7, \quad z = 45,$$

reducing by means of

$$x = 3, \quad y = 0, \quad z = 2,$$

$$x = 16 . 3 + 2, \quad y = 7, \quad z = 16 . 2 + 13 ;$$

then reducing

$$x' = 2, \quad y' = 7, \quad z' = 13$$

by means of

$$x = 0, \quad y = 3, \quad z = 5,$$

$$x' = 2, \quad y' = 2 . 3 + 1, \quad z' = 2 . 5 + 3,$$

and the remaining part

$$x'' = 2, \quad y'' = 1, \quad z'' = 3$$

is irreducible.

Hence in this case we have

$$\lambda = 16, \quad \mu = 2, \quad \nu = 0, \quad \rho = 1.$$

Of course if we substitute the expressions in (A) for x, y, z the equation is satisfied identically; the important point is that *every* positive integral solution can be written in the form there indicated.

97. The idea of reducibility can be extended at once to any number of linear homogeneous equations, for the sum of two solutions is always a solution, and we may enunciate our first lemma as follows:

The number of irreducible solutions in positive integers of a system of homogeneous linear equations is finite.

Consider first a single equation,

$$a_1 x_1 + a_2 x_2 + \ldots + a_m x_m = b_1 y_1 + b_2 y_2 + \ldots + b_n y_n,$$

connecting the x's and y's, where the coefficients a, b are positive integers.

If the two solutions

$$x_1 = \xi_1, \; x_2 = \xi_2, \; \ldots \; x_m = \xi_m, \; y_1 = \eta_1, \; y_2 = \eta_2, \; \ldots \; y_n = \eta_n ;$$

$$x_1 = \xi_1', \; x_2 = \xi_2', \; \ldots \; x_m = \xi_m', \; y_1 = \eta_1', \; y_2 = \eta_2', \; \ldots \; y_n = \eta_n',$$

be typified by

$$x = \xi, \quad y = \eta ;$$

$$x = \xi', \quad y = \eta',$$

respectively, then

$$x = \xi + \xi', \quad y = \eta + \eta'$$

also typifies a solution and this latter is reducible.

First the equation has mn solutions of the type

$$x_r = b_s, \quad y_s = a_r$$

with the rest of the variables zero.

Next suppose that in a solution one of the x's (say x_1) is greater than

$$b_1 + b_2 + \ldots + b_n,$$

then the right-hand side of the equation must be greater than

$$a_1(b_1 + b_2 + \ldots + b_n),$$

i.e. $$b_1(y_1 - a_1) + b_2(y_2 - a_1) + \ldots + b_n(y_n - a_1) > 0,$$

so that at least one y must be greater than a_1. Let $y_r > a_1$, then the solution in question is reducible by means of the solution

$$x_1 = b_r, \quad y_r = a_1$$

with the other variables zero.

Hence $(b_1 + b_2 + \ldots + b_n)$ is an upper limit to the value of any x in an irreducible solution, similarly $(a_1 + a_2 + \ldots + a_m)$ is an upper limit to the value of any y; but the number of solutions reducible or irreducible subject to these restrictions is manifestly finite, therefore *à fortiori* the number of irreducible solutions is finite.

If the irreducible solutions be typified by

$$x = \alpha_1, \quad y = \beta_1;$$
$$x = \alpha_2, \quad y = \beta_2;$$
$$\vdots \qquad \vdots$$
$$x = \alpha_p, \quad y = \beta_p,$$

then by continued reduction any solution can be expressed in the typical form

$$x = t_1\alpha_1 + t_2\alpha_2 + \ldots + t_p\alpha_p$$
$$y = t_1\beta_1 + t_2\beta_2 + \ldots + t_p\beta_p,$$

where the t's are all positive integers.

Suppose now we have a second equation of the same nature between the variables; then on replacing the x's and y's by their values in terms of the t's the first equation will be satisfied identically and the second equation will become a linear equation between the t's with integral coefficients.

Hence by the above reasoning every solution of the equation among the t's may be written in the typical form

$$t = T_1\gamma_1 + T_2\gamma_2 + \ldots + T_\sigma\gamma_\sigma,$$

where $$t = \gamma_1,\ t = \gamma_2,\ \ldots\ t = \gamma_\sigma$$

typify the irreducible solutions, and the T's are all positive integers.

Now substitute these values for the t's in the expressions for the x's and y's and we find at once that

$$x = \kappa_1 T_1 + \kappa_2 T_2 + \ldots + \kappa_\sigma T_\sigma$$

$$y = \lambda_1 T_1 + \lambda_2 T_2 + \ldots + \lambda_\sigma T_\sigma,$$

where the κ's and λ's are fixed positive integers.

Thus the only possible irreducible solutions of the two equations are those typified by

$$x = \kappa_1,\ y = \lambda_1;\ x = \kappa_2,\ y = \lambda_2;\ \ldots\ x = \kappa_\sigma,\ y = \lambda_\sigma,$$

for every other solution can be expressed as a linear combination of these.

If we had a third equation, on substituting for the x's and y's their values in terms of the T's the first two equations would be satisfied identically, and the third would become a linear equation among the T's. Then this equation in turn has only a finite number of irreducible solutions, and hence, reasoning exactly as before, we should find that the three equations given have only a finite number of irreducible solutions. The process can be manifestly extended to any number of equations, and hence our theorem is established. A formal proof by induction from $(r-1)$ equations to r equations could of course be easily given.

Ex. To find the irreducible solutions of the two equations

$$\left.\begin{aligned} 2x + 3y &= z + w \\ x + w &= y + z \end{aligned}\right\}.$$

The irreducible solutions of the second equation are easily found since no letter can exceed 2; they are

$$x=1,\ y=1;\ x=1,\ z=1;\ y=1,\ w=1;\ z=1,\ w=1;$$

variables not mentioned in a solution being zero.

Hence the general solution of the second equation is

$$x=a+b,\ y=a+c,\ z=b+d,\ w=c+a,$$

where a, b, c, d are positive integers.

The first equation now becomes

$$5a+b+2c=2d$$

and for an irreducible solution a, b, c cannot exceed 2.

On trial we find the following irreducible solution

$$a=2,\ d=5;\ \ b=2,\ d=1;\ \ c=1,\ d=1;\ \ a=1,\ b=1,\ d=3.$$

Hence the general solution is

$$a=2\alpha+\delta,\ b=2\beta+\delta,\ c=\gamma,\ d=5\alpha+\beta+\gamma+3\delta.$$

These values for a, b, c, d give

$$\left.\begin{aligned} x&=2\alpha+2\beta+2\delta\\ y&=2\alpha+\gamma+\delta\\ z&=5\alpha+3\beta+\gamma+4\delta\\ w&=5\alpha+\beta+2\gamma+3\delta\end{aligned}\right\}$$

as the general solution of the two equations.

The only possible irreducible solutions are accordingly

$$\left.\begin{array}{llll} x=2, & y=2, & z=5, & w=5\\ x=2, & y=0, & z=3, & w=1\\ x=0, & y=1, & z=1, & w=2\\ x=2, & y=1, & z=4, & w=3\end{array}\right\}$$

Of these the first is the sum of the third and fourth, while the fourth is the sum of the second and third, so the only irreducible solutions are the second and third. In other words any solution of the two equations may be written in the form

$$x=2p,\ y=q,\ z=3p+q,\ w=p+2q,$$

where p, q are positive integers.

Ex. (i). Prove that the equation $7x+4y=3z$ has four irreducible solutions and that every solution of the two equations $7x+4y=3z$, $z+5w=2y$ can be written in the form

$$x=2a+c,\ y=7a+15b+11c,\ z=14a+20b+17c,\ w=2b+c.$$

Ex. (ii). Find the number of irreducible solutions of the equation

$$a_1x_1+a_2x_2+\ldots+a_nx_n=2x,$$

the a's being positive integers.

Ans. If all the a's are even there are n irreducible solutions, if r of the a's are odd there are $n+\dfrac{r(r-1)}{2}$ solutions. In an irreducible solution at the most only two of the letters on the left are different from zero.

Ex. (iii). Prove directly that in an irreducible solution of the two equations

$$a_1x_1+a_2x_2+\ldots+a_mx_m=x+z$$

$$b_1y_1+b_2y_2+\ldots+b_ny_n=y+z$$

x is less than the greatest a, and y is less than the greatest b.

98. **System of forms derived by transvection from two given systems.** Consider two systems of binary forms in the same variables x_1, x_2, viz.

$A_1, A_2, \ldots A_m$, of orders $a_1, a_2, \ldots a_m$ respectively

and $B_1, B_2, \ldots B_n$, of orders $b_1, b_2, \ldots b_n$ respectively;

we suppose each form written symbolically and denote by U, V two products of the types

$$A_1^{\alpha_1}A_2^{\alpha_2}\ldots A_m^{\alpha_m}, \quad B_1^{\beta_1}B_2^{\beta_2}\ldots B_n^{\beta_n}$$

wherein all the exponents are either zero or positive integers.

The system C is said to be derived from the systems A and B by transvection when it includes all terms in all transvectants of the form

$$(U, V)^\gamma.$$

It is clear that some of the members of the system C are reducible, that is they can be expressed as rational integral functions of simpler members of that system—in fact, if

$$U=U_1U_2,\ V=V_1V_2,\ \gamma=\gamma_1+\gamma_2\,^*,$$

then there are many terms in the transvectant

$$(U, V)^\gamma$$

which are products of two terms, one belonging to the transvectant

$$(U_1, V_1)^{\gamma_1}$$

and the other to the transvectant

$$(U_2, V_2)^{\gamma_2}.$$

99. We can now enunciate and prove our first theorem, viz.

The number of transvectants of the form $(U, V)^\gamma$ *which do not contain reducible terms is finite.*

* It is of course assumed that γ_1, γ_2 are such that these transvectants are possible, *e.g.* γ_1 must not exceed the order of U_1.

For suppose that any term of the transvectant

$$(U, V)^{\gamma}$$

contains ρ symbols of the A's not in combination with a symbol of the B's and σ symbols of the B's not in combination with a symbol of the A's, then we have

$$\left.\begin{array}{l} a_1\alpha_1 + a_2\alpha_2 + \ldots + a_m\alpha_m = \rho + \gamma \\ b_1\beta_1 + b_2\beta_2 + \ldots + b_n\beta_n = \sigma + \gamma \end{array}\right\} \ldots\ldots\ldots\ldots\ldots (\mathrm{I})$$

because each side of the first equation, for example, represents the total number of the symbols of the forms A which occur in the product U.

Now to each positive integral solution of the above equations in α, β, ρ, σ, γ there correspond definite products U, V and a definite value of γ and hence a unique transvectant. But as we have already remarked if the solution corresponding to $(U, V)^{\gamma}$ be the sum of those corresponding to $(U_1, V_1)^{\gamma_1}$ and $(U_2, V_2)^{\gamma_2}$, then $(U, V)^{\gamma}$ certainly contains reducible terms. Hence transvectants corresponding to reducible solutions always contain reducible terms and inasmuch as the number of irreducible solutions has been proved to be finite it follows that the number of transvectants not containing reducible terms is finite.

100. In actually finding the transvectants which do not contain reducible terms we may use the equations (I), but it is generally easier to proceed directly.

Suppose that the system A contains the single form $f = a_x^5$ and the system B the single form $i = b_x^2$, then we have to consider transvectants

$$(f^{\alpha}, i^{\beta})^{\gamma}.$$

If $\gamma > 2\beta$ this transvectant vanishes, if $\gamma < 2\beta - 1$ it contains terms of $i(f^{\alpha}, i^{\beta-1})^{\gamma}$; hence for an irreducible transvectant we must have $\gamma = 2\beta - 1$ or 2β. In the same way

$$\gamma \ngtr 5\alpha \text{ and } \nless 5\alpha - 4.$$

Again if $\alpha > 2$, then for an irreducible transvectant $\gamma > 10$ and hence $\beta > 4$, so that some terms may be reduced by means of $(f^2, i^5)^{10}$. Thus we need only consider $\alpha = 0$, $\alpha = 1$ and $\alpha = 2$.

For $\alpha = 0$ we have i.

For $\alpha = 1$ we have f, (f, i), $(f, i)^2$, $(f, i^2)^3$, $(f, i^2)^4$, $(f, i^3)^5$.

For $\alpha = 2$ we have $(f^2, i^3)^6$, $(f^2, i^4)^7$, $(f^2, i^4)^8$, $(f^2, i^5)^9$, $(f^2, i^5)^{10}$.

Now $(f^2, i^3)^6$ contains terms which are products of a term of $(f, i^2)^4$ and $(f, i)^2$ and a like argument applies to

$$(f^2, i^4)^7,\ (f^2, i^4)^8,\ (f^2, i^5)^9,$$

so the only transvectants not containing reducible terms are

$$f,\ i,\ (f, i),\ (f, i)^2,\ (f, i^2)^3,\ (f, i^2)^4,\ (f, i^3)^5,\ (f^2, i^5)^{10}.$$

Ex. (i). If f be any form of order $2n+1$, then the transvectants

$$(f^\alpha, i^\beta)^r$$

which do not contain reducible terms are $2n+4$ in number.

Ex. (ii). Find the corresponding result when f is a form of even order.

Ex. (iii). The only transvectants

$$f_1^{\alpha_1} (f_2^{\alpha_2}, i^\beta)^r,$$

where $f_1 = a_x^4$ and $f_2 = b_x^3$, which do not contain reducible terms are

$$f_1,\ f_2,\ i,\ (f_1, i),\ (f_1, i)^2,\ (f_1, i^2)^3,\ (f_1, i^2)^4,$$
$$(f_2, i),\ (f_2, i)^2,\ (f_2, i^2)^3,\ (f_2^2, i^3)^6.$$

101. Definition. The system of forms A is said to be *complete* when any expression derived by convolution from a product U of powers of the forms A is itself a rational integral function of the A's.

Thus for example the system of forms

$$f = a_x^3 = b_x^3 = \dots$$
$$H = (ab)^2 a_x b_x$$
$$t = (ab)^2 (ca) b_x c_x^2$$
$$\Delta = (ab)^2 (cd)^2 (ac)(bd)$$

is complete because any expression derived in the above manner is a covariant of f and therefore a rational integral function of f, H, t and Δ. Again the system H and Δ included in the above is itself complete.

More generally the system A is said to be *relatively complete* for the modulus G consisting of a number of symbolical determinants when any expression derived by convolution from a product U is a rational integral function of the A's together with terms involving the modulus G.

Thus the system consisting of a single form

$$f = a_x^n = b_x^n = \dots$$

is relatively complete for the modulus $(ab)^2$, since any expression derived by convolution from a power of f can be transformed so that a factor $(ab)^2$ occurs in it.

Again for a quartic the system

$$f = a_x^4, \quad H = (ab)^2 a_x^2 b_x^2, \quad t = (ab)^2 (ca) a_x b_x^2 c_x^3,$$
$$j = (bc)^2 (ca)^2 (ab)^2$$

is relatively complete for the modulus $(ab)^4$, for all covariants of f are rational integral functions of

$$f, H, t, j, i,$$

where $i = (ab)^4$.

We may extend our definition of relative completeness still further: a system A is said to be complete relatively for several moduli G_1, G_2 ... when any expression derived by convolution from a product U is a rational integral function of the A's together with terms involving one at least of the moduli G_1, G_2 ....

It will be seen later (or it can be verified without difficulty) that in connection with any quantic $a_x^n = b_x^n$ ... the single form

$$H = (ab)^2 a_x^{n-2} b_x^{n-2}$$

is relatively complete with respect to the modulus $(ab)^4$ except when $n = 4$.

If $n = 4$ the complete system worked out for the form H shews that any expression derived by convolution from a power of H is a rational integral function of H together with terms involving i or j.

That is H is relatively complete for the two moduli $(ab)^4$ and $(bc)^2 (ca)^2 (ab)^2$.

It will be noticed that a complete system is relatively complete for any modulus or systems of moduli.

102. The system C derived by transvection from two given systems contains an infinite number of forms, but it is said to be a finite system when all its members can be expressed as rational integral functions of a certain finite number of them. More generally it is said to be relatively finite for a given modulus G

when every member of C can be expressed as a rational integral function of a certain finite number of them together with terms all of which involve the modulus G.

For example the number of covariants of a binary cubic is infinite but inasmuch as every one is a rational integral function of f, H, t and Δ the system of forms is said to be finite.

Again it will be seen later that every covariant of the binary n-ic

$$f = a_x^n = b_x^n = c_x^n$$

can be expressed in terms of f, H, t, where

$$H = (ab)^2 a_x^{n-2} b_x^{n-2}$$

$$t = (ab)^2 (ca) a_x^{n-3} b_x^{n-2} c_x^{n-1}$$

together with terms involving the factor $(ab)^4$. We should state this fact thus—The system of covariants of a binary n-ic is relatively finite for the modulus $(ab)^4$.

103. Theorem. *If the systems of forms A and B are both finite and complete, then the system derived from them by transvection is finite and complete.*

(*a*) The system is finite.

In the proof of this theorem we shall consider the transvectants

$$(U, V)^\gamma$$

in a certain order defined as follows:—

(i) Transvectants are taken in order of ascending total degree of the product UV in the coefficients of the forms involved in A and B.

(ii) Those for which the total degree is the same are taken in ascending order of indices.

Further than this the order is immaterial.

With this convention let T and T' be any two terms of the transvectant

$$(U, V)^\gamma,$$

then
$$(T - T') = \Sigma (\bar{U}, \bar{V})^{\gamma'}$$

where $\gamma' < \gamma$ and $\bar{U}$, $\bar{V}$ are derived by convolution from U, V respectively.

But since the systems A and B are complete

$$\bar{U} = F(A),$$

$$\bar{V} = \Phi(B),$$

where $F(A)$ is a rational integral function of the A's, that is, an aggregate of products of the type U, and $\Phi(B)$ a similar function of the B's.

Thus

$$(\bar{U}, \bar{V})^{\gamma'}$$

can be expressed as the sum of a number of transvectants in each of which the index is less than γ. By hypothesis all such transvectants have been examined before the one now under consideration and hence if all the C's derived from previously considered transvectants can be expressed in terms of

$$C_1, C_2, \dots C_r,$$

then all C's up to and including those derived from

$$(U, V)^{\gamma}$$

can be expressed in terms of

$$C_1, C_2, \dots C_r, T,$$

where T is any term of the last transvectant.

But if the transvectant

$$(U, V)^{\gamma}$$

contain a reducible term, say $T = T_1 T_2$, then inasmuch as T_1, T_2 must both arise from transvectants previously considered no term T need be added to the system

$$C_1, C_2, \dots C_r.$$

Thus in gradually building up a system of C's in terms of which all C's can be expressed we need only add a new member when we come to a transvectant containing no reducible term and then we need add only one new member. But the number of transvectants containing no reducible term is finite and hence a finite number of C's can be chosen such that every other is a rational integral function of these, that is the system C is finite.

Remark. A set of C's in terms of which all others can be expressed rationally and integrally can be chosen in various ways, for any term may be selected from each transvectant containing no reducible terms. Further since the difference of two terms of a transvectant can be expressed by means of terms of transvectants previously considered we may, instead of choosing a single term from any transvectant, take an aggregate of any number of such terms or even the transvectant itself, and it will still be true that every member of C can be expressed as a rational integral function of the members of our finite system.

(*b*) The finite system so constructed is complete.

Let
$$C_1, C_2, \ldots C_r$$
be the finite system, then we have to prove that an expression $\overline{W}$ derived by convolution from any product of the form

$$W = C_1^{\gamma_1} C_2^{\gamma_2} \ldots C_r^{\gamma_r}$$

is a rational integral function of $C_1, C_2, \ldots C_r$.

Suppose that $\overline{W}$ contains ρ determinantal factors in which a symbol belonging to a form A occurs in combination with a symbol belonging to a form B.

Then $\overline{W}$ is a term in a transvectant

$$(\overline{U}, \overline{V})^\rho,$$

where $\overline{U}$ contains only symbols of the A's and $\overline{V}$ only symbols of the B's, so that $\overline{U}$ is derived by convolution from a product U of the A's and $\overline{V}$ is derived by convolution from a product V of the B's.

Thus
$$\overline{W} = (\overline{U}, \overline{V})^\rho + \Sigma (\overline{\overline{U}}\overline{\overline{V}})^{\rho'},$$

where $\rho' < \rho$ and $\overline{\overline{U}}\overline{\overline{V}}$ are derived from $\overline{U}$, $\overline{V}$ by convolution and therefore ultimately from U, V.

Now
$$\overline{U} = F(A),$$
$$\overline{V} = \Phi(B),$$

accordingly $\overline{W}$ can be expressed as an aggregate of transvectants of the form

$$(U, V)^\nu.$$

But we have just proved that every term of such a transvectant is a rational integral function of the C's and consequently $\overline{W}$ is also a rational integral function of them.

Hence the system is not only finite but complete.

104. Theorem. *If a finite system of forms A, all the members of which are covariants of a binary form f, include f and be relatively complete for the modulus H; if, further, a finite system B be relatively complete for the modulus G, and include one form B_1 whose only determinantal factors are H, then the system C derived by transvection from A and B is relatively finite and complete for the modulus G.*

As an example of the theorem let A consist of

$$f = a_x^3 = b_x^3,$$

and B of the two forms

$$H = (ab)^2 a_x b_x, \quad \Delta = (ab)^2 (ac)(bd)(cd)^2.$$

Then A is relatively complete for the modulus $(ab)^2$, § 88, and B is absolutely complete, being the complete system of the Hessian of the cubic; hence according to the theorem the system derived by transvection should be absolutely complete. This is obviously true, for the new system contains f, H, t, Δ, where

$$t = (f, H) = -(ab)^2 (ac) b_x c_x^2,$$

and every possible member of the derived system is a covariant of f, therefore they are all rational integral functions of f, H, t, Δ, which constitute the complete system of the cubic.

105. Lemma. *If P be derived by convolution from a power of f any term in the transvectant*

$$(P, V)^\rho$$

can be expressed as an aggregate of transvectants of the type

$$(U, V)^\sigma$$

in which the degree of U is at most equal to that of P.

(Throughout we shall use U, V as typical symbols for products of powers of the forms of A and B respectively.)

This statement is manifestly true when the degree of P is zero; assuming it true when the degree of P is less than r we shall establish it when the degree is r.

In fact if T be a term in

$$(P, V)^\rho,$$

$$T = (P, V)^\rho + \Sigma (\overline{P}, \overline{V})^{\rho'};$$

and since P, $\bar{P}$ are derived by convolution from a power of the form f which is contained in A,

$$P = F(A) + HW^*,$$

$$\bar{P} = F'(A) + HW',$$

while $$\bar{V} = \Phi(B) + GZ \equiv \Phi(B), \text{ mod } G.$$

Hence T can be expressed as the sum of three parts;

(i) transvectants of the type $\{F(A), \Phi(B)\}^\sigma$ the degree of $F(A)$ being r;

(ii) transvectants of the type $(Q, V)^\tau$, where Q is of the same degree as P and contains the factor H;

(iii) terms containing the factor G.

Now Q can be derived by convolution from

$$B_1 f^s,$$

where s is less than r the degree of P; therefore any term in

$$(Q, V)^\tau$$

can be derived by convolution from

$$f^s B_1 V,$$

and is expressible in the form

$$\Sigma(P', \overline{B_1 V}),$$

where P' is derived by convolution from f^s and is of degree less than P. But by hypothesis every term in these transvectants can be expressed as an aggregate

$$\Sigma(U, V)^\sigma, \text{ mod } G,$$

for $$\overline{B_1 V} \equiv \Phi(B), \text{ mod } G,$$

where the degree of U is at most equal to s and therefore less than r.

On referring to the expression for T we see that T can be written in the form

$$\Sigma(U, V)^\sigma, \text{ mod } G;$$

consequently the statement in the lemma can be completely established by induction.

* HW simply means a symbolical product containing the factor H.

COR. *If the product P contain the factor H, then any term in*

$$(P, V)^\rho$$

can be expressed in the form

$$\Sigma (U, V)^\sigma,$$

where the degree of U is less than that of P.

For P is now of the form Q just discussed, and any term in a transvectant

$$(Q, V)^\rho$$

can be expressed as a sum

$$\Sigma (U, V)^\sigma$$

in which the degree of U is at most equal to s which is less than the degree of P.

106. The proof of the theorem is now the same in principle as that in § 103.

The transvectants are considered in the following order.

(i) In order of ascending degree of UV in the coefficients of f.

(ii) Those for which the degree of UV is the same are taken in order of ascending degree of U.

(iii) Transvectants for which these two degrees are the same are taken in order of ascending index.

Further than this the order is immaterial.

If T and T' be two terms in

$$(U, V)^\nu,$$

then

$$T' - T = \Sigma (\overline{U}, \overline{V})^{\nu'},$$

where

$$\nu' < \nu.$$

But

$$\overline{U} = F(A) + HW,$$

$$\overline{V} = \Phi(B) + GW';$$

therefore

$$T' - T \equiv \Sigma \{F(A), \Phi(B)\}^{\nu'} + \Sigma \{HW, \Phi(B)\}^{\nu'}, \text{ mod } G.$$

Transvectants of the type

$$\{F(A), \Phi(B)\}^{\nu'}$$

have been previously considered, for the degree of $F(A)$ is the same as that of U and $\nu' < \nu$; further by the lemma transvectants of the type

$$\{HW, \Phi(B)\}^{\nu'}$$

can be expressed in the form

$$\Sigma (U', V')^{\sigma}$$

where the degree of U' is less than that of HW, *i.e.* less than that of U.

Thus $T' - T$ can be written

$$\Sigma (U'', V'')^{\nu'} + \Sigma (U', V')^{\sigma}, \ \text{mod}\ G,$$

where the degree of U'' is the same as that of U and $\nu' < \nu$, while the degree of U' is less than that of U.

Hence if all terms of transvectants considered previously to

$$(U, V)^{\nu}$$

can be expressed rationally and integrally in terms of

$$C_1, C_2, \dots C_p$$

(except for terms involving G); then all terms of transvectants up to and including

$$(U, V)^{\nu}$$

can be expressed in the form

$$F(C_1, C_2, \dots C_p, T), \ \text{mod}\ G,$$

where T is any term of the last transvectant.

If the transvectant

$$(U, V)^{\nu}$$

contain a reducible term we may suppose it to be T, and since $T = T_1 T_2$ where T_1, T_2 are terms of former transvectants, there is no need to add the term T to $C_1, C_2, \dots C_p$.

It follows that in constructing a system of C's in terms of which all C's can be expressed we have to add a new member only when we come to a transvectant containing no reducible terms and then one only. The number of transvectants containing no irreducible terms is finite, § 99; hence if $C_1, C_2, \dots C_q$ be a series of terms one from each of this finite number of transvectants, any other member of the system C derived by transvection from A and B can be expressed as a rational integral function of $C_1, C_2, \dots C_q$ together with terms involving the factor G; in other words, the system C is relatively finite for the modulus G.

Next the system

$$C_1, C_2, \dots C_q$$

is relatively complete with respect to the modulus G.

For any term T derived by convolution from

$$W = C_1^{\gamma_1} C_2^{\gamma_2} \dots C_q^{\gamma_q}$$

may be regarded as a term in a transvectant

$$(\overline{U}, \overline{V})^\rho,$$

where $\overline{U}$ is derived by convolution from a product of the A's and $\overline{V}$ from a product of the B's.

Hence T can be expressed as an aggregate of transvectants

$$(\overline{U}, \overline{V})^\sigma;$$

while $\overline{U} = P$ can be derived by convolution from a power of f and

$$\overline{V} \equiv \Phi(B), \text{ mod } G;$$

therefore

$$\begin{aligned} T &\equiv \Sigma \{P, \Phi(B)\}^\sigma, \text{ mod } G, \\ &\equiv \Sigma (P, V)^\sigma, \text{ mod } G, \\ &\equiv \Sigma (U, V)^{\sigma'}, \text{ mod } G. \quad \text{(Lemma.)} \end{aligned}$$

Consequently, as has just been proved,

$$T = F(C_1, C_2, \dots C_q), \text{ mod } G,$$

and the system is complete.

107. Cor. I. *If the system B is absolutely complete, then the system derived by transvection from A and B is absolutely complete.*

Cor. II. If the system B is complete for two moduli G and G' and contains a form whose only determinantal factors are H, then the derived system is complete for the two moduli G and G'.

To prove this we have only to write

$$\overline{B} \equiv F(B), \text{ modd } (G, G')$$

instead of

$$\overline{B} \equiv F(B), \text{ mod } (G)$$

at every stage of the foregoing proof.

108. Gordan's Theorem. These long preliminary explanations are now at an end and the actual proof of the theorem does not present much difficulty.

Every covariant of a binary form

$$f = a_x^n = b_x^n = \text{etc.}$$

is either a power of f or else contains a factor $(ab)^2$, and hence the form f itself is a complete system with respect to the modulus $(ab)^2$.

Assuming now that a system of covariants containing f and relatively complete for the modulus $(ab)^{2k}$ can be found we shall shew how to construct a system also containing f and relatively complete for the modulus $(ab)^{2k+2}$. The system relatively complete mod $(ab)^{2k}$ is called A_{k-1}, and since every covariant can be derived from f by convolution it is a rational integral function of the forms in A_{k-1} except for terms involving the factor $(ab)^{2k}$.

To construct the system A_k when A_{k-1} is known we make use of the theorem of § 104.

We must therefore begin by constructing a system B_{k-1} possessing the following properties:

(i) it contains the form $(ab)^{2k} a_x^{n-2k} b_x^{n-2k}$,

(ii) it is relatively complete for the modulus $(ab)^{2k+2}$.

Then the system derived from A_{k-1} and B_{k-1} by transvection will be finite and complete with respect to the modulus $(ab)^{2k+2}$, and as it obviously contains f which is contained in A_{k-1} it is the system A_k required.

109. Accordingly we have now to shew how to construct the system B_{k-1}.

There are three cases.

I. If $2k < \frac{n}{2}$ then any form derived by convolution from a power of $H_k = (ab)^{2k} a_x^{n-2k} b_x^{n-2k}$ is of grade $(2k+1)$ at least and therefore of grade $(2k+2)$ since all symbols are now equivalent.

Hence H_k is itself relatively complete for the modulus $(ab)^{2k+2}$ and in this case the system B_k consists of the single form

$$(ab)^{2k} a_x^{n-2k} b_x^{n-2k}. \quad (\S\ 74.)$$

II. If $2k > \frac{n}{2}$ then $H_k = (ab)^{2k} a_x^{n-2k} b_x^{n-2k}$ is of order less than n, say m.

Now we suppose that the complete system of covariants for a form of order $< n$ is known and we derive a system from H_k on the model of the complete system of α_x^m as explained in §§ 79, 80.

Neglecting terms containing $(ab)^{2k+2}$ we can replace each copied form by a single term; the system so derived is complete for the modulus $(ab)^{2k+2}$ and is therefore the system B_{k-1} required.

III. If $2k = \frac{n}{2}$—a case which can only arise when n is a multiple of 4—we have a rather different state of things.

Here the form $H_k = (ab)^{2k} a_x^{n-2k} b_x^{n-2k}$ is relatively complete for the two moduli

$$(ab)^{2k+2}, \quad (ab)^{2k} (bc)^{2k} (ca)^{2k},$$

the latter being an invariant J, and hence by Cor. II. § 107 the system derived by transvection from A_{k-1} and B_{k-1} is relatively complete for the moduli $(ab)^{2k+2}$ and J; calling this system C_k for a moment we have

$$\bar{C}_k \equiv F(C_k) + J \, . \, P_1, \quad \text{mod } (ab)^{2k+2},$$

where P_1 is a covariant of degree less than $\bar{C}_k$.

Further since P_1 can be derived by convolution from f which is contained in C_k, we have

$$P_1 \equiv F_1(C_k) + J \, . \, P_2, \quad \text{mod } (ab)^{2k+2},$$

where P_2 is a covariant of degree less than P_1.

Proceeding in this way we see that $\bar{C}_k$ is a rational integral function of J and the forms in C_k together with terms involving the factor $(ab)^{2k+2}$.

Hence if we add J to the system C_k and call the total system A_k it follows at once that A_k is relatively complete for the modulus

$$(ab)^{2k+2}.$$

Therefore in every case, given the complete system mod $(ab)^{2k}$ we can construct that mod $(ab)^{2k+2}$; but the system A_0 is f, thence we find the system A_1, then from that the system A_2 and so on, in fact we can construct the system A_k relatively complete for the modulus $(ab)^{2k+2}$.

110. Consider now a little more closely what happens when we come to the end of the sequence of moduli $(ab)^2$, $(ab)^4$, $(ab)^6$ ..., and first let n be even and equal to $2g$.

Then the system A_{g-1} is relatively complete for the modulus $(ab)^{2g}$, and the system B_{g-1} consists of the single invariant $(ab)^{2g}$ so that it is of course absolutely complete.

Hence the system derived from A_{g-1} and B_{g-1} by transvection is absolutely complete and it contains f, therefore it is the complete system of invariants and covariants; further since B_{g-1} consists of a single invariant the complete system A_g consists of A_{g-1} and that invariant $(ab)^{2g}$.

Secondly let n be odd and equal to $2g+1$, then the system A_{g-1} can be constructed and it both contains f and is relatively complete for the modulus $(ab)^{2g}$.

The system B_{g-1} is derived from the quadratic

$$(ab)^{2g} a_x b_x$$

by the same convolutions as the complete system of the quadratic $\alpha_x^2 = \beta_x^2$ is found from this form. This complete system being α_x^2 and $(\alpha\beta)^2$ the system B_{g-1} consists of

$$(ab)^{2g} a_x b_x, \ (ab)^{2g} (ac)(bd)(cd)^{2g}.$$

This system is relatively complete for the modulus $(ab)^{2g+1}$ by § 109 II, and this being a vanishing invariant it follows that B_{g-1} is absolutely complete.

Hence the system derived from A_{g-1} and B_{g-1} contains f and is absolutely complete, that is it constitutes the complete system of f.

To recapitulate—the complete system mod $(ab)^2$ can be written down at once, then from that we deduce the complete system mod $(ab)^4$ and proceeding step by step we can finally construct an absolutely complete system as the last step in our series.

We have therefore proved that the complete system is finite, for all the systems $A_1, A_2, \dots$ are finite, and we have shewn how to construct it on the assumption that the systems for forms of lower orders are known—the proof is thus inductive in its nature.

111. We shall illustrate the above process by applying it to the quadratic, cubic, and quartic.

(i) *Quadratic.* The system A_0 is

$$f = a_x^2 = b_x^2$$

and the system B_0 is $(ab)^2$, hence the complete system is

$$a_x^2, \ (ab)^2.$$

(ii) *Cubic.* Here A_0 is

$$f = a_x^3 = b_x^3 = \text{etc.}$$

and B_1 is $$(ab)^2 a_x b_x, \ (ab)^2 (ac)(bd)(cd)^2,$$

in fact $$H, \ (H, H)^2.$$

This system B_1 is absolutely complete, therefore the system derived by transvection is the complete system.

It consists of

$$f, H, (H, H)^2 = \Delta \text{ and } (f^\alpha, H^\beta)^\gamma.$$

Proceeding as in § 88 we can shew that the only irreducible transvectant is (f, H).

(iii) *Quartic.* Here A_0 is

$$f = a_x^4 = b_x^4 \ldots,$$

B_0 is $$H = (ab)^2 a_x^2 b_x^2,$$

and this is complete modd $(ab)^4$ and $(ab)^2(bc)^2(ca)^2$.

The system derived by transvection is

$$(f^\alpha, H^\beta)^\gamma.$$

If $\gamma \geqslant 2$ this has a term containing the factor $(ab)^2(ac)^2$ which is congruent to zero modd $(ab)^4$, $(ab)^2(bc)^2(ca)^2$.

Hence we need take only $\gamma = 1$ and thence only $\alpha = 1$, $\beta = 1$, and we find that

$$f, H, (f, H)$$

is relatively complete modd $(ab)^4$, $(ab)^2(bc)^2(ca)^2$.

Therefore $$f, H, (f, H), (ab)^2(bc)^2(ca)^2$$

is complete mod $(ab)^4$ and is the system A_1.

Then B_1 being the invariant $i = (ab)^4$ we have for the complete system

$$f, H, t = (f, H), \ i = (ab)^4, \ j = (ab)^2(bc)^2(ca)^2.$$

112. We shall now apply the principles of §§ 73, 76 to the deduction of a complete system mod $(ab)^{\frac{n}{2}}$ for the binary form of order n.

The system A_0 consists of

$$f = a_x^n = b_x^n = \text{etc.}$$

and B_0 of $$H = (ab)^2 a_x^{n-2} b_x^{n-2} \ldots.$$

The system A_1 is derived by transvection from A_0 and B_0.

Now
$$(f^\alpha, H^\beta)^\gamma$$
has a term containing the factor $(ab)^2(ac)^2$ if $\gamma > 1$, and since such a term is
$$\equiv 0 \bmod (ab)^4$$
the transvectant may be rejected.

If $\gamma = 1$ the transvectant contains reducible terms unless $\alpha = \beta = 1$, and hence A_1 consists of
$$f, H, (f, H) = t.$$
The system B_1 is
$$(ab)^4 a_x^{n-4} b_x^{n-4}$$
and A_2 is derived by transvection from A_1 and B_1.

If the index of a transvectant be greater than two it contains a term having a factor $(ab)^4(ac)^3$ and this is
$$\equiv 0 \bmod (ab)^6. \quad (\S\ 70.)$$

We need only consider the cases in which the index is $\leqslant 2$, and since the order of each form in A_1 is certainly greater than 2 (in fact $\frac{n}{2} \geqslant 4$), products of forms may be rejected.

There remain transvectants of each form of A_1 taken simply with
$$H_2 = (ab)^4 a_x^{n-4} b_x^{n-4}.$$

For the future we shall only write down the determinantal factors of a covariant.

Transvectants with f give rise to
$$(ab)^4(bc),\ (ab)^4(bc)^2.$$
Those with H give
$$(ab)^4(bc)(cd)^2,\ (ab)^4(bc)^2(cd)^2,$$
and finally those with t give
$$(ab)^4(bc)(cd)^2(dc),\ (ab)^4(bc)^2(cd)^2(dc).$$

Now by § 76
$$(ab)^4(bc)^2(cd)^2 \equiv (ab)^4(cd)^4, \bmod (ab)^6;$$
hence
$$(ab)^4(bc)^2(cd)^2(de)$$
being a term of
$$\{(ab)^4(bc)^2(cd)^2, e_x^n\}$$

we have

$$(ab)^4 (bc)^2 (cd)^2 (de) \equiv \{(ab)^4 (cd)^4, e_x{}^n\}, \text{ mod } (ab)^6,$$

for all expressions derived by convolution from

$$(ab)^4 (bc)^2 (cd)^2$$

are $\equiv 0 \text{ mod } (ab)^6$. (§ 73.)

Now a term of the last transvectant is

$$(cd)^4 . (ab)^4 (ae),$$

$$\therefore \; (ab)^4 (bc)^2 (cd)^2 (de) \equiv (cd)^4 . (ab)^4 (ae), \text{ mod } (ab)^6$$

and accordingly may be rejected.

Finally $(ab)^4 (bc) (cd)^2 (de)$

is reducible as being the Jacobian of a Jacobian, and the system A_2 consists of

$$f, (ab)^2, (ab)^2 (bc),$$

$$(ab)^4, (ab)^4 (bc), (ab)^4 (bc)^2, (ab)^4 (bc) (cd)^2.$$

113. Before proceeding further we shall develope the results of § 76 by shewing that a symbolical product Γ containing the factor

$$(ab)^\lambda (bc)^\mu (cd)^\nu,$$

in which λ is even and equal to $\mu + \nu$, can in general be expressed in terms of covariants that are either reducible or of grade greater than λ.

The above reduction of

$$(ab)^4 (bc)^2 (cd)^2 (de)$$

is a case in point.

In fact Γ is a term of

$$\{(ab)^\lambda (bc)^\mu (cd)^\nu, \phi\}^\rho$$

which we write $(T, \phi)^\rho$.

Hence

$$\Gamma = (T, \phi)^\rho + \Sigma (\bar{T}, \bar{\phi})^{\rho'}, \quad \rho' < \rho$$

$$\equiv (T, \phi)^\rho + \Sigma (T, \bar{\phi})^{\rho'}, \text{ mod } (ab)^{\lambda+1},$$

since $\bar{T}$ derived by convolution from $(ab)^\lambda (bc)^\mu (ca)^\nu$ is of grade greater than λ, § 73.

Again

$$T = (ab)^\lambda . (cd)^\lambda + C_{\lambda+1} \quad (§ 76),$$

therefore

$$\Gamma \equiv \{(ab)^\lambda . (cd)^\lambda, \phi\}^\rho + \{(ab)^\lambda . (cd)^\lambda, \bar{\phi}\}^{\rho'} + C_{\lambda+1}.$$

Now if $2n - 2\lambda \geqslant \rho$ each of these transvectants contains terms having $(cd)^\lambda c_x^{n-\lambda} d_x^{n-\lambda}$ as a factor.

Hence

$$\{(ab)^\lambda . (cd)^\lambda, \phi\}^\rho$$
$$\equiv (cd)^\lambda \{(ab)^\lambda, \phi\}^\rho + \Sigma \{(ab)^\lambda . (cd)^\lambda, \bar{\phi}\}^\sigma, \text{ mod } (ab)^{\lambda+1}, \quad \sigma < \rho,$$

and by continuation of this process we can express Γ entirely in terms of reducible covariants and covariants of grade greater than λ; it suffices to remark that the index σ diminishes at every step.

It is quite easy to see that the condition

$$2n - 2\lambda \geqslant \rho$$

is satisfied in all our cases—at any rate it will be in the course of the subsequent work.

114. Returning now to the general form, B_2 consists of

$$H_3 = (ab)^6 a_x^{n-6} b_x^{n-6},$$

and A_3 is derived by transvection from A_2 and B_2.

The argument used in evolving A_1 and A_2 enables us to see

(i) that transvectants with index > 3 may be rejected,

(ii) thence that transvectants of products or powers of forms may be likewise rejected.

We are therefore left with transvectants of the forms of A_2 taken simply with H_2, the index being $\not> 3$.

Omitting Jacobians of Jacobians and forms having a factor

$$(ab)^6 (bc)^\mu (cd)^\nu, \text{ where } \mu + \nu \geqslant 6,$$

we have

from f,	$(ab)^6 (bc)$, $(ab)^6 (bc)^2$, $(ab)^6 (bc)^3$;
,, H,	$(ab)^6 (bc) (cd)^2$, $(ab)^6 (bc)^2 (cd)^2$, $(ab)^6 (bc)^3 (cd)^2$;
,, $(ab)^2 (bc)$,	$(ab)^6 (bc)^2 (cd)^2 (de)$, $(ab)^6 (bc)^3 (cd)^2 (de)$;
,, $(ab)^4$,	$(ab)^6 (bc) (cd)^4$;
,, $(ab)^4 (bc)$,	none;
,, $(ab)^4 (bc)^2$,	$(ab)^6 (bc) (cd)^4 (de)^2$;
,, $(ab)^4 (bc) (cd)^2$,	none.

Hence we have found for A_3 the above ten forms in addition to those of A_2. Putting aside the question as to whether any of these ten new forms are reducible, a continued repetition of the above process establishes the fact that all the forms of the system A_k, relatively complete for the modulus

$$(ab)^{2k+2}, \quad \left(k < \frac{n}{2}\right),$$

are included in the set

$$(ab)^{\lambda} (bc)^{\mu} (cd)^{\lambda'} (de)^{\mu'} (ef)^{\lambda''} (fg)^{\mu''} \ldots *$$

where the exponents satisfy the following conditions:

(i) $\lambda \not> 2k$,

(ii) $\lambda, \lambda', \lambda'', \ldots$ are all even,

(iii) $\lambda > \lambda' + \mu, \ \lambda' > \lambda'' + \mu', \ldots,$

(iv) no two of the exponents $\mu, \mu', \ldots$ are equal to unity.

In fact

(ii) follows immediately from the way in which the covariants are formed.

(iii) results from the application of §§ 73, 76.

(iv) is the expression of the fact that the Jacobian of a Jacobian is reducible.

Ex. (i). If the orders m, n, p of the forms f, ϕ, ψ, be each greater than two, then

$$(f\phi, \psi)^2 = \frac{m}{m+n} (f, \psi)^2 \phi + \frac{n}{m+n} (\phi, \psi)^2 f - \frac{mn}{(m+n)(m+n-1)} (f, \phi)^2 \psi.$$

Ex. (ii). For a form whose order is greater than four the covariants

$$(ab)^3 (bc) (cd), \quad (ab) (bc)^3 (cd), \quad (ab)^2 (bc)^2 (cd)$$

all vanish identically.

Ex. (iii). If $n > 5$, then

$$(bc)^2 (ca)^2 (ab)^2 a_x^{n-4} b_x^{n-4} c_x^{n-4}$$
$$= (ab)^4 (ac)^2 a_x^{n-6} b_x^{n-4} c_x^{n-2} - \tfrac{1}{2} (ab)^6 a_x^{n-6} b_x^{n-6} . c_x^{n}.$$

Ex. (iv). If $n > 4$, then

$$(H, f)^3 = \frac{n-4}{4n-10} (ab)^4 (bc) a_x^{n-4} b_x^{n-5} c_x^{n-1} = \frac{n-4}{4n-10} \{(f, f)^4, f\}$$

and if $n > 5$,

$$(H, f)^4 = -\frac{n-1}{2n-5} \{(f, f)^4, f\}^2 + \frac{3n^2 - 25n + 50}{4(2n-7)(n-4)} (f, f)^6 . f.$$

* Cf. Jordan, *Liouville's Journal*, 1876, 1879.

Ex. (v). For a form whose order is greater than three prove that

$$(H, H)^2 + \frac{(n-2)}{2(2n-5)} H(f, f)^4$$
$$= \{(f, H)^2, f\}^2 + \frac{2(2n-6)}{3n-8} \{(f, H)^3, f\} + \frac{2(2n-7)}{3(3n-8)} (f, H)^4 . f.$$

Hence replacing $(f, f)^4$ by i express $(H, H)^2$ as a linear combination of

$$Hi, f(i, f)^2, f^2(f, f)^6$$

and finally express t^2 in terms of the irreducible forms of the system.

Ex. (vi). Prove that all irreducible covariants of degree four and rank not greater than $\frac{n}{2}$ are included in

$$(ab)^{2\lambda}(bc)^{\mu}(cd)^{\nu}$$

where $$2\lambda \ngtr \frac{n}{2} \text{ and } \lambda > \mu > \nu.$$

Ex. (vii). In § 103 if no A be of order greater than m and no B be of order greater than n, then no form of the system C is of order greater than $m+n-2$.

CHAPTER VII.

THE QUINTIC.

115. To obtain the complete irreducible system of covariants of the quintic, we follow step by step Gordan's proof of the finiteness. Let us briefly recapitulate.

The complete system of forms, which are not expressible in terms of covariants having a symbolical factor $(ab)^2$, is first found; this is called A_0, it is the complete system mod $(ab)^2$. Generally A_k is used to denote the complete system mod $(ab)^{2k+2}$. To obtain the system A_{k+1} from the system A_k, a subsidiary system of forms B_k is used. This system is a system of forms having $\phi = (ab)^{2k+2} a_x^{n-2k-4} b_x^{n-2k-4}$ for ground-form.

When the order of this form is less than n, B_k consists of its complete irreducible system. Otherwise if the order of ϕ is greater than n we may take for B_k the single form ϕ; while when the order of ϕ is equal to n, the system B_k consists of ϕ and the invariant $(ab)^{\frac{n}{2}} (bc)^{\frac{n}{2}} (ca)^{\frac{n}{2}}$.

Then it has been proved that the system A_{k+1} may be obtained by taking transvectants of products and powers of forms from A_k with products and powers of forms from B_k.

116. The quintic will be written

$$f = a_x^5 = b_x^5 = \dots\dots$$

The system A_0 contains f only.

The system B_0 contains only

$$(ab)^2 a_x^3 b_x^3 = H.$$

The system A_1 is then obtained from the transvectants

$$(f^\alpha, H^\beta)^\gamma.$$

If $\gamma > 2$, this transvectant contains a term having a factor $(bc)^3$; such a term can be expressed as a sum of symbolical products each containing a factor $(ab)^4$, and is therefore $\equiv 0$, mod $(ab)^4$.

Hence we may reject these transvectants when $\gamma > 2$, for all transvectants which contain reducible terms may be rejected.

The transvectant $(H, f)^2$ contains the term

$$(ab)^2 (bc)^2 a_x^3 b_x c_x^3 = \tfrac{1}{2} f . (ab)^4 a_x b_x \qquad (\S\ 51, \text{Ex. (vi)}).$$

The system A_1 then consists of

$$f,\ H,\ (f,\ H) = t.$$

The system B_1 is built up from the form

$$(ab)^4 a_x b_x = i,$$

this is of order < 5, hence we must take the complete irreducible system of the quadratic i.

The system B_1 then consists of

$$i,\ (i,\ i)^2 = A.$$

The system A_2 is now the complete system of forms for the quintic, it is made up of the transvectants

$$U = (f^\alpha H^\beta t^\gamma,\ i^\delta A^\epsilon)^\eta.$$

Since A is an invariant we may suppose that $\epsilon = 0$ (if at the same time we remember that A belongs to the complete system).

Since H is a form of even order, and i is a quadratic, all transvectants are reducible except those which have

(i) $\beta = 0$,

(ii) $\alpha = 0,\ \gamma = 0,\ \beta = 1$.

Again t is the Jacobian of f and H, therefore

$$t^2 = -\tfrac{1}{2} \{(f,\ f)^2 H^2 - 2 (f,\ H)^2 f . H + (H,\ H)^2 . f^2\}$$
$$\equiv -\tfrac{1}{2} H^3 \text{ mod } (ab)^4. \qquad (\S\ 78.)$$

Hence any transvectant Γ, in which $\gamma > 1$, can be expressed in terms of transvectants Γ in which the degree of the product on the left has been decreased and that on the right has been increased together with reducible terms (§ 105, Cor.): for as we have just seen if $\beta > 1$ then U is reducible.

Accordingly we have the following cases to consider:

(i) $\alpha=1$ or 2, $\beta=0$, $\gamma=0$,

(ii) $\alpha=0$, $\beta=0$, $\gamma=1$,

(iii) $\alpha=1$, $\beta=0$, $\gamma=1$,

(iv) $\alpha=0$, $\beta=1$, $\gamma=0$.

All other transvectants are reducible or are expressible in terms of these.

(i) $\alpha=1, 2$, $\beta=0$, $\gamma=0$.

The irreducible transvectants are

$$(f, i), (f, i)^2, (f, i^2)^3, (f, i^2)^4, (f, i^3)^5, (f^2, i^5)^{10}.$$

To see that the other possibilities contain reducible terms we shall take one example. The transvectant $(f^2, i^5)^9$ contains the term $(f, i^3)^5 (f, i^2)^4$.

(ii) $\alpha=0$, $\beta=0$, $\gamma=1$.

Since $$t=-(ab)^2(bc)\,a_x^3b_x^2c_x^4,$$

and $$(bc)(bi)\,c_xi_x=\tfrac{1}{2}\{(bc)^2i_x^2+(bi)^2c_x^2-(ci)^2b_x^2\},$$

the term $$-(ab)^2(bc)(bi)\,a_x^3b_xc_x^4i_x$$

of $$(t, i)$$

is reducible. Similarly the transvectants

$$(t, i^2)^3, (t, i^3)^5, (t, i^4)^7$$

contain reducible terms.

Thus $(t, i^4)^7$ contains the term

$$-(ab)^2(bc)(ai')^2(ai'')(bi'')(ci''')^2(bi)\,c_x^2i_x$$

which is at once reduced by means of the above identity.

We are left with the forms

$$(t, i)^2, (t, i^2)^4, (t, i^3)^6, (t, i^4)^8, (t, i^5)^9.$$

(iii) $\alpha=1$, $\beta=0$, $\gamma=1$.

The only irreducible transvectant here, is

$$(f.t, i^7)^{14}.$$

To see that the other transvectants are reducible it is sufficient to remark that for example $(f.t, i^7)^{13}$ contains the term

$$(f, i^3)^5.(t, i^4)^8.$$

(iv) $\alpha = 0,\ \beta = 1,\ \gamma = 0.$

The transvectants

$$(H,\ i),\ (H,\ i)^2,\ (H,\ i^2)^3,\ (H,\ i^2)^4,\ (H,\ i^3)^5,\ (H,\ i^3)^6$$

prove to be all irreducible.

We are left with 23 forms which are as follows*:

Degree	Order								
	0	1	2	3	4	5	6	7	9
1						f			
2			i				H		
3				$(i, f)^2$		(i, f)			t
4	A				$(i, H)^2$		(i, H)		
5		$(i^2, f)^4$		$(i^2, f)^3$				$(i, t)^2$	
6			$(i^2, H)^4$		$(i^2, H)^3$				
7		$(i^3, f)^5$				$(i^2, t)^4$			
8	$(i^3, H)^6$		$(i^3, H)^5$						
9				$(i^3, t)^6$					
11		$(i^4, t)^8$							
12	$(i^5, f^2)^{10}$								
13		$(i^5, t)^9$							
18	$(i^7, ft)^{14}$								

* One very obvious remark is to be made regarding this, and all other complete systems obtained by the present methods. We are assured that every covariant can be expressed rationally and integrally in terms of those retained in the complete system, but there is nothing in the process to shew that the latter are all irreducible, except in so far as failure to reduce them may be taken as evidence in this direction. Theoretically then Gordan's process gives an upper limit to the irreducible system.

The enumerative method, depending on the generating function, introduced by Cayley and finally developed by Sylvester and Franklin (*Am. Jour.* vol. VII.) gives a lower limit to the system and when the two methods give the same result the irreducible set has been obtained. The results even when identical have to be received with some caution on account of the enormous labour involved.

117. It is found that for discussing the properties of covariants, it is convenient to have the indices of the transvectants which express them as low as possible. On this account it is usual to replace some of the forms in the irreducible system just given by others which differ from them by reducible terms only.

In the first place the covariant

$$j = -(f, i)^2$$

is of fundamental importance in the quintic system. It is a cubic, and the system of forms for which it is a ground-form are irreducible when considered as forms belonging to the quintic.

Now $(H, i^2)^4$ contains a term

$$(ab)^2 (ai)^2 (bi')^2 a_x b_x = (j, j)^2,$$

accordingly we shall take the irreducible form of degree 6 and order 2 to be

$$(j, j)^2 = \tau.$$

Similarly the form $(t, i^3)^6$ may be replaced by (j, τ), for $(t, i^3)^6$ contains a term

$$(ab)^2 (bc) (ai)^2 (bi')^2 (ci'')^2 a_x c_x^2$$
$$= ((ab)^2 (ai)^2 (bi')^2 a_x b_x,\ (ci'')^2 c_x^3) = (j, \tau).$$

And $(f^2, i^5)^{10}$ may be replaced by the invariant of j,

$$(\tau, \tau)^2.$$

This invariant will be denoted by C, the proof that it may be included in the system instead of $(f^2, i^5)^{10}$ will be given later (§ 121).

It will be found useful to denote the term

$$(ai)^2 (ai')^2 (bi'')^2 (bi''')^2 (ai^{\text{iv}}) (bi^{\text{iv}})$$

of this transvectant by M. Then M may be taken as the invariant of degree 12.

It may be recalled in fact in connection with the simultaneous system of a cubic and quartic (Gundelfinger, *Math. Ann.* Bd. IV.) that the two results originally agreed, but a revision of the generating function led to a reduction of the lower limit which it theoretically gives, and afterwards two forms included in the irreducible system as derived by the methods of Gordan and Clebsch were found to be reducible. The complete systems for the binary forms up to the octavic may be considered as accurately determined by the two methods combined.

Besides j and τ, there is one more quadratic covariant, given in the list as $(H, i^3)^5$. This is equal to $((H, i^2)^4, i)$; we may substitute τ for $(H, i^2)^4$, and hence take as the remaining quadratic covariant

$$(\tau, i) = -\vartheta.$$

118. *The linear covariants.*

$$(f, i^2)^4 = (ai)^2 (ai')^2 a_x = -(j, i)^2 = \alpha,$$

$$(f, i^3)^5 = (ai)^2 (ai')^2 (ai'') i_x'' = (\alpha, i) = -\beta.$$

$(t, i^4)^8$ contains the term

$$\begin{aligned}(ab)^2 (bc) (ai)^2 (bi')^2 (ci'')^2 (ci''')^2 a_x \\ = ((ab)^2 (ai)^2 (bi')^2 a_x b_x, \alpha) \\ = (\tau, \alpha) = \gamma.\end{aligned}$$

$(t, i^5)^9$ contains the term

$$\begin{aligned}&(ab)^2 (bc) (ai)^2 (bi')^2 (ai'') (ci''')^2 (ci^{\text{iv}})^2 i_x'' \\ &= ((ab)^2 (ai)^2 (bi')^2 (ai'') i_x'' b_x, \alpha) - \tfrac{1}{2} (ab)^2 (ai)^2 (bi')^2 (ai'') (bi'') . \alpha,\end{aligned}$$

of which the second term is reducible and the first

$$= ((\tau, i), \alpha) = -(\vartheta, \alpha) = -\delta^*.$$

119. *The invariants.*

$$(i, i)^2 = A,$$

$$(H, i^3)^6 = ((H, i^2)^4, i)^2.$$

The latter may be replaced by

$$(\tau, i)^2 = B.$$

$(f^2, i^5)^{10}$ has been replaced already by M.

$(ft, i^7)^{14}$ contains a term

$$((f, i^2)^4, (t, i^5)^9),$$

and hence may be replaced by $(\alpha, \delta) = -R$.

Taking the Jacobians of the 6 linear forms two and two, we obtain the 6 invariants

$$(\alpha\beta), (\beta\gamma), (\gamma\alpha), (\alpha\delta), (\beta\delta), (\gamma\delta).$$

* This is the definition of the linear covariant δ of degree 13 given by Clebsch. In Gordan's book $\delta = (\tau, \beta) = -(\vartheta, \alpha) - \frac{1}{2}(i, \tau)^2 . \alpha$. In other respects the letters common to the two books are identical in meaning.

The values of these invariants are*

$$(\alpha\beta) = -(ai)^2(ai')^2(ai'')(bi''')^2(bi^{\text{iv}})^2(bi'') = -M = -(i\alpha)^2,$$
$$(\beta\gamma) = (i\alpha)(\tau\alpha)(i\tau) = (\vartheta\alpha)^2 = (\delta\alpha) = R,$$
$$(\gamma\alpha) = (\tau\alpha)^2 = N,$$
$$(\alpha\delta) = -R.$$

Now $\delta = ((i, \tau), \alpha) = (i\tau)(\tau\alpha)\, i_x + \tfrac{1}{2}(i\tau)^2 . \alpha_x,$

hence

$$(\beta\delta) = (i\alpha)(ii')(i'\tau)(\tau\alpha) + \tfrac{1}{2}B . (i\alpha)^2$$
$$= \frac{-1}{2}AN + \frac{1}{2}BM,$$
$$(\gamma\delta) = ((\tau\alpha)\,\tau_x, (i\tau')(i\alpha)\,\tau_x' - \tfrac{1}{2}B . \alpha)$$
$$= \tfrac{1}{2}(CM - BN).$$

Also $N = (\gamma\alpha) = (\tau\alpha)^2 = (ji)^2(j'i')^2(\tau j)(\tau j')$
$$= (ji)(j'i')^2(\tau j)\{(\tau i)(jj') + (\tau j)(j'i)\}.$$

Now from the theory of the cubic we know that any symbolical product which contains a factor $(\tau j)^2$ is zero.

Hence

$$N = (jj')(ji)(j'i')^2(\tau j)(\tau i)$$
$$= \tfrac{1}{2}(jj')(\tau i)\{(ji)(j'i')^2(\tau j) - (j'i)(ji')^2(\tau j')\}$$
$$= \tfrac{1}{2}(jj')(\tau i)\{(jj')(ii')(j'i')(\tau j) + (ji')(j'i)(j'j)(\tau i')\}$$
$$= \tfrac{1}{4}(jj')^2(\tau j')(\tau j) . (ii')^2 - \tfrac{1}{2}((jj')^2 j_x j_x', (\tau i)(\tau i')\, i_x i_x')^2$$
$$= \tfrac{1}{4}AC - \tfrac{1}{2}(\tau, \tfrac{1}{2}[(\tau i)^2 . i_x'^2 + (\tau i')^2 . i_x^2 - (ii')^2\,\tau_x^2])^2$$
$$= \tfrac{1}{2}(AC - B^2).$$

120. The third transvectant of f with j is identically zero. For

$$(f, j)^3 = -(a_x^5, (bi)^2 b_x^3)^3 = -(ab)^3(bi)^2 a_x^2$$
$$= +\tfrac{1}{2}(ab)^3\{(ai)^2 b_x^2 - (bi)^2 a_x^2\} = \tfrac{1}{2}(ab)^4 i_x . \{(ai)\, b_x + (bi)\, a_x\}$$
$$= (i, i) = 0.$$

This property is sufficient to define j. For if ψ be an arbitrary cubic then $(f, \psi)^3$ is a quadratic. And in order that $(f, \psi)^3$ may

* In future when no confusion can arise the comma between the two forms in a transvectant will be omitted. This is the usual practice, cf. Stroh.

vanish identically the coefficients of x_1^2, x_1x_2, x_2^2 must be separately zero; giving three equations to determine the ratios of the four coefficients of ψ,—see Chap. XII.

Again

$$\begin{aligned}(f, \tau)^2 &= (aj)(aj')(jj')^2 a_x^3 \\ &= (aj)(aj') a_x \{(aj') j_x - (aj) j_x'\}^2 \\ &= -2 (aj)^2 (aj')^2 a_x j_x j_x', \text{ since } (aj)^3 a_x^2 = 0 \\ &= -2 (ab)^2 (ac)^2 (bi)^2 (ci')^2 a_x b_x c_x.\end{aligned}$$

Now
$$\begin{vmatrix} (ab)^2 & 0 & (i'b)^2 & (cb)^2 \\ (ai)^2 & (bi)^2 & A & (ci)^2 \\ (ac)^2 & (bc)^2 & (i'c)^2 & 0 \\ 0 & (ba)^2 & (i'a)^2 & (ca)^2 \end{vmatrix} = 0 \qquad (\S\ 77).$$

Hence, if Σ include all possible expressions obtained by interchanging a, b, c,

$$\begin{aligned}\Sigma (ab)^2 (bi)^2 (i'c)^2 (ca)^2 \\ = (ab)^4 (ci)^2 (ci')^2 + (bc)^4 (ai)^2 (ai')^2 + (ca)^4 (bi)^2 (bi')^2 \\ + 2A \,.\, (ab)^2 (bc)^2 (ca)^2.\end{aligned}$$

And therefore

$$\begin{aligned}(f, \tau)^2 &= -2 (ab)^2 (ac)^2 (bi)^2 (ci')^2 a_x b_x c_x \\ &= -(ab)^4 a_x b_x \,.\, (ci)^2 (ci')^2 c_x - \tfrac{2}{3} A \,.\, (ab)^2 (bc)^2 (ca)^2 a_x b_x c_x.\end{aligned}$$

But
$$\begin{aligned}&(ab)^2 (bc)^2 (ca)^2 a_x b_x c_x \\ &= (ab)^2 (bc)^2 (ca) [-(bc) a_x - (ab) c_x] a_x c_x \\ &= -(ab)^2 (bc)^3 (ca) a_x^2 c_x - (ab)^3 (bc)^2 (ca) a_x c_x^2 \\ &= \tfrac{1}{2} (bc)^4 (ca)(ab) a_x^3 + \tfrac{1}{2} (ab)^4 (bc)(ca) c_x^3 \\ &= -(i, f)^2 = j.\end{aligned}$$

Therefore
$$(f, \tau)^2 = -i \,.\, \alpha - \tfrac{2}{3} A \,.\, j.$$

121. To obtain the relation between the invariants C and M, we take the expression for M and introduce in it so far as possible symbols referring to the cubic j and its Hessian τ, for

$$C = (\tau\tau)^2.$$

Now

$$M = (i\alpha)^2$$
$$= (ji)^2 (j'i')^2 (ji'') (j'i'')$$
$$= (ji)^2 (j'i')^2 (ja) (j'b) (ab)^4$$
$$= (ja) (j'b) [(ja) (ib) - (jb) (ia)]^2 [(j'a) (i'b) - (j'b) (i'a)]^2$$
$$= (ja) (j'b) [-2 (ja) (jb) (ia) (ib) + (jb)^2 (ia)^2] [(j'a)^2 (i'b)^2 - 2 (j'a) (j'b) (i'a) (i'b)],$$

since $(ja)^3 a_x^2 = 0$, and $(j'b)^3 b_x^2 = 0$.

In this expression we will introduce, as far as possible, symbols referring to j; for

$$(ia)^2 a_x^3 = j = (i'b)^2 b_x^3.$$

M is then seen to be the sum of four terms, viz.

$$(ja) (j'b) (jb)^2 (ia)^2 (j'a)^2 (i'b)^2$$
$$= (jj'') (j'j''') (jj''')^2 (j'j'')^2 = -(\tau\tau)^2 = -C:$$

$$4 (ja)^2 (j'b)^2 (jb) (j'a) (ia) (i'b) (i'a) (ib)$$
$$= 2 (ja)^2 (j'b)^2 (jb) (j'a) \{(ia)^2 (i'b)^2 + (ib)^2 (i'a)^2 - (ii')^2 (ab)^2\}$$
$$= 4 (jj'')^2 (j'j''')^2 (jj''') (j'j'')$$
$$- 2A \,.\, (ja)^2 (j'b)^2 (ab)^2 \{(ja) (j'b) - (jj') (ab)\}$$
$$= -4C + A \,.\, (jj')^2 (ab)^4 \{(ja) (j'b) + (jb) (j'a)\},$$

since $(ja)^3 a_x^2 = 0$,

$$= -4C + 2A \,.\, (\tau i)^2 = -4C + 2AB:$$

$$-2 (ja)^2 (j'a)^2 (jb) (j'b) (i'b)^2 (ia) (ib)$$
$$= 2 (ja)^2 (j'a)^2 (jj'') (j'j'') (ia) (ij'')$$
$$= (ja) (j'a) (ia) (ij'') \{(ja)^2 (j'j'')^2 + (j'a)^2 (jj'')^2 - (j''a)^2 (jj')^2\}$$
$$= -(\tau a)^2 (j''a)^2 (ia) (ij'') = -(\tau a) (\tau j'') (j''a)^2 (ia)^2 = C:$$

the last term of M

$$-2 (jb)^2 (j'b)^2 (ja) (j'a) (ia)^2 (i'a) (i'b)$$

is obtained from the one just considered by interchanging i and i', a and b, its value is therefore C.

Hence
$$M = -C - 4C + 2AB + 2C$$
$$= -3C + 2AB.$$

122. The covariants of orders 0, 1, 2, with the exception of i, have been replaced by transvectants, of index not greater than 2, of simpler forms. We may simplify the expressions of the others in the same way, in fact this has been done already for two of the covariants degree 3, viz. j and (j, τ). The remaining cubic covariant $(f, i^2)^3 = ((f, i)^2, i)$

$$= -(j, i).$$

The transvectant $(H, i)^2$ may be replaced by its term

$$(ab)^2 (bi)^2 a_x^3 b_x = -(aj)^2 a_x^3 j_x;$$

it will be convenient to write this covariant

$$(f, j)^2 = p_x^4.$$

The remaining quartic covariant is

$$(H, i^2)^3 = ((H, i)^2, i),$$

and may then be replaced by (p, i).

Now since $(aj)^3 a_x^2 = 0$,

$$(aj)^2 a_x^3 j_y = (aj)^2 a_x^2 j_x a_y$$
$$= \tfrac{1}{4} \{(aj)^2 a_x^3 j_y + 3 (aj)^2 a_x^2 j_x a_y\}$$
$$= p_x^3 p_y.$$

Hence

$$(p, i) = (aj)^2 (ji) a_x^3 i_x = (aj)^2 (ai) a_x^2 j_x i_x.$$

Now
$$(f, \alpha) = (f, -(ji)^2 j_x)$$
$$= -(aj)(ji)^2 a_x^4$$
$$= -(aj) \{(ai) j_x - (aj) i_x\}^2 a_x^2$$
$$= 2 (aj)^2 (ai) i_x j_x a_x^2$$
$$= 2 (p, i),$$

which gives another expression for the same covariant.

For order 5 we have only to consider the transvectant $(t, i^2)^4$; this has a term

$$(ab)^2 (bc) (ci)^2 (ai')^2 a_x b_x^2 c_x^2,$$

hence this transvectant may be replaced by $((H, i)^2, (f, i)^2)$, and therefore by

$$(p, j).$$

Lastly the covariant order 7 may be replaced by the Jacobian (H, j).

123. To express any transvectant of two covariants of the quintic in terms of members of the irreducible system, it is in general advisable to use as far as possible symbolical letters referring to covariants such as j, i, τ, ϑ etc. instead of those belonging to the quintic itself. We give here the values of some of the transvectants, partly for the sake of reference, and partly as examples; we would recommend the student to verify a few of them*.

The following table gives the 2nd transvectants and Jacobians of the quadratic and cubic covariants.

2nd Trans-vectants	Jacobians					
	i	τ	ϑ	j	(j, i)	(j, τ)
i	A	ϑ	$\frac{1}{2}(Bi - A\tau)$	$-(j, i)$	$\frac{2}{3}ai + \frac{1}{2}Aj$	$\frac{1}{2}a\tau + \frac{1}{2}Bj$
τ	B	C	$\frac{1}{2}(Ci - B\tau)$	$-(j, \tau)$	$\frac{1}{6}a\tau + \frac{1}{2}Bj$	$\frac{1}{2}Cj$
ϑ	0	0	N	$-\frac{1}{2}a\tau$	$\frac{1}{6}a\vartheta + \frac{1}{2}\gamma i$	$\frac{1}{2}\gamma\tau$
j	$-a$	0	$-\gamma$	τ	$-\frac{1}{2}i\tau - \frac{1}{3}ja$	$-\frac{1}{2}\tau^2$
(j, i)	$\frac{1}{3}\beta$	$-\frac{2}{3}\gamma$	$\frac{1}{3}\delta - \frac{1}{2}Ba$	$-\frac{1}{3}\vartheta$	$-\frac{1}{9}\{2a^2 - \frac{3}{2}A\tau - 3Bi\}$	$-\frac{1}{6}\{3\gamma j + a(j, \tau)\}$
(j, τ)	γ	0	$-\frac{1}{2}Ca$	0	$\frac{1}{3}B\tau + \frac{1}{6}Ci$	$\frac{1}{2}C\tau$

The third transvectants of the cubic covariants are

$$(j, (j, i))^3 = B, \quad (j, (j, \tau))^3 = C, \quad ((j, i), (j, \tau))^3 = 0.$$

In obtaining the values of Jacobians it is well to remember the formula

$$((f, \phi), \psi) = \frac{m-n}{2(m+n-2)}(f, \phi)^2 \cdot \psi + \tfrac{1}{2}\{(f, \psi)^2 \phi - (\phi, \psi)^2 \cdot f\}$$

proved, Chap. IV. § 77; where f, ϕ are binary forms of orders m, n respectively; and the order of each of the forms f, ϕ, ψ is not less than 2. This will be of frequent assistance, since of the 15

* Should he experience any great difficulty he will find some of them worked out in Gordan's *Invariantentheorie* and Clebsch's *Binären Formen*.

	a	β	γ	δ
f	$2(p, i)$	$-Ap+aj-\frac{3}{2}i\tau$	$\frac{3}{2}\tau^2+Bp$	$2B(p, i)+j\gamma-\frac{3}{2}\tau\vartheta$
i	β	$-\frac{1}{2}Aa$	$\delta-\frac{1}{2}Ba$	$\frac{1}{2}(B\beta-A\gamma)$
τ	γ	$-\delta-\frac{1}{2}Ba$	$-\frac{1}{2}Ca$	$\frac{1}{2}(C\beta-B\gamma)$
ϑ	δ	$\frac{1}{2}(A\gamma-B\beta)$	$\frac{1}{2}(B\gamma-C\beta)$	$-\frac{1}{2}Na$
j	$-\vartheta$	$a^2-\frac{1}{2}(A\tau-Bi)$	$\frac{1}{2}(Ci-B\tau)$	$a\gamma$
(j, i)	$\frac{1}{3}a^2-\frac{1}{2}(A\tau-Bi)$	$-\frac{2}{3}a\beta+\frac{1}{2}A\vartheta$	$\frac{1}{3}a\gamma+\frac{1}{2}B\vartheta$	$-\frac{1}{3}(a\delta+\beta\gamma)+\frac{1}{3}Ba^2-\frac{1}{6}Ni$
(j, τ)	$\frac{1}{2}(Ci-B\tau)$	$\frac{1}{2}B\vartheta-a\gamma$	$\frac{1}{2}C\vartheta$	$\frac{1}{2}(Ca^2-N\tau)$
p	$-(j, \tau)+\frac{2}{3}A(j, i)-\frac{1}{3}\beta i$	$(\frac{1}{2}B-\frac{1}{3}A^2)j-\frac{1}{2}Aia+\frac{3}{2}\tau a$	$\frac{1}{2}Bia+\frac{1}{6}jM$	$\tau\gamma+\frac{1}{6}M(ji)-\frac{1}{6}Bi\beta+\frac{1}{2}B(pa)$
(p, i)	$\frac{3}{4}a\tau-\frac{1}{2}Aia+(\frac{1}{2}B-\frac{1}{3}A^2)j$	$\frac{1}{2}A(j\tau)-\frac{1}{3}A^2(ji)+\frac{1}{6}Ai\beta-\frac{3}{4}\tau\beta$	$\frac{1}{4}\tau\gamma+\frac{1}{6}M(ji)-\frac{1}{6}Bi\beta$	$(\frac{1}{4}B^2-\frac{1}{3}A^2B+\frac{1}{4}AC)j$ $-\frac{1}{2}ABia+\frac{3}{4}B\tau a-\frac{3}{4}\tau\delta$

irreducible covariants of the quintic for which the order is not less than 2, 9 have been expressed as Jacobians.

It is useful to know the values of the following transvectants

$$(p, f)^2 = \tfrac{1}{6} ij\,;\ (p, i)^2 = -\tau\,;\ (p, j)^2 = \tfrac{1}{2} i\alpha + \tfrac{1}{3} Aj.$$

124. For purposes to be presently explained the transvectants $(f, \alpha^5)^5$, $(f, \alpha^4\beta)^5$, ... will be required. Such transvectants are calculated step by step, first (f, α), then $(f, \alpha^2)^2 = ((f, \alpha), \alpha)$, and so on.

It will be useful then to know the values of the transvectants of certain of the covariants with α, β, γ, δ; when these are known the values of such transvectants as $(f, \alpha^5)^5$ may be obtained with great ease.

125. Syzygies. It has been proved (§ 77) that the product of two Jacobians, or the square of a Jacobian can be expressed as a sum of terms, each term being the product of at least three forms. Now nine of the covariants of the quintic are Jacobians,—we must exclude the forms $(i\alpha)$, $(\tau\alpha)$, $(\vartheta\alpha)$, $(\alpha\beta)$, $(\alpha\delta)$ for one at least of the quantics in each of these Jacobians is of order less than 2: hence we have 45 syzygies.

A more general method of obtaining syzygies is given by Stroh. He considers four different forms, and seeks to obtain the syzygies which are of unit degree in each of these forms. First for three forms there is evidently only one such relation, that for weight unity

$$f_1 (f_2, f_3) + f_2 (f_3, f_1) + f_3 (f_1, f_2) = 0,$$

this is written for short $(f_1 f_2 f_3) = 0$.

The other syzygies obtained by Stroh arise from the symbolical relation

$$(ab)\, c_x d_x + (cd)\, a_x b_x = (ad)\, b_x c_x + (cb)\, a_x d_x,$$

which may easily be verified. Raise both sides of this identity to the power i and expand by the binomial theorem: hence

$$\Sigma \binom{i}{\lambda} (ab)^\lambda (cd)^{i-\lambda} a_x^{i-\lambda} b_x^{i-\lambda} c_x^\lambda d_x^\lambda$$

$$= \Sigma \binom{i}{\lambda} (ad)^\lambda (cb)^{i-\lambda} a_x^{i-\lambda} d_x^{i-\lambda} c_x^\lambda b_x^\lambda.$$

Hence

$$\Sigma \binom{i}{\lambda} (f_1, f_2)^\lambda (f_3, f_4)^{i-\lambda} - \Sigma \binom{i}{\lambda} (f_1, f_4)^\lambda (f_3, f_2)^{i-\lambda} = 0.$$

This is written

$$(f_1 f_2 f_3 f_4)_i = 0.$$

Other syzygies may be obtained from it by interchange of the various quantics concerned. But it will be noticed that $(f_1 f_2 f_3 f_4)_i$ is unaltered if one pair of letters is interchanged and at the same time the other pair. Also this expression is only changed in sign if f_2 and f_4 are interchanged. Hence only three distinct syzygies of weight i are obtained in this manner.

If $i = 2$

$$(f_1 f_2 f_3 f_4)_2 \equiv (f_1 f_2)^2 f_3 f_4 + (f_3 f_4)^2 f_1 f_2 - (f_1 f_4)^2 f_2 f_3 - (f_3 f_2)^2 f_1 f_4$$
$$+ 2 (f_1 f_2)(f_3 f_4) - 2 (f_1 f_4)(f_3 f_2) = 0.$$

Whence

$$\{f_1 f_2 f_3 f_4\}_2 \equiv \tfrac{1}{4} [(f_1 f_2 f_3 f_4)_2 + (f_1 f_3 f_4 f_2)_2 + (f_1 f_4 f_2 f_3)_2]$$
$$= (f_1 f_2)(f_3 f_4) + (f_1 f_3)(f_4 f_2) + (f_1 f_4)(f_2 f_3) = 0,$$

a result already well known.

Also

$$[f_1 f_2 f_3 f_4]_2 \equiv \tfrac{1}{2} [(f_1 f_2 f_3 f_4)_2 - (f_1 f_3 f_4 f_2)_2 - (f_1 f_4 f_2 f_3)_2]$$
$$= (f_1 f_2)^2 f_3 f_4 + (f_3 f_4)^2 f_1 f_2 - (f_1 f_4)^2 f_2 f_3$$
$$- (f_3 f_2)^2 f_1 f_4 + 2 (f_1 f_3)(f_2 f_4) = 0.$$

This is the same relation as that given (§ 78) for the product of two Jacobians. It will be seen at once that the other syzygy of weight 2 is deducible from this.

The syzygies of higher weight may often be simplified, in the same way.

The syzygy

$$\{f_1 f_2 f_3 f_4\}_2 = 0$$

is remarkable from the fact that the forms in it need not be of order higher than unity; while in the syzygy

$$(f_1 f_2 f_3 f_4)_2 = 0$$

from which it was deduced, each of the forms is of necessity of order 2 at least.

In general in the syzygy

$$(f_1 f_2 f_3 f_4)_i = 0$$

each of the forms f_1, f_2, f_3, f_4 must be of order i at least; but from these syzygies others may be deduced which are true for forms of lower order. Stroh, in his papers on syzygies*, deduces many such.

We will obtain such a syzygy from that of weight 3,

$$(f_1 f_2 f_3 f_4)_3 = 0.$$

If λ, μ, ν be three quantities whose sum is zero, we know that

$$\lambda^3 + \mu^3 + \nu^3 - 3\lambda\mu\nu = 0,$$

hence

$$3\,(ab)\,(bc)\,(ca)\,a_x b_x c_x = (bc)^3\, a_x^3 + (ca)^3\, b_x^3 + (ab)^3\, c_x^3.$$

We thus obtain

$$\tfrac{1}{6}\left[(a_x^3 b_x^3 c_x^3 d_x^3)_3 - (a_x^3 c_x^3 d_x^3 b_x^3)_3 + (a_x^3 d_x^3 b_x^3 c_x^3)_3\right]$$

$$= (ab)\,(bc)\,(ca)\,a_x b_x c_x d_x^3$$

$$+ (b_x^3,\, c_x^3)\,(a_x^3,\, d_x^3)^2 + (c_x^3,\, a_x^3)\,(b_x^3,\, d_x^3)^2 + (a_x^3,\, b_x^3)\,(c_x^3,\, d_x^3)^2 = 0.$$

It will be seen at once that a_x, b_x, c_x, d_x are factors of this syzygy, we may then divide by any one or all of these factors; or else we may multiply the syzygy by a power of any one of them. The syzygy is then true whenever the order of each form is greater than unity. It is not difficult to see that this derived syzygy is just as general as the original one

$$(f_1 f_2 f_3 f_4)_3 = 0,$$

in other words, the original syzygy might be derived from it.

Stroh writes the syzygy just obtained in the form

$$\{f\phi q q'\}_3 = (fq)^2\,(\phi q') - (fq')^2\,(\phi q) - (f\phi)^2\,(qq') + f\,((qq')\,\phi)^2 = 0,$$

where q, q' are quadratics, and f, ϕ are forms of order 2 at least.

Again from the syzygy

$$(f_1 f_2 f_3 f_4)_4 = 0$$

* *Math. Ann.* Bd. 33, pp. 61–107 (§§ 18–22); Bd. 34, pp. 306–320, 354–370; Bd. 36, pp. 262–288; in § 3 of this latter paper he gives a list of syzygies of the kind just mentioned which he has deduced.

we obtain

$$(ff\phi\phi)_4$$
$$=f^2(\phi\phi)^4+6(ff)^2(\phi\phi)^2+\phi^2(ff)^4$$
$$-2f\phi(f\phi)^4+8(f\phi)^3(f\phi)-6[(f\phi)^2]^2=0.$$

In this write i^2 for ϕ, where i is a quadratic.

Now

$$(i^2, i^2)^4=\tfrac{2}{3}[(ii)^2]^2,$$
$$(i^2, i^2)^2=\tfrac{1}{2}(ii)^2 \,.\, i^2,$$
$$2(f, i^2)^3(f, i^2)=2i((f, i)^2, i)\,.\,(f, i)$$
$$=i^2f(f, i^2)^4+i^2[(f, i)^2]^2-if(i, i)^2(f, i)^2-i^3((f, i)^2, f)^2$$

—see § 78.

Hence substituting these values in the syzygy obtained by writing i^2 for ϕ, and then dividing the result by $2i^2$, we obtain

$$(ffii)_4$$
$$=f(fi^2)^4-2i((fi)^2f)^2-[(fi)^2]^2+(ii)^2(ff)^2+\tfrac{1}{2}i^2(ff)^4=0.$$

From this may be obtained a syzygy

$$(f\psi ii)_4=0$$

by means of the operator $\left(\psi\frac{\partial}{\partial f}\right)$. This operator requires that ψ and f should be of the same order, but when the syzygy is written symbolically it will be seen at once that factors of the form a_x and b_x may be introduced so that the relation is true whatever be the orders of f and ψ, provided that neither is less than 4.

In the same way we may obtain a syzygy

$$[f\psi i\tau]_4$$

where both i and τ are quadratics.

126. Application to the quintic. The forms of the quintic consist of f, H, i, j, τ, p, α, β, γ, δ and some of the Jacobians of these forms.

A large number of syzygies may be at once obtained by writing for the forms in the general syzygies of the last paragraph covariants of f.

We shall content ourselves with a few examples.

$$(fiH) = f.(iH) + H(fi) - it = 0.$$

$$(fjH) = f.(jH) - H(jf) - jt = 0.$$

$$[ffii]_2 = ((fi))^2 + Hi^2 + 2fij + f^2A = 0.$$

$$\{ij\tau i\}_3 = B(ji) - A(j\tau) - \alpha\vartheta + i\gamma = 0.$$

$$(ppii)_4 = -Bp + \tfrac{2}{3}i\alpha^2 - \tfrac{1}{6}Ai\tau - \tfrac{1}{6}Bi^2 - \tau^2 + \tfrac{1}{3}Aj\alpha = 0.$$

$$\{\alpha\beta\gamma\delta\}_2 = \tfrac{1}{2}M(BN - CM) - \tfrac{1}{2}N(AN - BM) - R^2 = 0.$$

From this last we deduce the relation connecting the irreducible invariants A, B, C, R.

It is easy to write down a great number of syzygies in this way. Stroh (*Math. Ann.* Bd. 34, pp. 354—370) has given a list of 168 syzygies of the quintic. The notation for the elementary syzygies given here is not quite the same as that used in Stroh's paper, but the notation is there explained; the notation here has been mainly taken from a later paper by the same author (Stroh, *Math. Ann.* Bd. 36, pp. 262—288).

127. Reducibility of syzygies. If

$$S_1 = 0, \ S_2 = 0, \ \ldots$$

be any syzygies, and if P_1, P_2, ... be any products of forms such that P_1S_1, P_2S_2, ... are expressions all of the same degree and order, then

$$P_1S_1 + P_2S_2 + \ldots = 0$$

is a syzygy. In this way it will be seen that an infinite number of syzygies may be built up.

A syzygy $$S = 0$$

is said to be *reducible*, when

$$S \equiv P_1S_1 + P_2S_2 + \ldots,$$

where $S_1 = 0$, $S_2 = 0$, ... are syzygies whose degree is less than the degree of $S = 0$ and the P's are covariants. Otherwise it is said to be *irreducible*.

It will be seen at once that any syzygy which contains a product of only two irreducible forms must be irreducible. All the irreducible syzygies which have yet been found for the quintic are of this nature.

128. It may happen that certain syzygies

$$S_1 = 0,\ S_2 = 0,\ \ldots\ S_i = 0$$

are such that certain products of forms $P_1, P_2, \ldots P_i$ may be found, for which

$$P_1S_1 + P_2S_2 + \ldots + P_iS_i \equiv 0,$$

the expressions P, S being regarded as functions of the concomitants—which for the moment are treated as independent variables.

Such a relation is called a syzygy of the second kind.

The following is an example,

$$(f_1f_2f_3f_4) \equiv f_1(f_2f_3f_4) - f_2(f_1f_3f_4) + f_3(f_1f_2f_4) - f_4(f_1f_2f_3) \equiv 0.$$

Thus for the quintic

$$(fHij) \equiv f(Hij) - H(fij) + i(fHj) - j(fHi) \equiv 0$$

is a syzygy of the second kind.

Syzygies of the second kind may clearly be reducible or irreducible. Between them may arise syzygies of the third kind, and so on.

The following questions at once present themselves. 'Is the number of syzygies finite when the system of forms is finite?' 'When the syzygies of the first kind are finite in number, are also those of the second and of higher kinds finite?' 'Is there any limit to the number of *kinds* of syzygies which arise from a finite system of forms?'

All these questions have been answered in the affirmative by Hilbert (*Math. Ann.* Bd. 36, pp. 473—534). They are partly considered in Chapter IX.

129. The typical representation of the binary quintic. For special purposes, some particular linear transformation of a binary quantic may have peculiar advantages. Thus any particular term of the quantic may by a special transformation be made to vanish. If the quantic has two linear covariants, such that the determinant formed by their coefficients does not vanish, these may be taken for the variables: the transformed quantic will then possess the property that every one of its coefficients is an invariant. We proceed to prove this. Let α_x, β_x be two linear covariants of the quantic a_x^n, which are such that $(\alpha\beta)$ is not zero. Then raising the identity

$$(\alpha\beta)\, a_x = (a\beta)\, \alpha_x - (a\alpha)\, \beta_x$$

to the nth power, we obtain

$$(\alpha\beta)^n \,.\, a_x^n = (a\beta)^n \,.\, \alpha_x^n - n\,(a\beta)^{n-1}\,(a\alpha)\,\alpha_x^{n-1}\beta_x + \dots$$

The expression on the right is the transformed quantic, and from the symbolical form of the coefficients, it follows that they are all invariants.

For the general quintic any pair of linear covariants may be chosen; for example those which we have written α and β.

The coefficients may be easily calculated with the help of the table given on p. 139; they are as follows:

$$(f,\, \alpha^5)^5 = -(B - \tfrac{2}{3}A^2)\,R = -\lambda R.$$

$$(f,\, \alpha^4\beta)^5 = -\left(\tfrac{1}{2}N - \tfrac{1}{3}AM + \lambda\frac{B}{2}\right)M + \lambda\tfrac{1}{2}\,(2BM - AN) = \mu.$$

$$(f,\, \alpha^3\beta^2)^5 = R\left(M + \frac{\lambda}{2}A\right).$$

$$(f,\, \alpha^2\beta^3)^5 = \tfrac{1}{2}M\,(AN - BM) - \mu\,\frac{A}{2}.$$

$$(f,\, \alpha\beta^4)^5 = -AR\left(M + \frac{\lambda A}{4}\right).$$

$$(f,\, \beta^5)^5 = -M^3 + \tfrac{1}{2}AM\,(AN - BM) + \mu\,\frac{A^2}{4}.$$

Further the invariant $(\alpha\beta) = -M = 3C - 2AB$ must not be zero.

To be more accurate the coefficients given above should be divided by $(-M)^5$. In the expression for any covariant in terms of the actual coefficients, the above transformed coefficients may be substituted, the covariant multiplied by a power of the determinant of transformation is then equal to the expression thus obtained. In this way any covariant may be expressed in terms of the invariants and two of the linear covariants.

To illustrate a different method of expressing any covariant in terms of the invariants and two linear covariants we shall obtain j in terms of the invariants and the covariants α and δ. Raising the identity

$$(\alpha\delta) j_x = (j\delta)\,\alpha_x - (j\alpha)\,\delta_x$$

to the third power

$$-R^3 j = (j,\ \delta^3)^3\,\alpha^3 - 3\,(j,\ \delta^2\alpha)^3\,\alpha^2\delta + 3\,(j,\ \delta\alpha^2)^3\,\alpha\delta^2 - (j,\ \alpha^3)^3\,\delta^3.$$

But by the method of § 124

$$(j,\ \alpha^3)^3 = -R,\quad (j,\ \alpha^2\delta)^3 = 0,$$
$$(j,\ \alpha\delta^3) = -\tfrac{1}{2}NR,\quad (j,\ \delta^3)^3 = -\tfrac{1}{2}R\,(CM - BN).$$

Therefore

$$R^2 j = \tfrac{1}{2}(CM - BN)\,\alpha^3 - \tfrac{3}{2}N\alpha^2\delta - \delta^3.$$

130. Given two binary forms of the same order, in particular two quintics, can one be linearly transformed into the other, and if so how?

The reply (in part) to the first question is that if the absolute invariants of the two quantics are equal to one another, and if they each possess a corresponding pair of linear covariants of which the determinants do not vanish, then the quantics are transformable into each other. The question will be found discussed in Clebsch, *Binären Formen*, § 92; and for the case where there are no linear covariants in § 105.

When two quintics have equal absolute invariants and one of the 6 invariants $(\alpha\beta)$, $(\alpha\gamma)$ ... is other than zero, say $(\alpha\beta)$, we may transform one quintic into the other thus:—

Let unaccented letters refer to one quintic, and accented letters to the other; we transform each quintic, so that the variables in the first are α, β, in the second are α', β'.

Thus
$$f = A_0\alpha^5 + 5A_1\alpha^4\beta + \ldots\ldots$$
$$f' = A_0'\alpha'^5 + 5A_1'\alpha'^4\beta' + \ldots\ldots.$$

Let the ratio $\dfrac{A}{A'} = r$, then since the absolute invariants for the two quintics are equal it follows that

$$\frac{B}{A^2} = \frac{B'}{A'^2},$$

and hence
$$\frac{B}{B'} = r^2.$$

Similarly

$$\frac{C}{C'} = r^3, \text{ and } \frac{R}{R'} = r^{\frac{9}{2}}.$$

Hence also

$$\frac{A_0}{A_0'} = r^{-6}, \quad \frac{A_1}{A_1'} = r^{-\frac{13}{2}}, \quad \frac{A_2}{A_2'} = r^{-7},$$

$$\frac{A_3}{A_3'} = r^{-\frac{15}{2}}, \quad \frac{A_4}{A_4'} = r^{-8}, \quad \frac{A_5}{A_5'} = r^{-\frac{17}{2}}.$$

The quintic f' may now be transformed into f by means of the transformation

$$\alpha = r^{\frac{6}{5}}\alpha',$$

$$\beta = r^{\frac{6}{5}+\frac{1}{2}}\beta'.$$

131. Associated forms. If y_1, y_2 is a pair of variables cogredient with x_1, x_2; then the two forms

$$\eta = (xy), \quad \xi = a_x^{n-1}a_y$$

are invariantive. Now regard x for the moment as a constant, and the two equations just written down as equations of linear transformation to transform from the variables y_1, y_2 to new variables ξ, η. The variables of the transformed form are covariants, hence its coefficients are invariants—or to be more accurate covariants, for they contain x but not y. Let us proceed exactly as in § 6.

The determinant of transformation is

$$(\xi\eta) = a_x^n = f.$$

From the identity

$$(\xi\eta)\, b_y = (b\eta)\,\xi - (b\xi)\,\eta$$

we obtain the transformed quantic

$$(\xi\eta)^n b_y^n = (b\eta)^n \xi^n - n\,(b\eta)^{n-1}(b\xi)\,\xi^{n-1}\eta + \dots \qquad \text{......(I).}$$

Let us calculate the coefficients of the transformed form for the case of the quintic.

$$(b\eta)^5 = b_x^5 = f,$$

$$(b\eta)^4(b\xi) = b_x^4(ba)\,a_x^4 = 0,$$

$$(b\eta)^3(b\xi)^2 = b_x^3(ba)\,a_x^4(ba')\,a'^4_x = \tfrac{1}{2}H\,.f,$$

$$(b\eta)^2(b\xi)^3 = b_x^2(ba)(ba')(ba'')\,a_x^4a'^4_xa''^4_x$$

$$= \tfrac{1}{2}b_x^2(ba)\,a_x^4a'^3_xa''^3_x\{(ba')^2a''^2_x + (ba'')^2a'^2_x - (a'a'')^2b_x^2\}$$

$$= -t\,.f - \tfrac{1}{2}H\,.(ba)\,b_x^4a_x^4 = -t\,.f,$$

$$(b\eta)(b\xi)^4 = b_x(ba)(ba')(ba'')(ba''')\,a_x^4 a'^4_x a''^4_x a'''^4_x$$

$$= (ba')^2(ba)(ba''')\,b_x a'^3_x a_x^4 a'''^4_x \,.\, a''^5_x$$

$$- \tfrac{1}{2}H\,.\,(ba)(ba''')\,b_x^3 a_x^4 a'''^4_x$$

$$= (ba')^2(ba)^2 b_x a'^3_x a_x^3 \,.\, f^2 - \tfrac{1}{2}H^2 f$$

$$- \tfrac{1}{4}H^2 f$$

$$= \tfrac{1}{2} i f^3 - \tfrac{3}{4}H^2 f,$$

$$(b\xi)^5 = (ba)(ba')(ba'')(ba''')(ba^{\text{iv}})\,a_x^4 a'^4_x a''^4_x a'''^4_x a^{\text{iv}\,4}_x$$

$$= (ba')^2(ba)^2(ba^{\text{iv}})\,a'^3_x a_x^3 a^{\text{iv}\,4}_x \,.\, f^2 + Htf.$$

Now $((ba)^2(ba')^2 b_x a_x^3 a'^3_x,\ f)$

$$= \tfrac{1}{7}(ba)^2(ba')^2\left\{(bc)\,a_x^3 a'^3_x + 3(ac)\,b_x a_x^2 a'^3_x + 3(a'c)\,b_x a_x^3 a'^2_x\right\} c_x^4$$

$$= (ba)^2(ba')^2(bc)\,a_x^3 a'^3_x c_x^4$$

$$+ \tfrac{6}{7}(ab)^3(ba')^2\,a_x^2 a'^3_x \,.\, f$$

$$= (ba)^2(ba')^2(bc)\,a_x^3 a'^3_x c_x^4 + \tfrac{3}{7}(ab)^4\left\{(a'b)\,a_x + (a'a)\,b_x\right\} a'^4_x \,.\, f$$

$$= (ba)^2(ba')^2(bc)\,a_x^3 a'^3_x c_x^4 + \tfrac{6}{7}(f,\ i)\,.\,f.$$

But the transvectant

$$((ba)^2(ba')^2 b_x a_x^3 a'^3_x, f) = (\tfrac{1}{2}if, f) = \frac{-1}{7}(f,\ i)\,.\,f.$$

Hence

$$(ba)^2(ba')^2(ba^{\text{iv}})\,a_x^3 a'^3_x c_x^4$$

$$= -(f,\ i)\,.\,f,$$

and therefore

$$(b\xi)^5 = -(f,\ i)\,.\,f^3 + Htf.$$

The transformation is then

$$f^4\,.\,f(y) = \xi^5 + 5H\xi^3\eta^2 + 10t\xi^2\eta^3$$

$$+ 5\,.\left(\frac{i}{2}f^2 - \frac{3}{4}H^2\right)\xi\eta^4 + ((f,\ i)\,f^2 - Ht)\,\eta^5.$$

Now let $\Phi(y)$ be any covariant of $f(y)$, then when the above transformation is made, the coefficients of $f(y)$ are replaced by the corresponding coefficients of the powers and products of ξ, η in the expression on the left. Let Φ thus transformed become $\Phi'(\xi,\ \eta)$, then Φ is equal to Φ' divided by a power of f the determinant of transformation; thus

$$\Phi(y) = \frac{\Phi'(\xi,\ \eta)}{f^\lambda}.$$

This equation is an identity. We may replace in it y by x; when this is done, ξ becomes f and η becomes zero, hence

$$\Phi(x) = \frac{\Phi'(f, 0)}{f^\lambda}.$$

Hence any covariant of the quintic is equal to its leading coefficient, when the original coefficients of the quintic are replaced by the corresponding coefficients in the form (I), divided by some power of f.

From this we see that all covariants of the quintic may be expressed rationally in terms of the covariants f, H, i, t, (f, i), in such a way that f alone occurs in the denominator. Such a system of covariants in terms of which all covariants of a system may be algebraically expressed is called a system of associated forms. We have confined ourselves to the case of the quintic, the results obtained are however true in general. The coefficients of the transformed quantic may always be expressed as rational integral functions of f, the covariants of degree 2, and the Jacobians of these latter with f. And this is in fact the simplest system of associated forms.

The matter will be found fully discussed in Clebsch, *Binären Formen*, ch. VII. The student who requires further information on the subject of typical representation will find it in the chapter just quoted and the two succeeding chapters of Clebsch's book.

The reduction of the quintic to a sum of three fifth powers will be discussed in Ch. XI., and so nothing need be said on the subject here, especially since it concerns the non-symbolical treatment of the subject rather than the symbolical treatment. The special canonical forms to which the quintic may be reduced, when one or other of its invariants vanishes, will be found in Prof. Elliott's *Algebra of Quantics*.

For a symbolical treatment of the subject the student is referred to Gordan's *Invariantentheorie*, or Clebsch, *Binären Formen*, §§ 93—96.

132. The Sextic. The difficulty in obtaining the complete irreducible system of concomitants of a binary form increases very much with the order. The system for the sextic is obtained here;

it affords examples of a method of reduction applicable to forms of a higher order, but not required when dealing with the quintic.

The arrangement in systems of forms whose grade does not exceed a certain number is followed as before.

The system A_0 contains only f; B_0 contains only

$$(f, f)^2 = H.$$

The system A_1 consists of

$$f,\ H,\ (f,\ H) = t.$$

For B_1 we must take the complete system of

$$i = (f, f)^4.$$

Now i is a quartic, and its complete system is

$$i,\ (i,\ i)^2 = \Delta,\quad (i,\ \Delta) = v,\quad (i,\ i)^4,\quad (i\Delta)^4.$$

To find the system A_2 we must take the transvectants of powers and products of forms of A_1 with powers and products of forms of B_1.

Now the form $(if)^3$ can be shewn to vanish, for

$$\begin{aligned}(if)^3 &= (ab)^4 (bc)^2 (ac)\, a_x c_x^3 \\ &= -\tfrac{1}{3}(ab)(bc)(ca)\,[(ab)^3 (bc)\, a_x c_x^3 + (bc)^3 (ca)\, b_x a_x^3 \\ &\qquad\qquad\qquad\qquad + (ca)^3 (ab)\, c_x b_x^3] \\ &= \tfrac{1}{6}(ab)(bc)(ca)\,[(ab)^4 c_x^4 + (bc)^4 a_x^4 + (ca)^4 b_x^4],\end{aligned}$$

on using Stroh's series

$$\begin{pmatrix} f & f & f \\ 1 & 1 & 1 \end{pmatrix}_7.$$

But $$(ab)^5 (bc)(ca)\, c_x^4 = 0,$$

since it changes sign when a and b are interchanged.

Hence $$(if)^3 = 0.$$

The quadratic covariant $(if)^4$ is of great importance; it is usually denoted by the symbol l.

If any covariant can be expressed as a symbolical product in which the factor $(ia)^3$ appears, it can be expressed as a sum of transvectants of l with other forms. For such a covariant

$$\begin{aligned}&= ((ia)^3\, i_x a_x^3,\ \Phi)^\rho + \Sigma\, ((ia)^4\, a_x^2,\ \Phi')^\rho \\ &= \Sigma\, (l,\ \Phi')^{\rho'}.\end{aligned}$$

Again
$$\Delta = (i, i)^2 = (ia)^2 (ab)^4 i_x^2 b_x^2 + \lambda (ab)^6 . i.$$
And
$$((ia)^3 i_x a_x^3, b_x^6)^3 = \tfrac{1}{4} (ia)^3 b_x^3 \{(ab)^3 i_x + 3 (ab)^2 (ib) a_x\}$$
$$= (ia)^3 (ab)^3 i_x b_x^3 + \tfrac{3}{4} (ia)^4 (ab)^2 b_x^4,$$
but
$$(ia)^3 i_x a_x^3 = 0,$$
hence
$$(ia)^3 (ab)^3 i_x b_x^3 = -\tfrac{3}{4} (ia)^4 (ab)^2 b_x^4 = -\tfrac{3}{4} (lf)^2.$$

Now since a and b are equivalent symbols
$$(ia)^3 (ab)^3 i_x b_x^3 = -\tfrac{1}{2} (ab)^4 i_x^2 \{(ia)^2 b_x^2 + (ia)(ib) a_x b_x + (ib)^2 a_x^2\}$$
$$= -\tfrac{3}{2} (ia)^2 (ab)^4 i_x^2 b_x^2 + \tfrac{1}{4} (ab)^6 . i,$$
and therefore
$$\Delta \equiv \tfrac{1}{2} (lf)^2 \text{ mod. } (ab)^6.$$

Thus every form of B_1 except i and the invariant $(i, i)^4$
$$\equiv 0 \text{ modd. } (ia)^4, (ab)^6.$$

133. We shall first find the system which is relatively complete with respect to the moduli $(ai)^4$ and $(ab)^6$ (§ 107, Cor. II.). This is obtained by taking the transvectants of powers of i, with powers and products of forms of the system A_1. Let us call this the system C.

First consider the forms
$$(i^\alpha, f^\beta)^\gamma.$$

We have (i, f), $(i, f)^2$. Every other one of these forms
$$\equiv 0 \text{ mod. } (ia)^4.$$

Next consider forms
$$(i^\alpha, H^\beta)^\gamma.$$

We retain only (i, H), for $(i, H)^2$ contains the term
$$(ia)^2 (ab)^2 i_x^2 a_x^2 b_x^4;$$
this can (§ 63) be linearly expressed in terms of covariants
$$(ia)^4 a_x^2 b_x^6,\ (ib)^4 b_x^2 a_x^6,\ (ia)^3 (ab) i_x a_x^2 b_x^5,$$
$$(ib)^3 (ab) i_x b_x^2 a_x^5,\ (ab)^4 a_x^2 b_x^2 . i_x^4.$$

Hence
$$(i, H)^2 \equiv \lambda i^2 + \mu l . f.$$

The form $(i, H)^3$ contains the term
$$(ia)^3 (ab)^2 i_x a_x b_x^4;$$

and the form $(i, H)^4$ contains the term

$$(ia)^4 (ab)^2 b_x^4;$$

hence these may both be rejected.

All other forms $(i^\alpha, H^\beta)^\gamma$ contain a term having a factor $(iH)^4$.

No one of the forms $(i^\alpha, t^\beta)^\gamma$ need be retained, for if $\gamma = 1$, the transvectant is the Jacobian of a Jacobian and another form and therefore reducible (§ 77).

If $\gamma > 2$, the transvectant always contains a term which involves a factor $(ia)^3$ or a factor $(ia)^4$, and therefore which

$$\equiv 0 \text{ mod. } (ai)^4.$$

If $\gamma = 2$, the transvectant $(i, t)^2$ contains the term

$$\begin{aligned}&(iH)^2 (Hf)\, i_x^2 H_x^5 f_x^5\\ &= ((i, H)^2, f) \text{ mod. } (iH)^3\\ &= \lambda (i^2, f) \text{ mod. } (ai)^4\\ &= \lambda i (i, f) \text{ mod. } (ai)^4.\end{aligned}$$

The general transvectant

$$(i^\alpha, f^\beta H^\gamma t^\delta)^\epsilon$$

may be treated in the same way. If $\epsilon > 2$, the transvectant contains a term which

$$\equiv 0 \text{ mod. } (ai)^4.$$

And if $\epsilon \not> 2$, it is certainly reducible, except for the cases already discussed.

The system C then contains the forms

$$f,\ H,\ t,\ i,\ (i, i)^4,\ (f, i),\ (H, i),\ (f, i)^2.$$

134. To find the system A_2 we must now take all possible transvectants of powers and products of the system C, with powers and products of the complete system of l.

Now l is a quadratic, its complete system must then consist of

$$l,\ (l, l)^2.$$

The invariant $(l, l)^2$ is the same as an invariant already found, viz.:

$$\begin{aligned}(i, \Delta)^4 &= \tfrac{1}{2} (i, (fl)^2)^4 \text{ mod. } (ab)^6\\ &= \tfrac{1}{2} (ll)^2 \text{ mod. } (ab)^6.\end{aligned}$$

Since l is a quadratic and all the covariants of C are of even degree, we need only consider the transvectants of powers of l with each form separately.

The forms

$$(l, f), \ (l, f)^2 = 2\Delta, \ (l^2, f)^3, \ (l^2, f)^4, \ (l^3, f)^5, \ (l^3, f)^6$$

are all irreducible; so also is

$$(l, H).$$

The covariant $(l, H)^2$ contains the term

$$\begin{aligned}(la)^2 (ab)^2 a_x^2 b_x^4 &= ((l, f)^2, f)^2 \\ &= 2((i, i)^2, f)^2 \\ &= \lambda (ii')^2 (i'a)^2 i_x^2 a_x^4 + \mu (i, i)^4 f.\end{aligned}$$

But the term

$$(ii')^2 (i'a)^2 i_x^2 a_x^4$$

is linearly expressible in terms of the covariants (§ 63)

$$(ii')^4 a_x^6, \ (ii')^3 (i'a) i_x a_x^5, \ (ia)^4 a_x^2 i_x'^4,$$
$$(ia)^3 (ii') i_x'^3 a_x^3, \ (i'a)^4 a_x^2 i_x^4,$$

each of which is reducible.

Hence $\qquad (l, H)^2 = \lambda_1 (ii)^4 f + \lambda_2 li.$

Now, if $\beta > 2$

$$(l^\alpha, H)^\beta$$
$$= ((l, H)^2, l^{\alpha-1})^{\beta-2} = (ii)^4 \Phi + \lambda_2 (li, l^{\alpha-1})^{\beta-2},$$

hence these transvectants are all reducible.

The covariant (l, t) is reducible, § 77.

The covariant $(l, t)^2$ contains the term

$$((l, H)^2, f),$$

and is therefore reducible.

The covariant, $\beta > 2$,

$$(l^\alpha, t)^\beta = ((l, t)^2, l^{\alpha-1})^{\beta-2}$$

and is reducible.

Hence all the covariants

$$(l^\alpha, t)^\beta$$

are reducible.

The covariants $(l, i),\ (l, i)^2,\ (l^2, i)^3$

are irreducible; but

$$(l^2, i)^4 = (l^2, (ab)^4 a_x^2 b_x^2)^4$$
$$\equiv ((l, f)^2, (l, f)^2)^4 \text{ mod. } (ab)^6$$
$$\equiv 4\,((i, i)^2, (i, i)^2)^4 \text{ mod. } (ab)^6,$$

which is reducible when considered as an invariant of the quartic i.

The form $(l, (f, i))$ is reducible, § 77;

$(l, (f, i))^2$ contains the term

$$((l, f)^2, i) = (\Delta, i) = -v;$$

$(l^2, (f, i))^3$ contains the reducible term

$$(l, ((l, f)^2, i)).$$

Similarly $(l^3, (f, i))^5,\ (l^4, (f, i))^7$

may be reduced.

The forms

$$(l^2, (f, i))^4,\ (l^3, (f, i))^6,\ (l^4, (f, i))^8$$

are however irreducible.

The form $(l, (f, i)^2)$ is not reducible.

Now $(l, (f, i)^2)^2$ contains the term

$$((l, f)^2, i)^2 = 2\,(\Delta, i)^2 = \tfrac{1}{3}\, i (ii)^4 \ldots$$

(see § 89).

Hence $(l^\alpha, (f, i)^2)^\beta$ is reducible when $\beta \geqslant 2$, for this

$$= (l^{\alpha-1}, (l, (f, i)^2)^2)^{\beta-2}$$
$$= \tfrac{1}{3}\,(ii)^4\,(l^{\alpha-1}, i)^{\beta-2} + \lambda\,(l^{\alpha-1}, l^2)^{\beta-2}.$$

Lastly $(l, (H, i))$ is reducible by § 77.

$(l, (H, i))^2$ contains a term

$$((l, H)^2, i) = \lambda_1\,(ii)^4\,(f, i) + \lambda_2\,(li, i)$$

and $(l^\alpha, (H, i))^\beta,\quad \beta > 2,$

$$= (l^{\alpha-1}, (l, (H, i))^2)^\beta$$

which is reducible.

Thus the system A_2 contains the forms of the system C together with l, $(l, l)^2$ and

$$(l, f), (l, f)^2, (l^2, f)^3, (l^2, f)^4, (l^3, f)^5, (l^3, f)^6,$$
$$(l, H), (l, i), (l, i)^2, (l^2, i)^3, (l, (f, i))^2, (l^2, (f, i))^4,$$
$$(l^3, (f, i))^6, (l^4, (f, i))^8, (l, (f, i)^2).$$

The system B_2 contains only the invariant $(f, f)^6$; we merely add this to the system A_2, and the result is the complete system for f.

We append the following table giving the 26 irreducible concomitants of the sextic.

Degree	Order 0	2	4	6	8	10	12
1				f			
2	$(f, f)^6$		$(f, f)^4 = i$		$(f, f)^2 = H$		
3		$(f, i)^4 = l$		$(f, i)^2$	(f, i)		$(f, H) = t$
4	$(i, i)^4$		$(f, l)^2$	(f, l)		(H, i)	
5		$(i, l)^2$	(i, l)		(H, l)		
6	$(l, l)^2$			$((f, i)^2, l)$ $((f, i), l)^2$			
7		$(f, l^2)^4$	$(f, l^2)^3$				
8		$(i, l^2)^3$					
9			$((f, i), l^2)^4$				
10	$(f, l^3)^6$	$(f, l^3)^5$					
12		$((f, i), l^3)^6$					
15	$((f, i), l^4)^8$						

Ex. (i). Prove that if the covariant a, of a quintic f, is identically equal to zero, then also

$$\beta \equiv 0, \quad \gamma \equiv 0, \quad \delta \equiv 0, \quad \vartheta \equiv 0, \quad (pi) \equiv 0, \quad M \equiv 0, \quad N \equiv 0, \quad R \equiv 0,$$
$$Bi \equiv A\tau, \quad (B - \tfrac{2}{3}A^2)j \equiv 0, \quad Bf + \tfrac{5}{2}j\tau \equiv 0.$$

Ex. (ii). In the last example either

$$j \equiv 0 \text{ or } B \equiv \tfrac{2}{3}A^2.$$

Prove that in the former case every covariant vanishes with the exception of

$$f,\ H,\ i,\ t,\ (fi),\ A\ ;$$

and that these are connected by the relations

$$AH+\tfrac{1}{2}i^3=0,$$
$$At+\tfrac{1}{2}i^2(fi)=0,$$
$$2[(f,i)]^2+Hi^2+f^2A=0.$$

Ex. (iii). Prove that if a vanishes identically and j is not zero, then

$$B\equiv\tfrac{2}{3}A^2,\quad t\equiv\tfrac{2}{3}Ai,\quad A\,(Af+\tfrac{15}{4}ij)\equiv 0.$$

The latter result gives an alternative, but if

$$Af=-\tfrac{15}{4}ij,$$

then $$Aj=\tfrac{15}{4}(ij,i)^2=\tfrac{3}{2}Aj.$$

Hence in either case since j is other than zero,

$$A\equiv 0,\quad B\equiv 0,\quad C\equiv 0.$$

Shew further that j is a perfect cube: that it is a factor of f: that i is a factor of j: that p is a perfect fourth power, having j for a factor: and that p is a factor of H.

Ex. (iv). If all the invariants of a quintic vanish shew that a must vanish and that j must be a perfect cube and a factor of f.

Ex. (v). If $$\eta=(xy),\quad \xi=a_x^{n-1}a_y,\quad f=a_x^n,$$

then

(i) $n=2$,

$$f\cdot f(y)=\xi^2+\tfrac{1}{2}(f,f)^2\cdot\eta.$$

(ii) $n=3$,

$$f^2\cdot(y)=\xi^3+\tfrac{3}{2}H\xi\eta^2+t\eta^3.$$

(iii) $n=4$,

$$f^3\cdot f(y)=\xi^4+3H\xi^2\eta^2+4t\xi\eta^3+\left(\frac{if^2}{2}-\frac{3}{4}H^2\right)\eta^4.$$

(iv) $n=6$,

$$f^5\cdot f(y)=\xi^6+\frac{15}{2}H\xi^4\eta^2+20t\xi^3\eta^3+15\left(\frac{i}{2}f^2-\frac{3}{4}H^2\right)\xi^2\eta^4$$
$$+6\,((fi)f^2-Ht)\,\xi\eta^5+\left(\frac{A}{2}f^4-\frac{15}{4}iHf^2+\frac{45}{8}H^3+10t^2\right)\eta^6$$

(*Clebsch*).

[The student, who wishes for information concerning special quintics—when some of the invariants vanish—is referred to Clebsch, *Binären Formen*, ch. VIII., and to Elliott, *Algebra of Quantics*, ch. XIII.]

CHAPTER VIII.

SIMULTANEOUS SYSTEMS.

135. It was proved in Chap. VI. § 103, that if S_1, S_2 be any two finite and complete systems of forms, then the system S formed by taking transvectants of powers and products of powers of forms of S_1 with powers and products of powers of forms of S_2 is both finite and complete. If S_1, S_2 be the complete systems for any two binary forms f_1, f_2; then S is the complete system of concomitants for the forms f_1, f_2 taken simultaneously; for S is complete and contains both f_1 and f_2. Hence the complete irreducible system of concomitants of a pair of binary forms is finite.

136. To make the matter clearer, let us briefly recapitulate the argument.

(i) Any concomitant of the simultaneous system can be expressed as a sum of symbolical products; the factors in which are all of the following types

$$(ab),\ (\alpha\beta),\ (a\alpha),\ a_x,\ \alpha_x:$$

where letters of the Roman alphabet refer to the quantic f_1, and letters of the Greek alphabet to the quantic f_2.

(ii) Any concomitant of the simultaneous system can be expressed linearly in terms of transvectants of products of forms belonging to the complete system for f_1, with products of forms belonging to the complete system for f_2.

For any symbolical product in which the letters are partly Roman and partly Greek is a term of a transvectant $(U, V)^\rho$, where U is a product containing only Roman letters and V a product containing only Greek letters. But by § 51 any term of the transvectant $(U, V)^\rho$ is equal to

$$(U, V)^\rho + \Sigma\lambda(\overline{U}, \overline{V})^{\rho'},$$

where $\overline{U}$, $\overline{V}$ are obtained by convolution from U, V respectively; λ is numerical and ρ' is less than ρ.

Now U, $\overline{U}$ are covariants of f_1 and hence may be expressed as a sum of products of the irreducible forms of f_1; similarly V, $\overline{V}$ may be expressed as a sum of products of the irreducible forms of f_2.

Hence the theorem is true for any symbolical product, the letters of which refer some to f_1 and some to f_2: and therefore it is true for any concomitant of the simultaneous system.

(iii) The system of transvectants $(U, V)^\rho$, where U is a product of concomitants of f_1 and V a product of the concomitants of f_2, is both finite and complete. This was proved in § 103.

137. *The complete irreducible system of concomitants of a finite number of quantics is finite.*

The proof of this theorem is inductive. Let us suppose that it has been proved that the complete system of concomitants of any n quantics is finite.

Consider a set of $n+1$ quantics,

$$f_1, f_2, \ldots f_{n+1}.$$

The n quantics $f_1, f_2, \ldots f_n$ possess, by hypothesis, a finite system of concomitants which may be called S_1. The single form f_{n+1} also possesses a finite system of concomitants, which may be called S_2. The complete system, S, of concomitants of the $n+1$ quantics is obtained by combining S_1 with S_2. And since the systems S_1 and S_2 are both finite and complete, it follows that the complete system S is finite. Hence if the complete system of concomitants of any n quantics is finite then that for any $n+1$ quantics is also finite. But the complete system for any one or any two quantics is finite. Hence the complete system of concomitants for any finite number of quantics is finite.

We proceed to find the complete systems, in a few of the simpler cases.

138. Linear form and quadratic. The complete system of concomitants of a quadratic f consists simply of the quadratic itself and the invariant $(f, f)^2$.

Thus the system S_1 is

$$f, \ (f, f)^2 = D.$$

The system S_2—of the linear form l—is simply l.

The combined system S is obtained by taking all possible transvectants,

$$(f^\alpha D^\beta, l^\gamma)^\delta.$$

But unless $\beta = 0$, this is equal to

$$D^\beta (f^\alpha, l^\gamma)^\delta,$$

and is certainly reducible.

Again $$(f^\alpha, l^\gamma)^\delta = (f^\alpha, l^\delta)^\delta . l^{\gamma-\delta},$$

which is reducible unless $\gamma = \delta$.

Further $(f^\alpha, l^\delta)^\delta$ contains a term

$$(f, l)^2 . (f^{\alpha-1}, l^{\delta-2})^{\delta-2},$$

and is reducible if $\delta > 2$.

The system S then consists of

$$f, \Delta, l, (f, l), (f, l)^2.$$

138 A. Linear form and any finite system. Let the finite system referred to be denoted by S_1. The system S_2 consists simply of the linear form l.

Let the system S_1 consist of the forms $C_1, C_2, \ldots C_\lambda$ which are of orders $s_1, s_2, \ldots s_\lambda$.

Then we have to consider all possible transvectants

$$(C_1^{\alpha_1} C_2^{\alpha_2} \ldots C_\lambda^{\alpha_\lambda}, l^\beta)^\gamma.$$

If $\beta > \gamma$ this transvectant contains a factor l, and is therefore reducible. It may then be supposed that

$$\beta = \gamma.$$

Let us suppose that $\alpha_r \neq 0$; then if $\gamma > s_r$ the transvectant contains the terms

$$(C_r, l^{s_r})^{s_r} (C_1^{\alpha_1} \ldots C_r^{\alpha_r - 1} \ldots C_\lambda^{\alpha_\lambda}, l^{\gamma - s_r})^{\gamma - s_r},$$

and is reducible. If $\gamma \not> s_r$ it contains the term

$$(C_r, l^\gamma)^\gamma . C_1^{a_1} C_2^{a_2} \ldots C_r^{a_r-1} \ldots C_\lambda^{a_\lambda},$$

and is therefore reducible.

Hence the system S contains the forms

$$C_1, C_2, \ldots C_\lambda; \ l; \ (C_r, l^\gamma)^\gamma,$$

$$\gamma = 1, 2, \ldots s_r,$$

$$r = 1, 2, \ldots \lambda;$$

and these forms only.

Ex. Prove that the complete system for a linear form l, and a given finite system of forms, consists of the linear form, the given system, and the forms obtained by operating with powers of

$$\left(l_2 \frac{\partial}{\partial x_1} - l_1 \frac{\partial}{\partial x_2}\right)$$

on the members of the given system.

139. Two quadratics. Let f_1, f_2 be the two quadratics. We have to combine the two systems S_1, S_2, where S_1 consists of

$$f_1, \ (f_1, f_1)^2 = D_1,$$

and S_2 consists of

$$f_2, \ (f_2, f_2)^2 = D_2.$$

Since D_1 and D_2 are invariants, they give rise to no new forms. Hence we have only to consider transvectants

$$(f_1^a, f_2^\beta)^\gamma.$$

The only irreducible transvectants which can be obtained are

$$J_{12} = (f_1, f_2),$$

and

$$D_{12} = (f_1, f_2)^2.$$

The required system is then

$$f_1, f_2, J_{12}, D_1, D_2, D_{12}.$$

139 A. Any number of quadratics. Consider first three quadratics f_1, f_2, f_3. To obtain their simultaneous system we combine the system S_1 for f_1, f_2 with the system S_2 for f_3.

Leaving invariants out of account we must consider all transvectants

$$(f_1^a f_2^\beta J^\gamma_{1,2}, f_3^\delta)^\epsilon.$$

Since all the forms are quadratics the only irreducible transvectants obtainable are

$$(f_1, f_3),\ (f_2, f_3),\ (J_{12}, f_3);$$

$$(f_1, f_3)^2,\ (f_2, f_3)^2,\ (J_{12}, f_3)^2.$$

Of these (J_{12}, f_3) is reducible, for it is the Jacobian of a Jacobian and another form.

The rest are

$$J_{13},\ J_{23},\ D_{13},\ D_{23},$$

and another invariant which may be called

$$E_{123}.$$

The complete system for the three quadratics is then seen to be

$$f_1, f_2, f_3, J_{12}, J_{13}, J_{23}, D_1, D_2, D_3, D_{12}, D_{23}, D_{31}, E_{123}.$$

There is only one form of a new kind, and this is an invariant. Hence in forming the system for four quadratics, we shall not meet with any new kind of concomitant. And in fact it is easy to see that every irreducible concomitant in the system for any number of quadratics belongs to one or other of the *types*

$$f,\ J,\ D,\ E.$$

139 B. It is easy to obtain the syzygies between the forms of the last paragraph. First J_{12} is a Jacobian, and therefore, § 78

$$2J^2_{12} = -D_1 f_2^2 - D_2 f_1^2 + 2D_{12} f_1 f_2 \text{ .... (1).}$$

It will be convenient to use the notation

$$f_1 = a_x^2,\ f_2 = b_x^2,\ f_3 = c_x^2, \ldots .$$

Then as in § 77 we obtain

$$2J_{12}J_{34} = 2(ab)\,a_x b_x \,.\, (cd)\,c_x d_x$$

$$= \begin{vmatrix} (ac)^2 & (ad)^2 & a_x^2 \\ (bc)^2 & (bd)^2 & b_x^2 \\ c_x^2 & d_x^2 & 0 \end{vmatrix}$$

$$= -D_{13} f_2 f_4 - D_{24} f_1 f_3 + D_{14} f_2 f_3 + D_{23} f_1 f_4 \text{ ...... (2).}$$

By replacing f_4 by f_1, a syzygy for $2J_{12}J_{31}$ is obtained.

Again $$E_{123} = -(ab)(bc)(ca),$$

hence, as in § 77,

$$2E_{123}\,.\,E_{456} = \begin{vmatrix} (ad)^2 & (ae)^2 & (af)^2 \\ (bd)^2 & (be)^2 & (bf)^2 \\ (cd)^2 & (ce)^2 & (cf)^2 \end{vmatrix}$$

$$= \begin{vmatrix} D_{14} & D_{15} & D_{16} \\ D_{24} & D_{25} & D_{26} \\ D_{34} & D_{35} & D_{36} \end{vmatrix} \quad \ldots\ldots\ldots\ldots (3).$$

From this may be obtained syzygies for

$$2E^2_{123},\ 2E_{123}\,.\,E_{124},\ 2E_{123}\,.\,E_{145}.$$

Similarly, the syzygy

$$2E_{123}J_{45} = \begin{vmatrix} D_{14} & D_{15} & f_1 \\ D_{24} & D_{25} & f_2 \\ D_{34} & D_{35} & f_3 \end{vmatrix} \quad \ldots\ldots\ldots\ldots\ldots (4)$$

may be obtained, and other particular cases may be deduced.

Again $$\begin{vmatrix} a_1^2 & a_1a_2 & a_2^2 & a_x^2 \\ b_1^2 & b_1b_2 & b_2^2 & b_x^2 \\ c_1^2 & c_1c_2 & c_2^2 & c_x^2 \\ d_1^2 & d_1d_2 & d_2^2 & d_x^2 \end{vmatrix} = 0,$$

for the last column of this determinant is a sum of multiples of the first three columns.

But it has been shewn, § 77, that

$$E_{123} = -(ab)(bc)(ca) = \begin{vmatrix} a_1^2 & a_1a_2 & a_2^2 \\ b_1^2 & b_1b_2 & b_2^2 \\ c_1^2 & c_1c_2 & c_2^2 \end{vmatrix}.$$

Hence

$$f_1E_{234} - f_2E_{134} + f_3E_{124} - f_4E_{123} = 0 \ldots\ldots\ldots\ldots (5).$$

If in the above determinant the elements of the last column are replaced by $(ae)^2$, $(be)^2$, $(ce)^2$, $(de)^2$ respectively, another syzygy is obtained,

$$D_{15}E_{234} - D_{25}E_{134} + D_{35}E_{124} - D_{45}E_{123} = 0 \ \ldots\ldots\ldots (6).$$

In § 77, it was proved that

$$\begin{vmatrix} (ae)^2 & (af)^2 & (ag)^2 & (ah)^2 \\ (be)^2 & (bf)^2 & (bg)^2 & (bh)^2 \\ (ce)^2 & (cf)^2 & (cg)^2 & (ch)^2 \\ (de)^2 & (df)^2 & (dg)^2 & (dh)^2 \end{vmatrix} = 0 \quad \ldots\ldots\ldots\ldots (7)$$

Similarly we obtain the syzygies

$$\begin{vmatrix} D_{15} & D_{16} & D_{17} & f_1 \\ D_{25} & D_{26} & D_{27} & f_2 \\ D_{35} & D_{36} & D_{37} & f_3 \\ D_{45} & D_{46} & D_{47} & f_4 \end{vmatrix} = 0 \quad \ldots\ldots\ldots\ldots (8),$$

and

$$\begin{vmatrix} D_{14} & D_{15} & D_{16} & f_1 \\ D_{24} & D_{25} & D_{26} & f_2 \\ D_{34} & D_{35} & D_{36} & f_3 \\ f_4 & f_5 & f_6 & 0 \end{vmatrix} = 0 \quad \ldots\ldots\ldots\ldots (9).$$

Every kind of syzygy which occurs in the irreducible system of concomitants for any number of quadratics has now been obtained.

Ex. (i). Shew that the last three syzygies just written down are not independent of those which come before, but may be obtained from them on multiplying by forms of the types E and J.

Ex. (ii). Obtain the syzygies (1), (2), (5) by means of Stroh's method.

Ex. (iii). Obtain (4) from (3), and (6) from (5) by transvection.

140. Quadratic and cubic. Let ϕ be the quadratic and f the cubic. Then we have to combine the systems of forms S_1 and S_2; where S_1 contains

$$\phi, \quad (\phi, \phi)^2 = D,$$

and S_2 contains

$$f, \quad (f, f)^2 = H, \quad (f, H) = T, \quad (H, H)^2 = \Delta.$$

All transvectants

$$(\phi^\alpha D^\beta, f^\gamma H^\delta T^\epsilon \Delta^\eta)^\zeta$$

must be considered. Any transvectant for which either β or η is other than zero is obviously reducible, it may then be supposed that

$$\beta = 0, \quad \eta = 0.$$

Again both ϕ and H are quadratics, hence if δ is not zero and $\zeta > 2$ the transvectant contains the reducible term

$$(\phi,\ H)^2\,(\phi^{\alpha-1},\ f^{\gamma}H^{\delta-1}T^{\epsilon})^{\zeta-2}.$$

We have then only to discuss the transvectants

$$(\phi,\ H),\quad (\phi,\ H)^2,\quad (\phi^{\alpha},\ f^{\gamma}T^{\epsilon})^{\zeta}.$$

Of the transvectants

$$(\phi^{\alpha},\ f^{\gamma})^{\zeta},$$

all contain reducible terms except

$$(\phi,\ f),\quad (\phi,\ f)^2,\quad (\phi^2,\ f)^3,\quad (\phi^3,\ f)^6.$$

Now since, by § 91,

$$T^2 = -\tfrac{1}{2}(H^3 + \Delta f^2),$$

those transvectants for which $\epsilon > 1$ are all reducible. Hence of the transvectants

$$(\phi^{\alpha},\ T^{\epsilon})^{\zeta}$$

all are reducible except

$$(\phi,\ T)^2,\quad (\phi^2,\ T)^3,$$

for $(\phi,\ T)$ is reducible by § 77.

The only other irreducible transvectant is readily seen to be

$$(\phi^3,\ fT)^6.$$

The simultaneous system for the quadratic ϕ and the cubic f then consists of:

five invariants

$$D,\quad \Delta,\quad (\phi,\ H)^2,\quad (\phi^3,\ f^2)^6,\quad (\phi^3,\ fT)^6;$$

four linear covariants

$$(\phi,\ f)^2,\quad (\phi^2,\ f)^3,\quad (\phi,\ T)^2,\quad (\phi^2,\ T)^3;$$

three quadratic covariants

$$\phi,\quad H,\quad (\phi,\ H);$$

three cubic covariants

$$f,\quad T,\quad (\phi,\ f).$$

141. Quadratic and any system of forms. Let the system of forms referred to be denoted by S_1, the system S_2 for the quadratic f consists of

$$f,\quad D = (f,\ f)^2.$$

The invariants of both systems may be left out of account as they produce no new forms.

All possible transvectants

$$U = (P, f^r)^\rho$$

must be discussed, where P is a product of the forms

$$C_1, C_2, \ldots C_\lambda$$

of the system S_1.

If $\rho < 2r - 1$, U contains the term

$$f \cdot (P, f^{r-1})^\rho,$$

and is therefore reducible. Since ρ cannot be greater than $2r$ we may confine ourselves to the cases

$$\rho = 2r, \quad 2r - 1.$$

If P be a product of two factors one of which is of even order, then U is reducible.

For let $P = P_1 P_2$, where P_1 is of order $2t$, then

$$(P_1 P_2, f^r)^\rho$$

contains the term

$$(P_1, f^t)^{2t} \cdot (P_2, f^{r-t})^{\rho-2t},$$

and is in consequence reducible.

Also if P is of order $> \rho$, and is a product of two factors, U is reducible. By what we have just proved, if one of the factors of P is of even order U is reducible; let then, $P = P_1 P_2$ where P_1 is of order $2t_1 + 1$ and P_2 of order $2t_2 + 1$, then

$$(P_1 P_2, f^r)^\rho$$

contains the term

$$(P_1, f^{t_1})^{2t_1} \cdot (P_2, f^{r-t_1})^{\rho-2t_1},$$

since $\rho < 2t_1 + 2t_2 + 2$, and therefore $\rho - 2t_1 < 2t_2 + 2$.

Thus U must always be reducible except when P consists of a single term C; or when P consists of a product of two terms C_i, C_j each of which is of odd order, their total order being $\rho = 2r$, so that

$$U = (C_i C_j, f^r)^{2r}.$$

Thus the irreducible forms belong to three classes:

$$\text{(i)}\quad (C, f^r)^{2r-1},$$
$$\text{(ii)}\quad (C, f^r)^{2r},$$
$$\text{(iii)}\quad (C_iC_j, f^r)^{2r},$$

where C_i is of order $2t+1$, and C_j of order $2r-2t-1$; this latter class furnishes invariants only.

It has not been proved that all transvectants belonging to these three classes are irreducible; on the contrary we proceed to examine a case in which certain of the transvectants thus retained are reducible.

142. Let C be a Jacobian $=(C_l, C_m)$.

(i) Let $$C_l = \phi_x^{2\sigma},\quad C_m = \psi_x^{2\tau},$$
$$f = a_x^2 = b_x^2 = \ldots,\quad C = (\phi\psi)\,\phi_x^{2\sigma-1}\psi_x^{2\tau-1}.$$

Then the form $$(C, f^r)^{2r-1}$$
is reducible. For if $2r > 2\sigma + 2\tau - 1$, this transvectant vanishes; and if $2r < 2\sigma + 2\tau - 1$ it contains the term

$$(\phi\psi)(\phi a)\,a_x\psi_x(\phi b^{(1)})^2(\phi b^{(2)})^2\ldots(\phi b^{(\lambda)})^2$$
$$(\psi c^{(1)})^2(\psi c^{(2)})^2\ldots(\psi c^{(r-\lambda-1)})^2\,\phi_x^{2\sigma-2\lambda-2}\,\psi_x^{2\tau+2\lambda-2r}.$$

But
$$(\phi\psi)(\phi a)\,a_x\psi_x = \tfrac{1}{2}\left[-(a\psi)^2\phi_x^2 + (\phi\psi)^2 a_x^2 + (\phi a)^2\psi_x^2\right].$$

And hence the term written down is

$$-\tfrac{1}{2}(\phi, f^\lambda)^{2\lambda}.(\psi, f^{\mu+1})^{2\mu+2} + \tfrac{1}{2}f.T + \tfrac{1}{2}(\phi, f^{\lambda+1})^{2\lambda+2}.(\psi, f^\mu)^{2\mu},$$

where T is a term of $((C_\lambda, C_\mu)^2, f^{r-1})^{2r-2}$.

(ii) If $$C_l = \phi_x^{2\sigma},\quad C_m = \psi_x^{2\tau+1},$$

then the transvectant
$$(C, f^r)^{2r-1}$$

vanishes if $2r > 2\sigma + 2\tau$, and if $2r < 2\sigma + 2\tau$ it contains the reducible term

$$(\phi\psi)(\phi a)\,a_x\psi_x(\phi b^{(1)})^2\ldots(\phi b^{(\lambda)})^2$$
$$(\psi c^{(1)})^2\ldots(\psi c^{(r-\lambda-1)})^2\,\phi_x^{2\sigma-2\lambda-2}\,\psi_x^{2\tau+2\lambda-2r+1}.$$

We are left with the case

$$2r = 2\sigma + 2\tau.$$

(iii) If $$C_l = \phi_x^{2\sigma+1}, \quad C_m = \psi_x^{2\tau+1},$$
then the transvectant
$$(C, f^r)^{2r-1}$$
vanishes if $2r > 2\sigma + 2\tau + 1$, and contains a reducible term if $2r < 2\sigma + 2\tau$; but not if $2r = 2\sigma + 2\tau$.

Hence: "*The transvectant* $(C, f^r)^{2r-1}$ *is reducible, if* C *is a Jacobian, except when one at least of the forms of which* C *is composed is of odd order, and the order of* C *itself is equal to* $2r$ *or* $2r-1$."

143. Quadratic and Quartic. The simultaneous system of irreducible concomitants when the ground-forms are a quadratic and a quartic may now be written down.

The complete system for the quartic ϕ is known to be
$$\phi, \quad H = (\phi, \phi)^2, \quad T = (\phi, H), \quad i = (\phi, \phi)^4, \quad j = (H, \phi)^4.$$
Since there are no forms here of odd order, there can arise no irreducible concomitants belonging to the third of the three classes mentioned above. The simultaneous forms are
$$(\phi, f), \quad (\phi, f)^2, \quad (\phi, f^2)^3, \quad (\phi, f^2)^4,$$
$$(H, f), \quad (H, f)^2, \quad (H, f^2)^3, \quad (H, f^2)^4,$$
$$(T, f)^2, \quad (T, f^2)^4, \quad (T, f^3)^6.$$
It follows from the theorem of § 142 that the forms (T, f), $(T, f^2)^3$, $(T, f^3)^5$ are reducible, T being a Jacobian.

To complete the simultaneous system we must take into account the forms which belong to the quartic and quadratic separately; thus in all we have 18 concomitants.

Ex. Prove that all the forms of the complete system for the two covariants j and i of a binary quintic f, considered as separate quantics, are irreducible when considered as concomitants of the quintic; with the single exception of one invariant of degree 18 in the coefficients of f.

CHAPTER IX.

HILBERT'S THEOREM.

Hilbert's Proof of Gordan's Theorem.

144. WE shall now give another proof of Gordan's theorem that the irreducible system of invariants and covariants of any number of binary forms is finite. The method, which is due to Hilbert*, is of more general application than that of Gordan, inasmuch as with slight and non-essential modifications it applies to forms with any number of variables; on the other hand, unlike Gordan's process it gives practically no information as to the actual determination of the finite system whose existence it establishes, in other words it proves that the problem always has a solution, while the other method, although only proving this for binary forms, gives much information as to the nature of the solution.

In the exposition of Hilbert's proof we shall confine ourselves to binary forms, and to save trouble we shall deal with pure invariants only; inasmuch as the complete system of invariants and covariants of any number of forms is really equivalent to the system of invariants of the set of forms obtained by adjoining an arbitrary linear form to the original set, the proof for invariants is sufficient for the most general case. Cf. § 139.

145. The proof may conveniently be divided into two parts of the following purport.

I. *Proof of the fact that any invariant I of the system may be written in the form*

$$I = A_1 I_1 + A_2 I_2 + \dots + A_n I_n,$$

* *Math. Ann.* XXXVI. Story, *Math. Ann.* XLII.

where I_1, $I_2 \dots I_n$ are a finite number of fixed invariants of the system, and the A's while not necessarily invariants are integral functions of the coefficients.

II. *The application to both sides of the equation just given of a differential operator which leaves an invariant unaltered except for a numerical multiplier, and changes a term*

$$A_r I_r$$

into one of the form $J_r I_r$, where J_r is an invariant.

As a result of I. and II. any invariant may be obtained in the form

$$I_1 J_1 + I_2 J_2 + \dots + I_n J_n.$$

Then by applying the same argument to the J's and so on it follows at once that the I's form the complete system.

146. As a matter of fact the result I. is a particular case of a much more general proposition which we shall first enunciate, then illustrate, and finally prove.

THEOREM. *If a homogeneous function of any number of variables be formed according to any definite laws, then, although there may be an infinite number of functions F satisfying the conditions laid down, nevertheless a finite number F_1, F_2, ... F_r can always be chosen so that any other F can be written in the form*

$$F = A_1 F_1 + A_2 F_2 + \dots + A_r F_r,$$

where the A's are homogeneous integral functions of the variables but do not necessarily satisfy the conditions for the F's.

Suppose for example that we have three variables x, y, z which we take to represent coordinates and that $F = 0$ represents a curve through the point $y = 0$, $z = 0$ (this being the law according to which F is formed), then F may be written in the form

$$yP + zQ,$$

as follows at once since the highest power of x must be wanting in the equation; $y = 0$, $z = 0$ being two curves of the system, this is the application of the theorem to this case.

As another example, if the curve pass through all the vertices of the fundamental triangle its equation may be written

$$yzP + zxQ + xyR = 0,$$

where P, Q, R are integral functions of the coordinates, and here $yz = 0$ etc. are curves of the system.

Again, we have the famous theorem that the equation of any curve through all the points common to $\phi = 0$ and $\psi = 0$ may be written

$$A\phi + B\psi = 0.$$

In each of these cases it will be noted that the system of forms $F_1, F_2, \dots F_r$ is determined; in the general case it is not actually determined, the essential point being that it is finite and that the A's are integral functions.

To establish the theorem in its general form we first remark that it is manifestly true when there is only one variable x, because in this case each F consists of a power of x and therefore all the F's are divisible by that which is of lowest degree; thus there is only one form in the system $F_1, F_2, \dots F_r$.

We now assume that the theorem is true when there are $n-1$ variables and deduce that it is true when there are n variables.

Let $x_1, x_2, \dots x_n$ be the variables and first suppose that the system contains a form H of order r in which the coefficient of $x_n{}^r$ does not vanish. Then we can divide any form in which x_n occurs to a power equal to or greater than r by H without introducing coefficients fractional in the x's, and we can continue the process until the remainder contains no power of x_n higher than the $(r-1)$th.

Hence we can write any form of the system thus

$$F = HP + Mx_n{}^{r-1} + N,$$

where P is the quotient, M is a function of $x_1, x_2, \dots x_{n-1}$, and N is a function of the variables but of degree $r-2$ at the most in x_n.

Now the functions M are formed according to definite laws if the F's are, because each M is deduced from the corresponding F by a definite process, and as they only contain $n-1$ variables the theorem is true by hypothesis for them.

Accordingly we can choose a finite number of M's, say $M_1, M_2, \dots M_k$, such that any other may be written in the form

$$M = B_1M_1 + B_2M_2 + \dots + B_kM_k,$$

where the B's are integral functions of $x_1, x_2, \dots x_{n-1}$.

But since

$$x_n^{r-1} M = F - HP - N, \quad x_n^{r-1} M_1 = F_1 - HP_1 - N_1, \text{ etc.,}$$

we have

$$F - HP - N = B_1 (F_1 - HP_1 - N_1) + B_2 (F_2 - HP_2 - N_2) + \ldots + B_k (F_k - HP_k - N_k),$$

or

$$F = H (P - B_1 P_1 - B_2 P_2 - \ldots - B_k P_k) + B_1 F_1 + \ldots + B_k F_k + N - B_1 N_1 - \ldots - B_k N_k.$$

Now the part of the right-hand side which does not contain one of the forms as a factor consists of B's and N's and therefore only contains x_n to the power $r-2$ at most. Hence we may write

$$F = HQ_1 + B_1 F_1 + B_2 F_2 + \ldots + B_k F_k + M^{(1)} x_n^{r-2} + N^{(1)},$$

and now $M^{(1)}$ is a function of $x_1, x_2, \ldots x_{n-1}$ formed according to definite laws and $N^{(1)}$ is of order $r-3$ at most in x_n.

Thus we can write F as the sum of a finite number of terms each containing a form of the system for factor together with expressions of order $r-2$ at most in the last variable.

Then applying precisely the same argument to the $M^{(1)}$'s as we applied to the M's we see that by adding a finite number of F's to

$$H, F_1, F_2, \ldots F_k$$

we can reduce the order of the remaining portion in x_n to $r-3$.

Proceeding in this way and adding only a finite number of F's at each step we can finally write F in the form

$$HQ_r + C_1 F_1 + C_2 F_2 + \ldots + C_m F_m + M^{(r)},$$

where $M^{(r)}$ only involves $x_1, x_2, \ldots x_{n-1}$ and in the nature of things is formed according to definite laws. Hence applying the same process to the $M^{(r)}$'s as we applied to the M's we finally have F in the form

$$A_1 F_1 + A_2 F_2 + \ldots + A_s F_s,$$

where the F's include H and the number s is finite.

Consequently if the theorem be true for $n-1$ variables it is true for n, but it is true for one variable, therefore by induction it is true universally.

We have now to remove the limitation imposed above, viz., that there exists a form of the system in which the coefficient of the highest power of x_n is not zero.

If there is no such form among the F's let F_t be one of the forms and apply to all a linear substitution

$$x_r = a_{r1}y_1 + a_{r2}y_2 + \ldots + a_{rn}y_n; \quad r = 1, 2, \ldots n.$$

Suppose that $F_t(x)$ becomes $G_t(y)$, then the coefficient of the highest power of y_n in G_t is $F_t(a_{1n}, a_{2n}, \ldots a_{nn})$, and therefore unless F_t is identically zero we can choose the linear substitution so that this coefficient is not zero*. Hence the theorem is true for the forms G in the variables y, and therefore by changing the variables back again from y to x we see that it is true for the F's.

Q. E. D.

147. Returning now to the consideration of invariants it is clear that such an expression regarded as a homogeneous function of the coefficients of the forms is formed according to definite laws; hence, if I be any invariant of the system, we have

$$I = A_1I_1 + A_2I_2 + \ldots + A_nI_n,$$

where $I_1, I_2, \ldots I_n$ are n fixed invariants and the A's are homogeneous integral functions of the coefficients but not necessarily invariants. As to the functions A a simple remark may be added. All that is asserted in the general statement of the foregoing theorem is that they are homogeneous in all the coefficients, but an invariant is homogeneous in each set of coefficients involved taken separately, and consequently since I and I_m are homogeneous in each set of coefficients, A_m is also homogeneous in each set. That this is so could of course be seen in the proof of the general theorem because at no point of the investigation is the homogeneity disturbed.

148. We now come to the second part of the proof, but before proceeding with it we must prove a necessary lemma on the properties of the operator Ω so often used in the course of this work.

If P be a function of $\xi_1, \xi_2, \eta_1, \eta_2$ which is homogeneous and of order λ in ξ_1, ξ_2 and homogeneous and of order μ in η_1, η_2, then

$$\Omega^m(D^nP) = C_0D^{n-m}P + C_1D^{n-m+1}\Omega(P) + \ldots + C_mD^n\Omega^m(P),$$

* We assume here that unless a form vanishes identically values of the variables can be found for which it is not zero. It is easy to give a formal proof of this theorem. Cf. Weber's *Algebra*, Vol. I. p. 457.

where $D = \xi_1\eta_2 - \xi_2\eta_1$, m and n are positive integers and the C's are either zero or constant.

The result can be readily proved by induction, for we have

$$\Omega(DP)$$
$$= P + \eta_2 \frac{\partial P}{\partial \eta_2} + \xi_1 \frac{\partial P}{\partial \xi_1} + D \frac{\partial^2 P}{\partial \xi_1 \partial \eta_2} - \left(-P - \xi_2 \frac{\partial P}{\partial \xi_2} - \eta_1 \frac{\partial P}{\partial \eta_1} + D \frac{\partial^2 P}{\partial \xi_2 \partial \eta_1}\right),$$

and by Euler's theorem for homogeneous functions the right-hand side becomes

$$(\lambda + \mu + 2) P + D\Omega P.$$

Now in this result change P into $D^{n-1}P$ so that λ and μ are increased by $n - 1$, and we have

$$\Omega(D^n P) = (\lambda + \mu + 2n) D^{n-1} P + D\Omega(D^{n-1}P).$$

Hence

$$D\Omega(D^{n-1}P) = (\lambda + \mu + 2n - 2) D^{n-2} P + D^2\Omega(D^{n-2}P),$$

$$D^2\Omega(D^{n-2}P) = (\lambda + \mu + 2n - 4) D^{n-1} P + D^3\Omega(D^{n-3}P),$$

....

$$D^{n-2}\Omega(D^2P) = (\lambda + \mu + 4) D^{n-1} P + D^{n-1}\Omega(DP),$$

$$D^{n-1}\Omega(DP) = (\lambda + \mu + 2) D^{n-1} P + D^n\Omega(P).$$

By adding these results together we obtain

$$\Omega(D^n P) = \{n(\lambda + \mu) + n(n+1)\} D^{n-1} P + D^n\Omega(P),$$

which establishes the result when $m = 1$ for all values of n.

Assume that the result is true for any value of m so that

$$\Omega^m(D^n P) = C_0 D^{n-m} P + C_1 D^{n-m+1}\Omega(P) + \ldots + C_m D^m \Omega^m(P),$$

then operating again with Ω we have

$$\Omega^{m+1}(D^n P) = \sum_{r=1}^{r=m} C_r \Omega\{D^{n-m+r}\Omega^r(P)\}.$$

But

$$\Omega(D^{n-m+r}P)$$
$$= (n - m + r)(\lambda + \mu + n - m + r + 1) D^{n-m+r-1} P + D^{n-m+r}\Omega(P),$$

and changing P into $\Omega^r(P)$ so that λ, μ are each diminished by r, we deduce

$$\Omega\left\{D^{n-m+r}\Omega^r(P)\right\}$$

$$= (n-m+r)(\lambda+\mu-r-m+n+1)\,D^{n-m+r-1}\Omega^r(P)+D^{n-m+r}\Omega^{r+1}(P)$$

$$= \alpha_r D^{n-m+r-1}\Omega^r(P)+D^{n-m+r}\Omega^{r+1}P$$

when α_r is numerical.

Thus we have

$$\Omega^{m+1}(D^nP)=\overset{r=m}{\underset{r=1}{\Sigma}}\,(C_r\alpha_r+C_{r-1})\,D^{n-m+r-1}\Omega^r(P),$$

in other words, if the result be true for m it is true for $m+1$, for the right-hand side is of the stipulated form. Hence by induction the theorem is true universally.

Ex. Prove that

$$\Omega^n(D^mP)=\underset{r}{\Sigma}\binom{m}{r}\frac{n!\,(\lambda+\mu+n+1-r)!}{(\lambda+\mu+n-m+1)!\,(n-m+r)!}\,D^{n-m+r}\Omega^rP.$$

Cor. It clearly follows that if in the formal statement any exponent of D on the right-hand side be negative the corresponding coefficient C is zero because only integral functions can appear in the process.

149. With the aid of the above lemma the proof of Gordan's theorem may be easily completed.

For the sake of convenience we shall regard x_1, x_2 as the variables in the fundamental binary forms, despite the fact that in the general theorem proved above they play the rôle that the coefficients do in the remaining portion of the investigation.

Suppose the variables are changed by the linear transformation

$$\left.\begin{aligned}x_1 &= \xi_1 x_1 + \eta_1 x_2\\ x_2 &= \xi_2 x_1 + \eta_2 x_2\end{aligned}\right\},$$

then an invariant I of the forms becomes $(\xi\eta)^\mu I$.

Further we have

$$I = A_1 I_1 + A_2 I_2 + \ldots + A_n I_n$$

and an invariant I_m on the right becomes

$$(\xi\eta)^{\mu_m} I_m.$$

We now write down the identity which is the transformation of

$$I = A_1 I_1 + A_2 I_2 + \dots + A_n I_n,$$

i.e. the same identity for the transformed quantics; we suppose a coefficient A_m written in symbolical letters entirely, so that it is the sum of a number of terms each of which contains only factors of the types a_ξ and a_η, where a is a symbol belonging to one of the quantics.

If after transformation A_m become B_m we have

$$I(\xi\eta)^\mu = \sum_1^n B_m (\xi\eta)^{\mu_m} I_m$$

and the equation shews at once that B_m is of order $\mu - \mu_m$ in both ξ and η.

Now operate on both sides of this identity with

$$\Omega^\mu = \left(\frac{\partial^2}{\partial\xi_1 \partial\eta_2} - \frac{\partial^2}{\partial\xi_2 \partial\eta_1}\right)^\mu.$$

The left-hand side becomes a numerical multiple of I, viz. $(\mu + 1)(\mu\,!)^2 I$, and on the right-hand side we have

$$\Omega^\mu \{(\xi\eta)^{\mu_m} B_m\} I_m$$

$$= I_m \{C_0 (\xi\eta)^{\mu_m - \mu} B_m + C_1 (\xi\eta)^{\mu_m - \mu + 1} \Omega B_m + \dots + C_\mu (\xi\eta)^{\mu_m} \Omega^\mu B_m\}$$

by the lemma, since I_m does not involve ξ or η.

But if $\mu - \mu_m = \nu$, then

$$\mu_m - \mu,\ \mu_m - \mu + 1, \dots \mu_m - \mu + (\nu - 1)$$

are all negative.

Consequently

$$C_0, C_1, \dots C_{\nu-1}$$

are all zero.

Again B_m is of order $\mu - \mu_m = \nu$ in both ξ and η,

hence $$\Omega^{\nu+1}(B_m),\ \Omega^{\nu+2}(B_m), \dots \Omega^\mu (B_m)$$

are all zero, and the effect of the operator on

$$(\xi\eta)^{\mu_m} B_m$$

therefore reduces to a single term, namely

$$C_\nu \Omega^\nu (B_m).$$

Now B_m is the sum of a number of terms each containing ν factors of the type a_ξ and r factors of the type a_η, hence by a fundamental theorem,

$$\Omega^\nu (B_m)$$

is an invariant of the system.

Therefore after operating with Ω^μ on both sides of the equation we are left with

$$I = \Sigma J_m I_m, \text{ where } J_m = \frac{C_r}{(\mu + 1)(\mu!)^2} \Omega^\nu (B_m)$$

and is an invariant.

Since J_m is an invariant we can express it also as the sum of a number of terms each containing an I_m as a factor, hence by continual reduction we can ultimately express I as a rational integral function of $I_1, I_2, \ldots I_m$, that is to say, these invariants constitute a complete system and, as we have seen, their number is finite; Gordan's theorem is thus completely established.

150. Syzygies between the irreducible invariants.

Examples of relations between the members of an irreducible system of invariants or covariants have already been given, and in fact a very large number were obtained for the quintic.

It can be deduced from Hilbert's Lemma that the system of syzygies is finite, that is to say if $S = 0$ be any syzygy we can find a finite number of syzygies

$$S_1 = 0, \quad S_2 = 0, \ldots S_r = 0,$$

such that $$S = C_1 S_1 + C_2 S_2 + \ldots + C_r S_r,$$

where $C_1, \ldots C_r$ are invariants.

If there be such a relation, then of course all other syzygies are necessary consequences of

$$S_1 = 0, \quad S_2 = 0, \ldots S_r = 0,$$

and these constitute the finite system.

The proof is very simple. Let $I_1, I_2, \ldots I_m$ be the members of the irreducible system of invariants, then S is a function of $I_1, I_2, \ldots I_m$ formed according to the law that it must vanish when for the I's we substitute their actual values in terms of the coefficients.

Hence we have

$$S = C_1 S_1 + C_2 S_2 + \dots + C_r S_r,$$

where $S_1 = 0$, etc. are a finite number of syzygies and the C's, being functions of the I's, are invariants. (Cf. § 127.)

151. Gordan's Proof of Hilbert's Lemma.

Many versions have been given of the fundamental lemma of Hilbert on functions formed according to given laws, but the majority of them do not differ materially from the original proof due to Hilbert. Nevertheless Gordan has recently given a demonstration* which is so interesting and depends on such simple principles that we cannot refrain from giving an account of it here. We shall state it in the form of two theorems.

Theorem I. *If a simple product of positive integral powers of n letters*

$$x_1^{k_1} x_2^{k_2} \dots x_n^{k_n},$$

be formed in such a way that the exponents $k_1, k_2, \dots k_n$ satisfy certain prescribed conditions, then, although the number of products satisfying the conditions may be infinite, yet a finite number of them can be chosen so that every other is divisible by one at least of this finite number.

To illustrate the scope of this theorem take the case of products of three letters and suppose the conditions are

$$k_1 \equiv 0 \text{ (mod. 3)},$$

$$k_2^2 - k_3 = 7.$$

The simple products satisfying the conditions are

$$x_2^3 x_3^2, \quad x_2^4 x_3^9, \dots$$

$$x_1^3 x_2^3 x_3^2, \quad x_1^3 x_2^4 x_3^9 \dots,$$

and it is evident that all such products are divisible by $x_2^3 x_3^2$.

Again, suppose the sole condition is

$$k_1 - k_2 + k_3 > 0;$$

the products are

$$x_1, x_3; \quad x_1^2, x_3^2, x_1 x_3; \quad x_1^2 x_2, x_1 x_2 x_3, x_2 x_3^2, x_1^3, \dots$$

and all the products are divisible by x_1 or x_3.

* *Liouville's Journal*, 1900.

Other examples could be given, but the above will suffice to shew the nature of the theorem which we now proceed to prove.

If $n = 1$ the truth of the theorem is evident because all the products are powers of a single letter and are therefore divisible by that having the least exponent.

We shall now assume that the result is true for $n - 1$ letters and prove that it is true for n letters.

Let
$$x_1^{a_1} x_2^{a_2} \dots x_n^{a_n}$$
be a definite product P satisfying the given conditions and let
$$x_1^{k_1} x_2^{k_2} \dots x_n^{k_n}$$
be a typical product K of the system.

If K be not divisible by P one of the k's must be less than the corresponding a.

Suppose that $k_r < a_r$, then, consistently with this, k_r must have one of the values
$$0, 1, 2, \dots a_r - 1.$$

Hence if K be not divisible by P one of a number
$$a_1 + a_2 + \dots + a_n = N \text{ contingencies arises, viz.}$$
either
$$k_1 \text{ has one of the values } 0, 1, 2, \dots a_1 - 1,$$
or
$$k_2 \text{ has one of the values } 0, 1, 2, \dots a_2 - 1, \text{ etc.}$$

Suppose that $k_r = m$, and that this is the pth of the possible cases; then the remaining exponents $k_1, k_2, \dots k_{r-1}, k_{r+1} \dots k_n$ satisfy definite conditions which are obtained by making $k_r = m$ in the original conditions.

Let
$$K_p = x_1^{k_1} x_2^{k_2} \dots x_r^{m} \dots x_n^{k_m}$$
be a product of the system for which $k_r = m$ and write
$$K_p = x_r^m K'_p.$$

Then K'_p contains only $n - 1$ letters and the exponents satisfy definite conditions, and when these are satisfied the exponents of K_p satisfy the original conditions. Hence by hypothesis a finite number of products of the type K'_p can be found such that every other such product is divisible by one at least of these.

Denote this finite system by

$$L_1, L_2, \dots L_{a_p}$$

so that K_p is divisible by one at least of the L's.

Thus $K_p = x_r{}^m K'_p$ is divisible by one at least of the products

$$x_r{}^m L_1,\ x_r{}^m L_2, \dots x_r{}^m L_{a_p},$$

which all belong to the original system of products because every L belongs to the subsidiary system.

Denote these latter products by

$$M_p{}^{(1)},\ M_p{}^{(2)}, \dots M_p{}^{(a_p)};$$

then in the pth of the N possible contingencies K is divisible by one of the products

$$M_p{}^{(1)},\ M_p{}^{(2)}, \dots M_p{}^{(a_p)}.$$

Now one of these N contingencies certainly does arise when K is not divisible by P, and hence K must be divisible by one of the products

$$M_1{}^{(1)},\ M_1{}^{(2)}, \dots M_1{}^{(a_1)};\quad M_2{}^{(1)}, \dots M_2{}^{(a_2)}; \dots M_N{}^{(a_N)},$$

or else by P.

The exponents of the M's all satisfy the prescribed conditions and they are finite in number, hence if the theorem be true for $n-1$ letters it is true for n letters, but it is true for one letter and hence by induction it is true universally.

152. Theorem II. *If a system of homogeneous forms be constructed according to given laws, then a finite number of definite forms of the system can be chosen such that every other form of the system is an aggregate of terms each of which involves one of the finite number of forms as a factor, and the coefficients are integral in the variables.* (Hilbert's Lemma.)

Suppose in fact that $x_1, x_2, \dots x_n$ are the variables and that ϕ is a typical form of the system. Now construct an auxiliary system of functions η of the same variables according to the law that a function is an η function when it can be written in the form

$$\eta = \Sigma A\phi,$$

the A's being integral functions of the variables which make the right-hand side homogeneous, but otherwise unrestricted except that the number of terms on the right-hand side must be finite.

The class of functions η is infinitely more comprehensive than the class ϕ, and it possesses the important property that a function of the form $\Sigma B\eta$ which is homogeneous in the variables is also an η function.

Now in examining the functions η we arrange the terms of one of them of order r in such a way that x_1^r comes first and, generally, a term

$$S = x_1^{a_1}x_2^{a_2} \dots x_n^{a_n}$$

comes before a term

$$T = x_1^{b_1}x_2^{b_2} \dots x_n^{b_n},$$

when the first of the quantities

$$a_1 - b_1, \quad a_2 - b_2, \dots a_n - b_n$$

which does not vanish is positive.

In such a case we say that the term S is simpler than the term T and T is more complex than S, so that any function η is arranged with its terms in ascending order of complexity. Now the functions η being formed according to fixed laws, their first terms satisfy given conditions relating to the exponents, and hence by Theorem I. a finite number of η's, say $\eta_1, \eta_2, \dots \eta_p$, can be chosen such that the first term of any other η is divisible by the first term of one of these.

Take any function η of the auxiliary system, and suppose its first term is divisible by the first term of η_{m_1}, and that P_1 is the quotient.

Then $\eta - P_1\eta_{m_1}$ is an η function with a more complex first term than η because η and $P_1\eta_{m_1}$ have the same first term; if we denote this new function by $\eta^{(1)}$ we have

$$\eta = P_1\eta_{m_1} + \eta^{(1)}.$$

Next, if the first term of $\eta^{(1)}$ be divisible by that of η_{m_2} we have

$$\eta^{(1)} = P_2\eta_{m_2} + \eta^{(2)},$$

and the first term of $\eta^{(2)}$ is more complex than that of $\eta^{(1)}$.

Continuing this process of reduction we find

$$\eta^{(r-1)} = P_r\eta_{m_r} + \eta^{(r)} \text{ and so on.}$$

Now the first terms of

$$\eta,\ \eta^{(1)},\ \eta^{(2)}, \ldots \eta^{(r)} \ldots$$

are in ascending order of complexity, and hence the time must come when there is no η function of the same order as η with a more complex first term than $\eta^{(r)}$; in that case we have

$$\eta^{(r)} = P_{r+1}\eta_{m_{r+1}}.$$

Hence

$$\eta = P_1\eta_{m_1} + P_2\eta_{m_2} + \ldots + P_{r+1}\eta_{m_{r+1}}\,^*$$

where the η's on the right-hand side are all members of the finite system $\eta_1, \eta_2, \ldots \eta_p$.

Now the η system includes all the ϕ's, moreover each η contains only a finite number of ϕ's and hence every ϕ can be expressed in the form

$$A_1\phi_1 + A_2\phi_2 + \ldots + A_r\phi_r,$$

where $\phi_1, \phi_2, \ldots \phi_r$ are the ϕ's contained in the expression for $\eta_1, \eta_2, \ldots \eta_p$ and the A's are integral functions of the variables.

153. Remark. If all the conditions satisfied by $k_1, k_2, \ldots k_n$ in Theorem I. be linear homogeneous equations, then the theorem establishes the existence of a finite number of solutions by means of which any other solution can be reduced. The difference of two solutions being now a solution, it follows that by continual reduction we can express all solutions of the linear equations in terms of a finite number—this is the result otherwise proved in § 97.

* The η's on the right being members of a finite system are finite in number; hence even though the number of steps in the reduction be infinite, there can only be a finite number of terms on the right-hand side. The same η may of course occur in more than one term, but in that case we should add all such terms together.

CHAPTER X.

THE GEOMETRICAL INTERPRETATION OF BINARY FORMS.

154. GIVEN two points of reference A, B on any straight line, the position of any other point P may be determined by the value of the ratio $\frac{AP}{PB}$ of the distances of P from A and B.

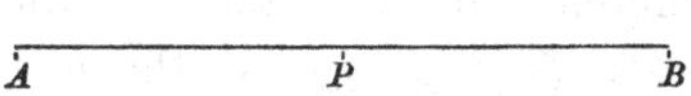

A convention as regards *sign* is necessary to complete the definition; it is convenient to regard the ratio as positive if P lie between A and B, otherwise as negative.

When a binary form of order n is equated to zero, the ratio $\frac{x_1}{x_2}$ may have any one of n values. These determine n points on the straight line AB, such that the ratio $\frac{AP}{PB}$ for each of these points is equal to one of the roots of the equation for $\frac{x_1}{x_2}$. The coordinates (x_1, x_2) then define the position of P on the straight line by means of the equation

$$\frac{AP}{PB} = \frac{x_1}{x_2}.$$

It is found advisable, as will be evident immediately, to define the position of the point P whose coordinates are (x_1, x_2) by means of the equation

$$\frac{AP}{PB} = \lambda \frac{x_1}{x_2},$$

where λ is a fixed numerical multiplier.

A further convention will be useful, viz. the positive direction of measurement from A is towards B and that from B is towards A.

To find the distance between (x_1, x_2) *and* (y_1, y_2), *in terms of the length* l *of* AB.

Denoting the two points by P and Q, we have

$$\frac{AP}{\lambda x_1} = \frac{PB}{x_2} = \frac{AB}{\lambda x_1 + x_2} = \frac{l}{\lambda x_1 + x_2},$$

and similarly

$$\frac{QB}{y_2} = \frac{l}{\lambda y_1 + y_2}.$$

Hence

$$PQ = (PB - QB) = l\frac{x_2(\lambda y_1 + y_2) - y_2(\lambda x_1 + x_2)}{(\lambda y_1 + y_2)(\lambda x_1 + x_2)}$$

$$= \frac{\lambda l (yx)}{(\lambda y_1 + y_2)(\lambda x_1 + x_2)}.$$

155. Let us consider the effect of a change in the points of reference. Let the new points of reference A', B' in terms of the original system of coordinates be (ξ_1, ξ_2), (η_1, η_2); if the new multiplier be μ and (X_1, X_2) be the new coordinates of P, then

$$\frac{X_1}{X_2} = \mu\frac{A'P}{PB'}.$$

Hence
$$\frac{X_1}{X_2} = \mu\frac{(x\xi)}{\lambda\xi_1 + \xi_2}\cdot\frac{\lambda\eta_1 + \eta_2}{(\eta x)}.$$

The change in coordinates is thus equivalent to the linear transformation

$$X_1 = \rho_1(x\xi),$$
$$X_2 = \rho_2(\eta x),$$

where
$$\frac{\rho_1}{\rho_2} = \mu\frac{\lambda\eta_1 + \eta_2}{\lambda\xi_1 + \xi_2}.$$

156. A linear transformation

$$x_1 = \xi_1 X_1 + \eta_1 X_2,$$
$$x_2 = \xi_2 X_1 + \eta_2 X_2,$$

may be regarded geometrically from two different points of view:

(i) As changing the points of reference and the constant multiplier, but leaving the other points on the straight line in their original position.

(ii) If the points of reference are regarded as fixed, the

transformation alters the positions of all the points defined by the algebraic forms under discussion.

Consider the first of these points of view. When $x_1 = 0$, the point $P(x_1, x_2)$ coincides with one of the points of reference A. Similarly, if $x_2 = 0$, P coincides with B. Hence to find the new points of reference in the original system of coordinates, it is only necessary to write $X_1 = 0$, and $X_2 = 0$. We obtain them at once as (η_1, η_2) and (ξ_1, ξ_2).

The distances of P from these new points of reference A', B' are

$$\lambda l \frac{(x\eta)}{(\lambda x_1 + x_2)(\lambda \eta_1 + \eta_2)} = \lambda l \frac{(\xi\eta) X_1}{(\lambda x_1 + x_2)(\lambda \eta_1 + \eta_2)},$$

and $$\lambda l \frac{(\xi x)}{(\lambda x_1 + x_2)(\lambda \xi_1 + \xi_2)} = \lambda l \frac{(\xi\eta) X_2}{(\lambda x_1 + x_2)(\lambda \xi_1 + \xi_2)}.$$

The ratio of these distances is

$$\frac{A'P}{PB'} = \frac{\lambda\xi_1 + \xi_2}{\lambda\eta_1 + \eta_2} \cdot \frac{X_1}{X_2}.$$

Hence the new multiplier is $\frac{\lambda\xi_1 + \xi_2}{\lambda\eta_1 + \eta_2}$; and the coordinates (X_1, X_2) define the same point as that defined by the coordinates (x_1, x_2). It should be observed that the sign of the expression $\frac{\lambda\xi_1 + \xi_2}{\lambda\eta_1 + \eta_2} \cdot \frac{X_1}{X_2}$ is positive if X lie between A' and B', otherwise it is negative.

Ex. (i). Shew that by properly choosing quantities a_1, a_2 the distance between the points (x_1, x_2), (y_1, y_2) may be written $\frac{(yx)}{(ax)(ay)}$. And that in this case the constant multiplier after transformation becomes $\frac{(a\xi)}{(a\eta)}$.

Ans. $a_1 = -\sqrt{\frac{1}{l\lambda}}$, $a_2 = \sqrt{\frac{\lambda}{l}}$.

Ex. (ii). The point (a_1, a_2) of the last example is the point at infinity on the range.

When an invariant of a binary quantic is zero, there exists some relation between its roots which is unaffected by any linear transformation. Hence when the binary quantic is regarded as the analytical expression of n points on a range, the vanishing of an invariant is the condition that there may be some definite geometrical relation between the points, independent of the points of reference and of the constant multiplier.

For example if two of the points coincide an invariant—the discriminant—vanishes. Again, as will be shewn later, if four points on a straight line which form a harmonic range are represented analytically by a quartic, then the invariant j of that quartic is zero.

157. In the second point of view stated in the last paragraph, the points of reference and multiplier are regarded as fixed, the point P takes a new position P' given by the coordinates (X_1, X_2).

Let the points Q, R, S, viz. (y_1, y_2), (z_1, z_2), (w_1, w_2) become Q', R', S'.

Then
$$PQ = \lambda l \frac{(yx)}{(\lambda x_1 + x_2)(\lambda y_1 + y_2)}$$

and
$$(yx) = (\xi\eta)(YX).$$

Hence
$$\frac{PQ \,.\, RS}{PS \,.\, RQ} = \frac{(yx)(wz)}{(wx)(yz)}$$
$$= \frac{(YX)(WZ)}{(WX)(YZ)} = \frac{P'Q' \,.\, R'S'}{P'S' \,.\, R'Q'}.$$

The expression $\frac{PQ \,.\, RS}{PS \,.\, RQ}$ is called the cross (or anharmonic) ratio of the four points P, Q, R, S; it is usually denoted by $\{PQRS\}$*. The result just proved may be written
$$\{PQRS\} = \{P'Q'R'S'\}.$$

* By rearranging the letters P, Q, R, S we obtain 24 such cross-ratios. It is easy to see, however, that only six of these are different. Then if λ, μ, ν are written for the three products $PQ \,.\, RS$, $PR \,.\, SQ$, $PS \,.\, QR$, the six different cross-ratios are
$$-\frac{\lambda}{\mu},\ -\frac{\mu}{\nu},\ -\frac{\nu}{\lambda},\ -\frac{\mu}{\lambda},\ -\frac{\nu}{\mu},\ -\frac{\lambda}{\nu}.$$
Since the four points are collinear,
$$PQ \,.\, RS + PR \,.\, SQ + PS \,.\, QR = 0$$
or
$$\lambda + \mu + \nu = 0;$$
by means of this relation all six cross-ratios of four points may be expressed in terms of one of them. This mode of presenting the subject is due to Mr R. R. Webb.

That is, the cross-ratio of four points on the range is unaltered by any linear transformation. Hence the transformed range is homographic with the original range.

Further any range homographic with the original range may be obtained from it by a linear transformation. To prove this, it is only necessary to prove that the coefficients of transformation may be so chosen that three non-coincident points P, Q, R of the original range are changed to any three non-coincident points P', Q', R' chosen at random on the straight line. For when P', Q', R' are known, the point S' of the transformed range corresponding to S is given by

$$\{P'Q'R'S'\} = \{PQRS\}\text{*}.$$

Let us suppose then that the values of the ratios $\dfrac{X_1}{X_2}, \dfrac{Y_1}{Y_2}, \dfrac{Z_1}{Z_2}$ are given.

Then $\quad x_1 = \left(\xi_1 \dfrac{X_1}{X_2} + \eta_1\right) X_2, \quad x_2 = \left(\xi_2 \dfrac{X_1}{X_2} + \eta_2\right) X_2$

and
$$\frac{x_1}{x_2} = \frac{\xi_1 \dfrac{X_1}{X_2} + \eta_1}{\xi_2 \dfrac{X_1}{X_2} + \eta_2}$$

or
$$\xi_2 \frac{X_1}{X_2} \cdot \frac{x_1}{x_2} - \xi_1 \frac{X_1}{X_2} + \eta_2 \frac{x_1}{x_2} - \eta_1 = 0.$$

Similarly
$$\xi_2 \frac{Y_1}{Y_2} \cdot \frac{y_1}{y_2} - \xi_1 \frac{Y_1}{Y_2} + \eta_2 \frac{y_1}{y_2} - \eta_1 = 0$$

and
$$\xi_2 \frac{Z_1}{Z_2} \cdot \frac{z_1}{z_2} - \xi_1 \frac{Z_1}{Z_2} + \eta_2 \frac{z_1}{z_2} - \eta_1 = 0.$$

We have three equations to determine the ratios of the coefficients $\xi_1, \xi_2, \eta_1, \eta_2$.

These ratios are thus determined uniquely.

Hence a range of points on a straight line may be transformed into any other range homographic with itself by a linear transformation.

If an invariant of a binary quantic representing a range of n points is zero, these points must possess some special property,

* This must give a unique position for S' since it is equivalent to a linear relation between its coordinates.

which is also a property of all ranges homographic with the original range. Such a property is said to be projective, and thus the vanishing of an invariant must be the condition for the existence of some projective property of the points which the quantic represents.

Conversely, if a system of n points on a straight line possesses some projective property, there will exist a corresponding analytical relation between the coefficients of the quantic represented by these points, which is unaltered by any linear transformation. It does not necessarily follow that the condition is represented by the vanishing of an invariant; it sometimes happens that a projective property necessitates the vanishing of all the coefficients of a covariant.

Again a covariant of a quantic will define a certain number of points on the straight line. These points are related to the original points of the range in the same way as their homologues are related to the homologues of the original points on a homographic range. It is usual to denote this by saying that the points are projectively related to the points of the original range.

158. A binary form is homogeneous in two variables, we are then—in such forms—only concerned with the ratio of the variables. Let $f(x_1, x_2)$ be any binary form of order n, then the equation

$$f(x_1, x_2) = 0$$

defines n values of the ratio $\frac{x_1}{x_2}$. Hence in any geometrical figure in which the geometric element is completely defined in position by a single parameter, the form $f(x_1, x_2)$ may be considered as defining n of these elements. For example a point which lies on a *unicursal* curve is such an element. If x, y, z are its Cartesian coordinates, it is well known that we may express x, y, z as rational algebraic functions of a single parameter. Again the tangent to a fixed unicursal curve may be taken to be the element. Or else the element might be the osculating plane of a twisted unicursal curve.

Now the two simplest figures of the kind, are a range of points on a fixed straight line, and a *pencil* of straight lines passing through a fixed point and lying in a fixed plane. We may deduce the properties of the latter from the former. For if any straight line be drawn to cut the pencil there is a one-to-one correspondence between the points on the range thus formed and the rays of the pencil. In fact any ray of the pencil may be defined by the coordinates (x_1, x_2) of the point in which it intersects the straight line. With this definition it appears that everything that has been said for the range applies equally well to the pencil.

159. A binary form may be expressed as a product of n linear factors. A covariant of the binary form is necessarily a covariant of the system of linear forms of which it is a product. Thus let

$$a_x^n = (xx^{(1)})(xx^{(2)}) \dots (xx^{(n)}) \dots\dots\dots\dots\dots\dots \text{(I)},$$

where $\dfrac{x_1^{(1)}}{x_2^{(1)}}, \dfrac{x_1^{(2)}}{x_2^{(2)}}, \dots$ are the roots of the equation $a_x^n = 0$.

Any invariant may be written in the form

$$I = \Sigma (x^{(1)}x^{(2)})^{\alpha_{12}} (x^{(1)}x^{(3)})^{\alpha_{13}} \dots (x^{(2)}x^{(3)})^{\alpha_{23}} \dots\dots\dots \text{(II)},$$

since it is an invariant of the linear forms $(xx^{(1)})$, $(xx^{(2)}) \dots$.

The coefficients of the quantic are given in terms of the roots by equating the different powers of x in (I). Two things are at once apparent.

(i) The coefficients are symmetric functions of the quantities $x^{(1)}, x^{(2)} \dots$.

(ii) The coefficients are functions homogeneous and linear in each of the n sets of variables $x_1^{(1)}, x_2^{(1)}$; $x_1^{(2)}, x_2^{(2)}$; ... $x_1^{(n)}, x_2^{(n)}$.

It follows that any function of the coefficients must, when expressed in terms of the quantities $x^{(1)}, x^{(2)} \dots$, be symmetrical in them. And further such a function must be homogeneous and of the same order in each of the sets of variables $x_1^{(1)}, x_2^{(1)}$; ... $x_1^{(n)}, x_2^{(n)}$.

Hence in the expression for an invariant (II) it is necessary that

$$\alpha_{12} + \alpha_{13} + \dots + \alpha_{1n} = p,$$

$$\alpha_{21} + \alpha_{23} + \dots + \alpha_{2n} = p,$$

where p is a quantity which is the same for each term of the sum representing the invariant.

Let T be any one term of this sum, then let

$$\frac{I}{T} = \Sigma\,(x^{(1)}x^{(2)})^{\beta_{12}}\,(x^{(1)}x^{(3)})^{\beta_{13}}\,\ldots\,(x^{(2)}x^{(3)})^{\beta_{23}}\ldots.$$

Then
$$\beta_{12} + \beta_{13} + \ldots + \beta_{1n} = 0,$$
$$\beta_{21} + \beta_{23} + \ldots + \beta_{2n} = 0.$$

We are going to prove that $\frac{I}{T}$ is a function of anharmonic ratios of the roots. It will be assumed that when the number of quantities $x^{(1)}$, $x^{(2)}\ldots$ is less than n, then the term

$$(x^{(1)}x^{(2)})^{\beta_{12}}\,(x^{(1)}x^{(3)})^{\beta_{13}}\,\ldots\,(x^{(2)}x^{(3)})^{\beta_{23}}\ldots,$$

where
$$\sum_r \beta_{1r} = 0,\ \sum_r \beta_{2r} = 0,$$

may be expressed as a function of the anharmonic ratios.

Now the ratio

$$\{x^{(1)}x^{(2)}x^{(4)}x^{(t)}\} = \frac{(x^{(1)}x^{(2)})\,(x^{(4)}x^{(t)})}{(x^{(1)}x^{(t)})\,(x^{(4)}x^{(2)})}.$$

Hence

$$(x^{(1)}x^{(t)}) = (x^{(1)}x^{(2)})\,.\,\frac{(x^{(4)}x^{(t)})}{(x^{(4)}x^{(2)})}\,.\,\frac{1}{\{x^{(1)}x^{(2)}x^{(4)}x^{(t)}\}}.$$

On replacing $(x^{(1)}x^{(t)})$ wherever possible by the value just found

$$(x^{(1)}x^{(2)})^{\beta_{12}}\,(x^{(1)}x^{(3)})^{\beta_{13}}\,\ldots\,(x^{(2)}x^{(3)})^{\beta_{23}}\ldots$$

becomes

$$(x^{(1)}x^{(2)})^{\beta_{12}+\beta'_{13}+\ldots}\,PQ = PQ,$$

where P is a function of anharmonic ratios, and Q is of the form

$$(x^{(2)}x^{(3)})^{\beta_{23}}\,(x^{(2)}x^{(4)})^{\beta_{24}}\ldots,$$

where
$$\sum_r \beta_{2r} = 0,\ \sum_r \beta_{3r} = 0.$$

The theorem has been assumed true for Q, hence with this assumption it is true when there are n quantities $x^{(1)}$, $x^{(2)}$,.... If there are only three quantities, then

$$\beta_{12} + \beta_{13} = 0,$$
$$\beta_{21} + \beta_{23} = 0,$$
$$\beta_{31} + \beta_{32} = 0,$$

and
$$\beta_{12} = 0 = \beta_{23} = \beta_{31}.$$

Hence it is true when there are 4 quantities x, and therefore also when there are 5, and so on universally. Thus, *any invariant of a binary form is the numerator of a rational function of the anharmonic ratios of the roots.* If the invariant contains only one term, there is an apparent exception. The invariant equated to zero then represents the condition for the equality of a pair of roots, it can only be the discriminant.

160. So far our remarks have been confined to the case of a single quantic; a slight alteration in the wording of the previous paragraphs is all that is necessary to make them applicable to any system of binary forms. Each binary form of a system is geometrically represented by a set of points on a range, or of rays of a pencil. Points belonging to the same quantic must be regarded as indistinguishable from one another. Thus if we have a set of n points on a straight line, we may regard them as given by a single quantic of order n; by two quantics of orders r and $n-r$ respectively, or even by n separate linear forms.

Now let two of these points coincide; then, if the n points are regarded as a single quantic, the discriminant is zero; but there is nothing to tell us which roots coincide. We may regard the n points as two quantics, in this case either the discriminant of one of the quantics or else the resultant of the two is zero.

161. We shall now discuss the geometrical representation of the invariants and covariants of the binary forms of lowest order.

A quadratic has only one invariant, this vanishes when the points representing the quadratic are coincident; it is the discriminant.

A pair of quadratics a_x^2, b_x^2 have a simultaneous invariant $(ab)^2$.

Then if $x^{(1)}$, $x^{(2)}$ are the roots of a_x^2 and $y^{(1)}$, $y^{(2)}$ those of b_x^2,

$(ab)^2 = (ay^{(1)})(ay^{(2)}) = \frac{1}{2}[(x^{(1)}y^{(1)})(x^{(2)}y^{(2)}) + (x^{(1)}y^{(2)})(x^{(2)}y^{(1)})]$.

If $(ab)^2 = 0$, it follows that

$$\frac{(x^{(1)}y^{(1)})(x^{(2)}y^{(2)})}{(x^{(1)}y^{(2)})(x^{(2)}y^{(1)})} = -1.$$

Hence the pair of points $y^{(1)}$, $y^{(2)}$ is harmonic with the pair $x^{(1)}$, $x^{(2)}$.

The quadratics have a covariant, their Jacobian, $\vartheta = (ab)\, a_x b_x$.

Now $$(a\vartheta)^2 = 0,\ (b\vartheta)^2 = 0,$$
hence ϑ represents the pair of points which is at the same time harmonic with a_x^2 and with b_x^2. In other words ϑ represents the pair of double points of the involution defined by the two pairs a_x^2, b_x^2.

The discriminant of ϑ is
$$(\vartheta\vartheta')^2 = \tfrac{1}{2}\{(aa')^2(bb')^2 - [(ab)^2]^2\};$$
if this is zero, ϑ is a perfect square; the double points of the involution coincide. Hence, as may be verified either geometrically or algebraically, one of each of the pairs a_x^2, b_x^2 coincides with the point represented by ϑ, and the other two points may be anywhere on the range. Thus a_x^2, b_x^2 have in this case a common point; hence $(\vartheta\vartheta')^2$ may be taken to be the resultant of the two quadratics. If there are more than two quadratics, there is only one more type of concomitant to be discussed; viz. the invariant
$$(ab)(bc)(ca).$$
This is equal to $-(\vartheta, c_x^2)^2$.

If this is zero then the pair of points ϑ is harmonic with the pair c_x^2. Hence
$$(ab)(bc)(ca) = 0$$
represents the condition that the three pairs of points a_x^2, b_x^2, c_x^2 should be pairs in involution.

To find the anharmonic ratio of the four points defined by a_x^2, b_x^2, we have
$$(x^{(1)}y^{(1)})(x^{(2)}y^{(2)}) + (x^{(1)}y^{(2)})(x^{(2)}y^{(1)}) = (ab)^2,$$
$$(x^{(1)}y^{(1)})(x^{(2)}y^{(2)}) - (x^{(1)}y^{(2)})(x^{(2)}y^{(1)}) = (x^{(1)}x^{(2)})(y^{(1)}y^{(2)})$$
$$= \sqrt{(aa')^2}\,.\,\sqrt{(bb')^2}.$$
Hence
$$2(x^{(1)}y^{(1)})(x^{(2)}y^{(2)}) = (ab)^2 + \sqrt{(aa')^2\,.\,(bb')^2},$$
$$2(x^{(1)}y^{(2)})(x^{(2)}y^{(1)}) = (ab)^2 - \sqrt{(aa')^2\,.\,(bb')^2},$$
and therefore
$$\{x^{(1)}y^{(1)}x^{(2)}y^{(2)}\} = \frac{(x^{(1)}y^{(1)})(x^{(2)}y^{(2)})}{(x^{(1)}y^{(2)})(x^{(2)}y^{(1)})}$$
$$= \frac{(ab)^2 + \sqrt{(aa')^2\,.\,(bb')^2}}{(ab)^2 - \sqrt{(aa')^2\,.\,(bb')^2}}.$$

Denoting the anharmonic ratio by ρ, and squaring, this equation becomes

$$D'^2(\rho-1)^2 = D.D''(\rho+1)^2,$$

or

$$\rho^2 - 2\rho\frac{D'^2+DD''}{D'^2-DD''}+1=0,$$

where $D=(aa')^2,\ D'=(ab)^2,\ D''=(bb')^2.$

The two values of ρ correspond to the two anharmonic ratios $\{x^{(1)}y^{(1)}x^{(2)}y^{(2)}\}$ and $\{x^{(1)}y^{(2)}x^{(2)}y^{(1)}\}$.

If $\vartheta\equiv 0$, then the two quadratics are such that one is a multiple of the other. This is merely a particular case of the general property of the Jacobian; it is not necessary to do more than mention it here.

162. When a range possesses geometrical peculiarities which are unaffected by projection, there exist analytical relations of an invariant nature among the coefficients of the corresponding binary form; but it must not be supposed that these relations can always be expressed in terms of the pure invariants of the form. If there is only one such relation

$$A=0,$$

which expresses the necessary and sufficient condition that the range may possess a certain projective property, then it will be found that A is an invariant, for it is unaltered by linear transformation. On the other hand, when the condition is expressed by a set of algebraical relations

$$A=0,\ B=0,\dots$$

$A, B,\dots$ will not, in general, be invariants. Thus the condition that a binary form of order n may be a perfect nth power is that *all* the coefficients of its Hessian vanish*.

163. The Cubic. Any three collinear non-coincident points can be projected into any other three collinear non-coincident points; it is not to be expected then, that the geometry of a binary cubic will be of much interest from a projective point of view. But in respect of the associated points furnished by covariants, the geometry of the binary cubic is highly interesting.

* See later, Chap. XI.

The single invariant Δ is its discriminant and

$$\Delta = 0$$

is the necessary and sufficient condition that two of the three points represented by the cubic should coincide.

If all three points coincide, then Δ, H and t are all identically zero, as may be easily verified; but

$$H \equiv 0$$

represents the necessary and *sufficient* condition*.

164. Let us consider the *pencil*

$$\kappa f + \lambda t \equiv f_{\kappa,\lambda},$$

where κ and λ are new constants which determine the particular members of the pencil of cubics. Then $f_{\kappa,\lambda}$ represents three points, which are called a *triad*; by varying κ and λ a pencil of triads is obtained. The covariants of $f_{\kappa,\lambda}$ will be denoted by the symbols $H_{\kappa,\lambda}$, $t_{\kappa,\lambda}$, $\Delta_{\kappa,\lambda}$. These may be at once calculated; the following table, most of the results of which are proved in Chap. V., will be found useful for this purpose.

Index	Transvectant					
	(f, f)	(f, H)	(f, t)	(H, H)	(H, t)	(t, t)
1	0	t	$-\frac{1}{2}H^2$	0	$\frac{1}{2}\Delta f$	0
2	H	0	0	Δ	0	$\frac{1}{2}\Delta H$
3	0		Δ			0

The fundamental forms, it will be remembered, are connected by one syzygy,

$$t^2 = -\tfrac{1}{2}\{H^3 + \Delta f^2\}.$$

To obtain $H_{\kappa,\lambda}$ we have

$$\begin{aligned} H_{\kappa,\lambda} &= (\kappa f + \lambda t,\ \kappa f + \lambda t)^2 \\ &= \kappa^2 (f, f)^2 + 2\kappa\lambda\,(f, t)^2 + \lambda^2\,(t, t)^2 \\ &= (\kappa^2 + \tfrac{1}{2}\,\Delta\lambda^2)\,H. \end{aligned}$$

* Chap. XI.

Hence, if we use the notation

$$\Theta = (\kappa^2 + \tfrac{1}{2}\Delta\lambda^2),$$

$$H_{\kappa,\lambda} = \Theta H.$$

In the same way, we obtain

$$t_{\kappa,\lambda} = (f_{\kappa,\lambda}, H_{\kappa,\lambda}) = \Theta\,(f_{\kappa,\lambda}, H)$$
$$= \Theta\,(\kappa t - \tfrac{1}{2}\Delta\lambda f)$$
$$= \tfrac{1}{2}\,\Theta\left(t\frac{\partial\Theta}{\partial\kappa} - f\frac{\partial\Theta}{\partial\lambda}\right).$$

And lastly $$\Delta_{\kappa,\lambda} = \Theta^2\Delta.$$

It is worth noticing that if we introduce the arguments κ, λ of Θ as suffixes, thus

$$\Theta_{\kappa,\lambda} = \kappa^2 + \tfrac{1}{2}\Delta\lambda^2,$$

the syzygy may be written

$$H^3 = -2\Theta_{t,f}.$$

165. Consider the relation

$$H^3 = -2\,(t^2 + \tfrac{1}{2}\Delta f^2),$$

if any pair of the three forms

$$f,\quad H,\quad t$$

have a common factor, then all three must have this factor. Let us suppose that such a factor exists, and let us change the variables so that the common factor is x_2. Then f is of the form

$$(0,\ a_1,\ a_2,\ a_3 \between x_1,\ x_2)^3,$$

and the coefficient of x_1^2 in H is $-a_1^2$; hence if x_2 is a factor of H, we must have $a_1 = 0$. This means that f has a double factor, and therefore

$$\Delta = 0.$$

The syzygy then becomes

$$H^3 = -2t^2,$$

whence it is easy to deduce that H is a perfect square and t a perfect cube. In fact if

$$f = \zeta^2\theta,$$

then $$H = A\zeta^2,\ t = \sqrt{\frac{-A^3}{2}}\,\zeta^3,\ \Delta = 0,$$

where $$A = -\tfrac{2}{9}\,[(\zeta\theta)]^2.$$

166. It will now be supposed that f, H, t have no common factor. The syzygy may be written

$$H^3 = -2\left(t + f\sqrt{\frac{-\Delta}{2}}\right)\left(t - f\sqrt{\frac{-\Delta}{2}}\right);$$

hence if ξ, η are the factors of H, so determined that

$$H = -2\xi\eta,$$

we may take
$$2\xi^3 = \left(t + f\sqrt{\frac{-\Delta}{2}}\right),$$

and
$$2\eta^3 = \left(t - f\sqrt{\frac{-\Delta}{2}}\right),$$

and therefore
$$\sqrt{\frac{-\Delta}{2}}\, f = \xi^3 - \eta^3,$$

$$t = \xi^3 + \eta^3.$$

Hence also

$$\sqrt{-\frac{\Delta}{2}} \cdot f_{\kappa,\lambda} = \left(\kappa + \lambda\sqrt{-\frac{\Delta}{2}}\right)\xi^3 - \left(\kappa - \lambda\sqrt{-\frac{\Delta}{2}}\right)\eta^3,$$

$$t_{\kappa,\lambda} = \Theta\left[\left(\kappa - \lambda\sqrt{-\frac{\Delta}{2}}\right)\xi^3 + \left(\kappa + \lambda\sqrt{-\frac{\Delta}{2}}\right)\eta^3\right],$$

$$H_{\kappa,\lambda} = -2\Theta\xi\eta.$$

It is at once apparent that the only members of the pencil which possess double factors are ξ^3 and η^3.

Let P_1, P_2, P_3 be the three points determined by any one member of the pencil, and A, B the two points determined by the Hessian (which are the same for every member of the pencil). Then if ω be a cube root of unity, the points P_1, P_2, P_3 correspond to linear forms

$$\xi - a\eta,\ \xi - \omega a\eta,\ \xi - \omega^2 a\eta.$$

The ranges

$$A,\ P_1,\ P_2,\ P_3,\ B$$
$$A,\ P_2,\ P_3,\ P_1,\ B$$
$$A,\ P_3,\ P_1,\ P_2,\ B$$

are projective. A set of points such as P_1, P_2, P_3 are said to be cyclically projective.

The simplest way to find the anharmonic ratio $\{AP_1P_2P_3\}$ is to transform the variables to ξ, η. This transformation may be regarded as merely changing the points of reference, and the constant multiplier. Then

$$\{AP_1P_2P_3\} = \frac{(-a)(-\omega^2 a + \omega a)}{(-\omega^2 a)(-a+\omega a)} = -\omega^2 = \{AP_2P_3P_1\} = \{AP_3P_1P_2\}.$$

Hence the range formed by a triad and one of its Hessian points is equianharmonic. The six distinct cross ratios of such a range are each $-\omega$ or $-\omega^2$.

167. The Quartic. Just as in the case of the cubic we considered a pencil of cubics instead of the single one, so here we shall find it convenient to consider the pencil

$$f_{\kappa,\lambda} = \kappa f + \lambda H,$$

instead of the single quartic f. Each member of the pencil will define four points, one of these points may be chosen at will on the range considered, but when this is done the ratio $\kappa : \lambda$ is fixed, and the remaining three points are uniquely determined. The calculation of the covariants of $f_{\kappa,\lambda}$ presents no serious difficulty. For convenience a table of the transvectants of the quartic is appended; most of the calculations were effected in Chap. v.

Index	Transvectant					
	(f, f)	(f, H)	(H, H)	(f, t)	(H, t)	(t, t)
1	0	t	0	$\frac{1}{12}if^2 - \frac{1}{2}H^2$	$\frac{1}{6}f(jf - iH)$	0
2	H	$\frac{1}{6}if$	$\frac{1}{3}jf - \frac{1}{6}iH$	0	0	$\frac{1}{6}jHf - \frac{1}{12}iH^2 - \frac{1}{72}i^2f^2$
3	0	0	0	$\frac{1}{4}(jf - iH)$	$\frac{1}{24}i^2f - \frac{1}{4}jH$	0
4	i	j	$\frac{1}{6}i^2$	0	0	0

The only transvectants which are not contained in this table are

$$(t, t)^5 = 0, \quad (t, t)^6 = \tfrac{1}{24}i^3 - \tfrac{1}{4}j^2.$$

The syzygy between the forms is

$$t^2 = -\tfrac{1}{2}(H^3 - \tfrac{1}{2}iHf^2 + \tfrac{1}{3}jf^3).$$

In connection with the system for $f_{\kappa,\lambda}$ the expression

$$\Omega_{\kappa,\lambda} = \kappa^3 - \frac{i}{2}\kappa\lambda^2 - \frac{j}{3}\lambda^3,$$

or more briefly Ω, will be found of great importance. We observe at once that the syzygy may be written

$$2t^2 = -\Omega_{H,-f}.$$

Now

$$\begin{aligned} H_{\kappa,\lambda} &= (\kappa f + \lambda H, \kappa f + \lambda H)^2 \\ &= \kappa^2 H + \tfrac{1}{3} i\kappa\lambda f + \lambda^2(\tfrac{1}{3}jf - \tfrac{1}{6}iH) \\ &= \tfrac{1}{3}\left(H\frac{\partial\Omega}{\partial\kappa} - f\frac{\partial\Omega}{\partial\lambda}\right), \\ t_{\kappa,\lambda} &= \left(\kappa f + \lambda H,\ \tfrac{1}{3}H\frac{\partial\Omega}{\partial\kappa} - \tfrac{1}{3}f\frac{\partial\Omega}{\partial\lambda}\right) \\ &= \tfrac{1}{3}t\left(\kappa\frac{\partial\Omega}{\partial\kappa} + \lambda\frac{\partial\Omega}{\partial\lambda}\right) = \Omega t, \\ i_{\kappa,\lambda} &= \kappa^2 i + 2\kappa\lambda j + \lambda^2\tfrac{1}{6}i^2 = -3(\Omega, \Omega)^2 \\ &= -3H_\Omega, \end{aligned}$$

where Ω is regarded as a binary cubic in κ and λ.

Lastly

$$\begin{aligned} j_{\kappa,\lambda} &= \tfrac{1}{3}\left\{-i\kappa\frac{\partial\Omega}{\partial\lambda} + j\left(\kappa\frac{\partial\Omega}{\partial\kappa} - \lambda\frac{\partial\Omega}{\partial\lambda}\right) + \tfrac{1}{6}i^2\lambda\frac{\partial\Omega}{\partial\lambda}\right\} \\ &= -3t_\Omega. \end{aligned}$$

168. The invariant of the cubic Ω is

$$\Delta_\Omega = \tfrac{1}{27}(i^3 - 6j^2).$$

This, as we proceed to shew, may be taken to be the discriminant of the quartic. The discriminant is the condition that the equation

$$f = 0$$

may have a pair of equal roots.

It is an invariant; for if the range represented by f be projected, the pair of coincident points project into a pair of coincident points. Further it is well known to be of degree

$2(n-1)$ for the quantic of order n; hence it is of degree 6 for the quartic. It is then a linear function of i^3 and j^2.

Let us suppose that f has a double linear factor α_x, the remaining quadratic factor being p_x^2, and then find what relation exists between i and j for the quartic

$$a_x^4 = f = \alpha_x^2 . p_x^2.$$

In the first place

$$H = (\alpha_x^2 . p_x^2,\ a_x^4)^2$$

$$= \tfrac{1}{6}\{(\alpha a)^2 p_x^2 + (pa)^2 \alpha_x^2 + 4(\alpha a)(pa)\alpha_x p_x\} a_x^2$$

$$= \tfrac{1}{6}\{3(\alpha a)^2 p_x^2 + 3(pa)^2 \alpha_x^2 - 2(\alpha p)^2 a_x^2\} a_x^2 \text{ ....(I).}$$

But

$$(\alpha a)^2 a_x^2 = (f,\ \alpha_x^2)^2 = \tfrac{1}{6}(\alpha p)^2 . \alpha_x^2,$$

$$(pa)^2 a_x^2 = (f,\ p_x^2)^2 = \tfrac{1}{6}\{(\alpha p)^2 p_x^2 + 3(p,\ p)^2 \alpha_x^2\}.$$

Therefore

$$H = -\tfrac{1}{6}(\alpha p)^2 f + \tfrac{1}{4}(p,\ p)^2 \alpha^4.$$

Similarly

$$i = (a\alpha)^2 (ap)^2$$

$$= ((a\alpha)^2 a_x^2,\ p_x^2)^2 = \tfrac{1}{6}[(\alpha p)^2]^2,$$

$$j = (f,\ H)^4 = -\tfrac{1}{6}(\alpha p)^2 . i$$

$$= -\tfrac{1}{36}[(\alpha p)^2]^3.$$

Therefore

$$i^3 - 6j^2 = 0.$$

Hence Δ_Ω may be taken to be the discriminant of the quartic f.

From the above form for H, it is evident that H contains the factor α_x twice over; hence if f contains a repeated linear factor then every form of the pencil $\kappa f + \lambda H$ contains the same repeated factor. This leads us to expect that the discriminant of $\kappa f + \lambda H$ is a multiple of the discriminant of f. This is so; for from the syzygy for the cubic we obtain

$$-H_\Omega^3 - 2t_\Omega^2 = \Delta_\Omega . \Omega^2,$$

or

$$(i_{\kappa,\lambda}^3 - 6j_{\kappa,\lambda}^2) = \Omega^2 (i^3 - 6j^2).$$

It is easy to obtain in the same way the condition that f may have two pairs of repeated factors. For writing as before

$$f = \alpha_x^2 . p_x^2,$$

we obtain

$$t = (f,\ H) = \tfrac{1}{4}(p,\ p)^2 (f,\ \alpha)\alpha_x^3.$$

Now if f contains two pairs of repeated factors, $(p, p)^2$ must vanish, for p_x^2 is a perfect square; hence in this case

$$t \equiv 0.$$

This is the necessary and sufficient condition, for if it is satisfied

$$(t, t)^6 = \tfrac{1}{4}\left(\tfrac{1}{6} i^3 - j^2\right) = 0,$$

and we are at liberty to assume that f has one repeated factor; then using the relation

$$(p, p)^2 (f, \alpha)\, \alpha_x^3 \equiv 0$$

we see that either

$$(p, p)^2 = 0,$$

in which case p is a perfect square; or

$$(f, \alpha) \equiv 0,$$

in which case

$$f = \alpha_x^4.$$

This furnishes another illustration of the remarks in § 162.

Ex. (i). Shew that the necessary and sufficient condition that f may have a three times repeated factor is

$$i = 0, \; j = 0.$$

Ex. (ii). There are in general three different members of the pencil $f_{\kappa, \lambda}$ which are perfect squares.

They are given by solving the equation

$$\Omega = 0.$$

169. As has been already pointed out, the syzygy for the quartic may be written

$$2t^2 = -\,\Omega_{H, -f}.$$

If m_1, m_2, m_3 be the roots of the cubic

$$\Omega = 0,$$

then
$$\Omega_{\kappa, \lambda} = (\kappa - m_1\lambda)(\kappa - m_2\lambda)(\kappa - m_3\lambda).$$

Hence also

$$2t^2 = -(H + m_1 f)(H + m_2 f)(H + m_3 f).$$

If H and f have a common factor, by transformation this may be made x_2. Then f is of the form

$$(0,\ a_1,\ a_2,\ a_3,\ a_4 \between x_1,\ x_2)^4.$$

In order that x_2 may be a factor of H, we must have

$$-a_1^2=0.$$

Hence x_2^2 is a factor of f; and therefore

$$\Delta_\Omega=0.$$

Excluding this exceptional case, it is evident that no pair of the expressions

$$H+m_1f,\ H+m_2f,\ H+m_3f$$

have a common factor (m_1, m_2, m_3 are distinct) for Δ_Ω is the discriminant of Ω. Hence the above relation shews that each of the expressions $H+mf$ must be a perfect square—since t^2 is a perfect square.

Let

$$H+m_1f=-2\phi^2$$

$$H+m_2f=-2\psi^2$$

$$H+m_3f=-2\chi^2,$$

where ϕ, ψ, χ are binary quadratics.

Then

$$t=2\phi\psi\chi.$$

As an example it may be verified that for the quartic

$$x_1^4+6a\,x_1^2x_2^2+x_2^4,$$

$$H=2ax_1^4+2(1-3a^2)\,x_1^2x_2^2+2ax_2^4,$$

$$\Omega=\kappa^3-(3a^2+1)\,\kappa\lambda^2-2a(1-a^2)\,\lambda^3,$$

and that the roots of

$$\Omega=0$$

are $a-1$, $a+1$, $-2a$; which are identical with the values of m which make

$$H+mf$$

a perfect square.

Now

$$(H+m_1f,\ H+m_2f)=(m_1-m_2)\,t$$

$$=(-2\phi^2,\ -2\psi^2)=4(\phi,\ \psi)\,\phi\psi;$$

putting in the value of t we obtain

$$2(\phi,\ \psi)=(m_1-m_2)\,\chi.$$

Similarly

$$2(\psi,\ \chi)=(m_2-m_3)\,\phi,$$

$$2(\chi,\ \phi)=(m_3-m_1)\,\psi.$$

Now from the expression for the Jacobian of a Jacobian we obtain

$$((\phi,\ \psi),\ \chi) = -\tfrac{1}{2}\{\phi(\psi,\ \chi)^2 - \psi(\phi,\ \chi)^2\}.$$

Hence by repeated use of this formula

$$\begin{aligned}
0 &= (\phi\chi)^2\psi - (\phi\psi)^2\chi,\\
-\frac{(m_1-m_2)(m_1-m_3)}{2}\chi &= (\phi\phi)^2\chi - (\phi\chi)^2\phi,\\
-\frac{(m_1-m_2)(m_1-m_3)}{2}\psi &= (\phi\phi)^2\psi - (\phi\psi)^2\phi,\\
-\frac{(m_2-m_3)(m_2-m_1)}{2}\chi &= (\psi\psi)^2\chi - (\psi\chi)^2\psi,\\
0 &= (\psi\phi)^2\chi - (\psi\chi)^2\phi,\\
-\frac{(m_2-m_3)(m_2-m_1)}{2}\phi &= (\psi\psi)^2\phi - (\psi\phi)^2\psi,\\
-\frac{(m_3-m_1)(m_3-m_2)}{2}\psi &= (\chi\chi)^2\psi - (\chi\psi)^2\chi,\\
-\frac{(m_3-m_1)(m_3-m_2)}{2}\phi &= (\chi\chi)^2\phi - (\chi\phi)^2\chi,\\
0 &= (\chi\psi)^2\phi - (\chi\phi)^2\psi.
\end{aligned}$$

Since ϕ, ψ, χ have no common factors these equations give the following six relations

$$\begin{aligned}
(\phi\phi)^2 &= -\tfrac{1}{2}(m_1-m_2)(m_1-m_3), & (\psi\chi)^2 &= 0,\\
(\psi\psi)^2 &= -\tfrac{1}{2}(m_2-m_3)(m_2-m_1), & (\chi\phi)^2 &= 0,\\
(\chi\chi)^2 &= -\tfrac{1}{2}(m_3-m_1)(m_3-m_2), & (\phi\psi)^2 &= 0.
\end{aligned}$$

The remaining invariant of these three quadratics is

$$\begin{aligned}
(\phi\psi)(\psi\chi)(\chi\phi) &= -((\phi\psi)\phi_x\psi_x,\ \chi^2_x)^2\\
&= -\tfrac{1}{2}(m_1-m_2)\,.\,(\chi\chi)^2 = -\tfrac{1}{4}(m_1-m_2)(m_2-m_3)(m_3-m_1).
\end{aligned}$$

170. By means of the equations

$$\begin{aligned}
H + m_1 f &= -2\phi^2,\\
H + m_2 f &= -2\psi^2,\\
H + m_3 f &= -2\chi^2 \quad \ldots\ldots\ldots\ldots\ \ldots\ldots\ldots\text{(II)},
\end{aligned}$$

the quartic f, or more generally $\kappa f + \lambda H$, may be separated into quadratic factors.

Thus
$$f = \frac{2}{m_2 - m_3} (\chi^2 - \psi^2),$$

$$H = \frac{2}{m_2 - m_3} (m_3 \psi^2 - m_2 \chi^2).$$

and
$$\kappa f + \lambda H = \frac{2}{m_2 - m_3} \{(\kappa - \lambda m_2) \chi^2 - (\kappa - \lambda m_3) \psi^2\}$$

$$= \frac{2}{m_2 - m_3} (\sqrt{\kappa - \lambda m_2}\, \chi + \sqrt{\kappa - \lambda m_3}\, \psi)(\sqrt{\kappa - \lambda m_2}\, \chi - \sqrt{\kappa - \lambda m_3}\, \psi).$$

The second transvectant of either of these quadratic factors with ϕ is zero. Hence the two points determined by ϕ are the harmonic conjugates of the two pairs of points represented by the above quadratic factors of $\kappa f + \lambda H$. Now we have only used the last two of equations (II) to find the quadratic factors of $\kappa f + \lambda H$. Any pair of the three equations might be taken. The three results represent the three ways into which the quartic $\kappa f + \lambda H$ may be separated into quadratic factors. Then the three quadratic factors of t are the three pairs of points harmonically conjugate with respect to the four points $\kappa f + \lambda H$, when divided into two pairs of points.

Now
$$(\phi\psi)^2 = 0, \ (\psi\chi)^2 = 0, \ (\chi\phi)^2 = 0,$$

hence the pair of points ϕ is harmonically conjugate with respect to each of the pairs ψ and χ. If the points $\kappa f + \lambda H$ are divided into two pairs in any way, these pairs determine an involution, one of the quadratic factors of t represents the pair of double points of the involution. The other two quadratic factors of t represent pairs of points belonging to the involution.

Now the points determined by t are independent of κ and λ, hence the pencil $\kappa f + \lambda H$ represents sets of four points such that when any set is separated into two pairs of points, these are pairs of one of three fixed involutions.

The quartic f is arbitrary, it may represent any four points. Hence the pairs of double points of the three involutions determined by four points on a line are harmonically conjugate two and two.

171. To determine the anharmonic ratio ρ of the four points f. We have obtained the quadratic factors of f, one pair is

$$\chi - \psi, \ \chi + \psi.$$

The anharmonic ratio of the four points determined by a pair of quadratics has been obtained in § 161, as a root of the equation

$$\rho^2 - 2\rho \frac{D'^2 + DD''}{D'^2 - DD''} + 1 = 0.$$

In our case

$$D = (\chi - \psi, \ \chi - \psi)^2$$

$$= -\tfrac{1}{2}(m_3 - m_1)(m_3 - m_2) - \tfrac{1}{2}(m_2 - m_3)(m_2 - m_1)$$

$$= -\tfrac{1}{2}(m_3 - m_2)^2 = D''.$$

$$D' = -\tfrac{1}{2}(m_3 - m_1)(m_3 - m_2) + \tfrac{1}{2}(m_2 - m_3)(m_2 - m_1).$$

Hence $$\rho^2 - \frac{(m_3 - m_1)^2 + (m_2 - m_1)^2}{(m_3 - m_1)(m_2 - m_1)}\rho + 1 = 0.$$

As there are six different values for the anharmonic ratio of four points, a sextic for ρ is to be expected. This will be obtained by multiplying together the three equations similar to the above.

It will be more convenient to write these equations in the form

$$\rho^2 - 2\rho + 1 - \frac{(m_3 - m_2)^2}{(m_3 - m_1)(m_2 - m_1)}\rho = 0.$$

Now m_1, m_2, m_3 are the roots of the cubic

$$\kappa^3 - \frac{i}{2}\kappa\lambda^2 - \frac{j}{3}\lambda^3 = 0.$$

The discriminant of this is

$$(m_1 - m_2)^2(m_2 - m_3)^2(m_3 - m_1)^2 = \tfrac{1}{2}(i^3 - 6j^2) = \Delta,$$

the exact expression is most quickly obtained by using the equation of the squared differences of the cubic.

The equation whose roots are $m_1 - m_2$, $m_2 - m_3$, $m_3 - m_1$ is obtained thus

$$\Sigma(m_1 - m_2) = 0,$$

$$\Sigma(m_3 - m_1)(m_1 - m_2) = -\Sigma m_1^2 + \Sigma m_2 m_3$$

$$= 3\Sigma m_2 m_3 = -3\frac{i}{2},$$

$$(m_1 - m_2)(m_2 - m_3)(m_3 - m_1) = \sqrt{\Delta}.$$

The equation is $\quad y^3 - 3\frac{i}{2}y - \sqrt{\Delta} = 0.$

The equation whose roots are $(m_1 - m_2)^3, \ldots$ is

$$z - 3\frac{i}{2}z^{\frac{1}{3}} - \sqrt{\Delta} = 0,$$

or

$$(z - \sqrt{\Delta})^3 = 27\frac{i^3}{8}z.$$

But

$$(m_2 - m_3)^3 = -\sqrt{\Delta}\,\frac{\rho^2 - 2\rho + 1}{\rho}.$$

Hence

$$\Delta^{\frac{3}{2}}\left\{-\frac{\rho^2 - 2\rho + 1}{\rho} - 1\right\}^3 = -27\frac{i^3}{8}\frac{\rho^2 - 2\rho + 1}{\rho}\Delta^{\frac{1}{2}},$$

or

$$\left(1 - 6\frac{j^2}{i^3}\right)\left\{1 + \frac{\rho^2 - 2\rho + 1}{\rho}\right\}^3 = \frac{\rho^2 - 2\rho + 1}{\rho}\cdot\frac{27}{4}.$$

Therefore

$$\frac{i^3}{j^2} = \frac{24(\rho^2 - \rho + 1)^3}{4(\rho^2 - \rho + 1)^3 - 27\rho^2(\rho - 1)^2}$$

$$= \frac{24(\rho^2 - \rho + 1)^3}{(\rho + 1)^2(\rho - 2)^2(2\rho - 1)^2}.^*$$

* When the quartic is not treated by the symbolical method it is usual to define the invariants as follows:—the quartic itself is

$$f = (a,\ b,\ c,\ d,\ e\between x_1,\ x_2)^4,$$

$$I = ae - 4bd + 3c^2,$$

$$J = \begin{vmatrix} a & b & c \\ b & c & d \\ c & d & e \end{vmatrix}.$$

The invariants i, j in the text above differ from these by numerical factors only. Thus

$$i = (ab)^4 = 2I, \quad j = (ab)^2(bc)^2(ca)^2 = 6J.$$

In connection with the calculation of the values of invariants given symbolically in terms of the actual coefficients the reader may find it interesting to discover the fallacy in the following:—

$$j = (bc)^2(ca)^2(ab)^2$$

$$= \{(bc)(ca)(ab)\}^2$$

$$= \begin{vmatrix} a_1^2 & b_1^2 & c_1^2 \\ a_1a_2 & b_1b_2 & c_1c_2 \\ a_2^2 & b_2^2 & c_2^2 \end{vmatrix}\begin{vmatrix} a_1^2 & b_1^2 & c_1^2 \\ a_1a_2 & b_1b_2 & c_1c_2 \\ a_2^2 & b_2^2 & c_2^2 \end{vmatrix}$$

$$= \begin{vmatrix} a_1^4 + b_1^4 + c_1^4 & \ldots & \ldots & \ldots \\ a_1^3a_2 + b_1^3b_2 + c_1^3c_2 & \ldots & \ldots & \ldots \\ a_1^2a_2^2 + b_1^2b_2^2 + c_1^2c_2^2\ldots & \ldots & \ldots & \ldots \end{vmatrix}$$

$$= 27J.$$

Thus the anharmonic ratio is expressed by means of a sextic equation in terms of the absolute invariant $\frac{i^3}{j^2}$.

We see from this equation at a glance that if $i=0$, the points represented by the quartic form an equianharmonic range, for then,

$$\rho^2-\rho+1=0.$$

Similarly if $j=0$, the four points form a harmonic range.

Again, if two of the points of the range are coincident, one value of ρ is unity; hence

$$j^2=6i^3,$$

as it should be.

172. The anharmonic ratio for the four points

$$\kappa f+\lambda H$$

may be obtained at once by writing $i_{\kappa,\lambda}$ for i, and $j_{\kappa,\lambda}$ for j.

To determine those values of $\kappa:\lambda$ for which the four points have any definite anharmonic ratio ρ; let

$$a=24\frac{(1-\rho+\rho^2)^3}{(\rho+1)^2(\rho-2)^2(2\rho-1)^2}.$$

Then $$i^3_{\kappa,\lambda}-aj^2_{\kappa,\lambda}=0,$$

or $$3H_\Omega^3+at_\Omega^2=0,$$

this is a sextic for $\kappa:\lambda$.

Now $$H_\Omega^3=-2t_\Omega^2-\Omega^2\Delta_\Omega.$$

Hence $$(a-6)\,t_\Omega^2=\Omega^2\Delta_\Omega,$$

or $$t_\Omega=\pm\Omega\sqrt{\frac{\Delta_\Omega}{a-6}}.$$

The sextic thus reduces to two cubics.

If $a=6$, it is easy to see that $\rho=0, 1, \infty$, hence two points must coincide. In this case $\Omega=0$, and $\frac{\kappa}{\lambda}$ has one of the three values

$$m_1,\ m_2,\ m_3.$$

Hence the three members of the pencil for which $a = 6$ are

$$H + m_1 f = \phi^2,$$

$$H + m_2 f = \psi^2,$$

$$H + m_3 f = \chi^2,$$

shewing that if one pair of points coincide, the other pair must also coincide.

If $a = \infty$, the four points form a harmonic range, and $t_\Omega = 0$. There are three members of the pencil for which the range is harmonic. If $a = 0$, then $H_\Omega = 0$; hence there are only two members of the pencil which form equianharmonic ranges.

In all other cases, there are six members of the pencil having a definite anharmonic ratio.

Ex. If $l_x l_x'$ is the Hessian of the cubic a_x^3, prove that the quartic $a_x^3 l_x$ is equianharmonic.

173. Case when $\Delta_\Omega = 0$. The discussion was limited in § 169 to the case when Δ_Ω is other than zero. Now Δ_Ω is the discriminant of the quartic f, and hence when Δ_Ω vanishes, two of the roots of $f = 0$ are the same. We may, as in § 168, write

$$f = \alpha_x^2 p_x^2,$$

where α_x^2 is the square of a linear form α_x, and p_x^2 is a quadratic. Then as before

$$H = \alpha_x^2 \{-\tfrac{1}{6}(\alpha p)^2 p_x^2 + \tfrac{1}{4}(pp)^2 \alpha_x^2\},$$

$$i = \tfrac{1}{6}\{(\alpha p)^2\}^2, \quad j = -\tfrac{1}{36}\{(\alpha p)^2\}^3.$$

The invariant Δ_Ω is also the discriminant of the cubic

$$\Omega = \kappa^3 - \frac{i}{2}\kappa\lambda^2 - \frac{j}{3}\lambda^3,$$

hence, in the present case, Ω has a repeated root. Let this be m_2, and let the other root be m_1. Then

$$2m_2 + m_1 = 0, \quad m_2^2 + 2m_1 m_2 = -\frac{i}{2}, \quad m_2^2 m_1 = \frac{j}{3}.$$

Hence

$$m_2 = -\frac{j}{i} = \tfrac{1}{6}(\alpha p)^2, \quad m_1 = \frac{2j}{i} = -\tfrac{1}{3}(\alpha p)^2,$$

and therefore

$$H + m_2 f = \tfrac{1}{4}(pp)^2 \cdot \alpha_x^4,$$

$$H + m_1 f = \alpha_x^2 \{-\tfrac{1}{2}(\alpha p)^2 p_x^2 + \tfrac{1}{4}(pp)^2 \alpha_x^2\}.$$

Again, since

$$2t^2 = -(H + m_1 f)(H + m_2 f)^2,$$

we obtain as before

$$H + m_1 f = -2\phi^2,$$

where ϕ is a quadratic. One of the factors of ϕ must be α_x, and if the other is β_x, then

$$2\beta_x^2 = \tfrac{1}{2}(\alpha p)^2 p_x^2 - \tfrac{1}{4}(pp)^2 \alpha_x^2.$$

The value of t is then seen to be

$$t = \tfrac{3}{4}(pp)^2 \alpha_x^5 \beta_x.$$

174. We shall now briefly explain another interesting method of representing invariant properties of binary forms geometrically.

If we put $x_1 = z$ and $x_2 = 1$ throughout the work on binary quantics the general linear substitution may be written

$$z = \frac{az' + b}{cz' + d}, \quad \text{since } z = \frac{x_1}{x_2}.$$

Now put $z = x + iy$, and represent z as the real point x, y in the Argand diagram in the usual way; then the substitution

$$z = \frac{az' + b}{cz' + d}$$

is a point transformation.

Unless $c = 0$ the relation between z and z' may be reduced to the form

$$(z - \alpha)(z' - \alpha') = k,$$

wherein α, α', k are constants.

Suppose z, z', α, α' are the points P, P', A, A' respectively, then the geometrical meaning of the above is

$$AP \cdot A'P' = \text{mod.}\ (k),$$

and the sum of the angles that AP and $A'P'$ make with any fixed line is constant. Hence the general linear substitution is equivalent to an inversion together with a change of origin and a reflexion of inclination of the line AP with respect to a fixed line.

If $c = 0$ the equation can be written

$$z' - \beta = m(z - \beta),$$

indicating that P' is derived from P by turning BP through a fixed angle and increasing it in a given ratio, B being the point which represents β.

Hence a binary form of order n represents n real points A in the plane, and a covariant of the form represents a group of points C whose relation to the points A is unaltered by a geometrical transformation of the types indicated.

In particular, the relation of the points C to the points A is unaltered by any inversion, because in the particular case in which A and A' coincide and k is real, the transformation is equivalent to an inversion with respect to A, and a reflexion with respect to the real axis through A; but the properties of the derived figure are evidently unaltered by a reflexion alone, and hence they are unaltered by an inversion alone.

174 A. Some of the simpler invariants and covariants can now be interpreted.

If $$az^2 + 2bz + c = 0; \quad a'z^2 + 2b'z + c' = 0$$

represent the points A, B and C, D respectively, then when

$$ac' + a'c - 2bb' = 0$$

A, B, C, D are four harmonic points on a circle.

In fact on changing the origin to O, the middle point of AB, the first quadratic becomes

$$z^2 - k^2 = 0,$$

and if the second be $$(z - z_1)(z - z_2) = 0,$$

then since the relation is invariantive we have

$$z_1 z_2 = k^2,$$

therefore $$OC \,.\, OD = OA^2 = OB^2$$

and OC, OD are equally inclined to OA.

If we produce CO to D' making $OD' = OD$, we have

$$OC \,.\, OD' = OA^2 = OA \,.\, OB;$$

therefore $CAD'B$ are concyclic, and by symmetry D is on the circle.

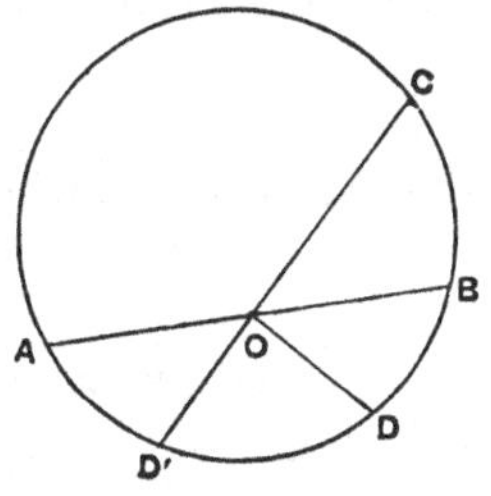

Further as the pencil $D'\{ACBD\}$ is harmonic the four points AB and CD form two harmonic pairs on the circle. We shall call them harmonically concyclic.

Ex. Shew that there is one pair of points (P, P') harmonically concyclic to each of two given pairs (A, B) and (C, D).

Further if (Q, Q') be harmonically concyclic to (A, C), (B, D) and (R, R') to (A, D), (B, C), then any two of the three pairs P, P'; Q, Q'; R, R' are harmonically concyclic.

175. We shall now apply the complex variable to prove certain properties of the foci of conics.

If the tangential equation of a conic be

$$Al^2 + 2Hlm + Bm^2 + 2Gl + 2Fm + C = 0,$$

the axes of coordinates being rectangular, and x_1y_1 be a focus, then the line

$$x + y\iota = x_1 + y_1\iota \text{ touches the curve.}$$

Hence the above equation is satisfied by $l = -\dfrac{1}{z_1}$, $m = -\dfrac{\iota}{z_1}$, and we have

$$A + 2H\iota - B - 2(G + F\iota)z_1 + Cz_1^2 = 0.$$

Consequently the two real foci z_1 and z_2 are given by the quadratic

$$Cz^2 - 2(G + F\iota)z + (A - B + 2H\iota) = 0.$$

Since $\dfrac{z_1 + z_2}{2} = \dfrac{G}{C} + \iota\dfrac{F}{C}$, we see that the centre is the point $\dfrac{G}{C}, \dfrac{F}{C}$.

Again, if O be the origin and S_1, S_2 the foci, we have

$$OS_1 . OS_2 = \frac{\sqrt{(A - B)^2 + 4H^2}}{C},$$

so that the origin can only be a focus when $A = B$ and $H = 0$.

If a system of conics be inscribed in the same quadrilateral, their tangential equations are of the form

$$\lambda\Sigma + \lambda'\Sigma' = 0,$$

and thus the real foci are given by the pencil of quadratics

$$\lambda f + \lambda' f' = 0.$$

All the quadratics are harmonic to the Jacobian (f, f'), and accordingly we have the theorem that the real foci of any conic inscribed in a quadrilateral are harmonically concyclic to a fixed pair of points J_1, J_2.

We leave the reader to prove that if T_1, T_2 be harmonically concyclic to S_1 and S_2, then the points of contact of the tangents drawn from T_1 and T_2 to any conic whose foci are S_1, S_2 lie on a circle through T_1 and T_2. Hence if tangents T_1P_1, T_1Q_1 be drawn to any one of these conics, the circles $T_1P_1Q_1$ all pass through another fixed point.

Thus the points J_1, J_2 are such that if tangents be drawn from them to any confocal to a conic inscribed in the quadrilateral, then the points of contact lie on a circle through J_1 and J_2.

176. *The binary cubic.*

Suppose the form is

$$az^3 + 3bz^2 + 3cz + d = 0,$$

and that the points A, B, C representing it are z_1, z_2, z_3.

The cubic covariant ϕ represents three points A', B', C' on the circle ABC such that A, A' are harmonically conjugate to B, C and so on. For these three points must be represented by a cubic covariant which is not f and therefore must be ϕ.

We have next to interpret the Hessian.

Let H_1, H_2 be the points representing h_1 and h_2 the roots of the Hessian, then we know that for real variables the range $\{ABCH_1\}$ is equianharmonic,

$$\text{i.e.} \qquad \frac{(z_1-z_2)(z_3-h_1)}{1} = \frac{(z_2-z_3)(z_1-h_1)}{\omega} = \frac{(z_3-z_1)(z_2-h_1)}{\omega^2}.$$

Hence since mod $(\omega) = 1$, we have on equating moduli

$$AB \,.\, CH_1 = BC \,.\, AH_1 = CA \,.\, BH_1,$$

therefore the points H_1, H_2 are the points whose distances from the vertices are inversely proportional to the opposite sides.

To construct them we draw a circle having BC for inverse points, and passing through A, with analogous circles for CA and AB; then these three circles meet in the points H_1 and H_2.

It follows in addition that H_1 and H_2 are inverse points with respect to the circle ABC.

It is interesting to notice that the Hessian points of $A'B'C'$ are the same as those of ABC, a fact which gives rise to a curious geometrical theorem.

Another easily proved property of the Hessian points is that if H_1L, H_1M, H_1N be drawn perpendicular to the sides, then the triangle LMN is equilateral.

CHAPTER XI.

APOLARITY AND ELEMENTARY GEOMETRY OF RATIONAL CURVES.

177. Two binary forms of the same order are said to be *apolar* when the joint invariant which is linear in the coefficients of both is zero.

Suppose that the two forms in question are

$$f \equiv a_0 x_1^n + n a_1 x_1^{n-1} x_2 + \dots + a_n x_2^n = a_x^n,$$

$$\phi \equiv b_0 x_1^n + n b_1 x_1^{n-1} x_2 + \dots + b_n x_2^n = b_x^n,$$

then the only lineo-linear invariant is

$$(ab)^n = a_0 b_n - n a_1 b_{n-1} + \dots + (-1)^n a_n b_0$$

and this vanishes when the forms are apolar. An immediate consequence is that a form of odd order is always apolar to itself.

Thus the discussion of apolar forms may be regarded as the development of the theory of the simplest type of invariant; the fact that each set of coefficients only occur to the first degree in the invariant renders such a discussion simple and accounts for the relative importance of the allied geometrical properties. If two quadratic forms are apolar they are harmonic so that we may regard apolarity as being, in a certain way, the generalisation of harmonic properties.

The condition for apolarity may be written in other useful forms.

In fact, if the linear factors of ϕ are $\beta_x^{(1)}, \beta_x^{(2)}, \dots \beta_x^{(n)}$, we have

$$0 = \{a_x^n, \beta_x^{(1)} \beta_x^{(2)} \dots \beta_x^{(n)}\}^n = (a\beta^{(1)})(a\beta^{(2)}) \dots (a\beta^{(n)}).$$

Again, if ϕ vanishes for the values

$$x_1 = y_1^{(r)},\ x_2 = y_2^{(r)},\ r = 1, 2, \dots n,$$

we have as far as ratios are concerned

$$\beta_1^{(r)} = -y_2^{(r)},\ \beta_2^{(r)} = +y_1^{(r)},$$

and hence the condition is

$$a_{y^{(1)}} a_{y^{(2)}} \dots a_{y^{(n)}} = 0.$$

This form of the condition at once shews the connection with polar forms; given the form f, $n-1$ of the vanishing points of ϕ can be chosen arbitrarily and the remaining one is given by polarizing f with respect to each of the $n-1$ given values successively and equating the final result to zero.

We can at once find all perfect nth powers apolar to f, for if all the y's are the same we have

$$a_y^n = 0,$$

so that y must be a vanishing point of f. Hence f is apolar to the nth power of any one of its linear factors, and these determine the only nth powers apolar to f.

If f be apolar to each of the forms $\phi_1, \phi_2, \dots \phi_n$ it is apolar to $\lambda_1\phi_1 + \lambda_2\phi_2 + \dots + \lambda_k\phi_k$, where the λ's are any constants; for

$$\{f, \lambda_1\phi_1 + \lambda_2\phi_2 + \dots + \lambda_k\phi_k\}^n = \lambda_1(f\phi_1)^n + \lambda_2(f\phi_2)^n + \dots + \lambda_k(f\phi_k)^n = 0,$$

which establishes the result.

This result also follows at once from the fact that the equations of condition are all linear.

Again if f be apolar to each of the $(n+1)$ forms

$$\phi_1, \phi_2, \dots \phi_{n+1},$$

then on elimination of the a's from the equations which they satisfy, we find that the determinant of the coefficients of the ϕ's is zero; but this is precisely the condition that there should be an identical relation of the form

$$\lambda_1\phi_1 + \lambda_2\phi_2 + \dots + \lambda_{n+1}\phi_{n+1} = 0 \quad \dots\dots\dots\dots\dots\dots(A),$$

and hence being given n linearly independent forms apolar to f any other apolar form is a linear combination of these. It is easy to prove directly that n linearly independent forms can be found which are apolar to a given form of order n, and this fact will appear independently from the special system of apolar forms to be constructed in the next article.

178. *To determine n linearly independent forms apolar to a given one.*

I. Suppose that the factors of the given form f are all different and that except for numerical multiples they are

$$\alpha_1^{(r)}x_1 + \alpha_2^{(r)}x_2, \quad r = 1, 2, \dots n,$$

then the nth power of each factor is an apolar form, and they are linearly independent.

For if there be a relation of the type

$$\sum_1^n \lambda_r (\alpha_1^{(r)}x_1 + \alpha_2^{(r)}x_2)^n = 0,$$

where λ_r is a constant, then the $(n+1)$ determinants of the array

$$| \alpha_1^{(r)n}, \ \alpha_1^{(r)n-1}\alpha_2^{(r)}, \ \alpha_1^{(r)n-2}\alpha_2^{(r)}, \dots \alpha_2^{(r)n} | \quad r = 1, 2, \dots n,$$

vanish identically.

Hence the determinant

$$\begin{vmatrix} \alpha_1^{(1)n}, & \alpha_1^{(1)n-1}\alpha_2^{(1)}, & \alpha_1^{(1)n-2}\alpha_2^{(1)2}, & \dots & \alpha_2^{(1)n} \\ \dots & \dots & \dots & \dots & \dots \\ \dots & \dots & \dots & \dots & \dots \\ \alpha_1^{(n)n}, & \alpha_1^{(n)n-1}\alpha_2^{(n)}, & \alpha_1^{(n)n-2}\alpha_2^{(n)2}, & \dots & \alpha_2^{(n)n} \\ p_1^n, & p_1^{n-1}p_2, & p_1^{n-2}p_2^2, & \dots & p_2^n \end{vmatrix}$$

vanishes for all values of p_1 and p_2.

But this determinant being homogeneous in each set of symbols is equal to

$$\pm \Pi (\alpha_1^{(r)}\alpha_2^{(s)} - \alpha_1^{(s)}\alpha_2^{(r)}) \prod_{r=1}^{r=n} (\alpha_1^{(r)}p_2 - \alpha_2^{(r)}p_1),$$

where in the first product r and s have all the values $1, 2, \dots n$, but are different.

Thus if we choose p_1, p_2 such that $\dfrac{p_1}{p_2} \neq \dfrac{\alpha_1^{(r)}}{\alpha_2^{(r)}}$, $r = 1, 2, \dots n$, the determinant can only vanish when for some pair of values of r and s

$$\alpha_1^{(r)}\alpha_2^{(s)} = \alpha_2^{(r)}\alpha_1^{(s)},$$

in which case the two factors $\alpha_1^{(r)}x_1 + \alpha_2^{(r)}x_2$ and $\alpha_1^{(s)}x_1 + \alpha_2^{(s)}x_2$ only differ by a numerical multiplier, contrary to the hypothesis that the factors are all different.

Consequently the n forms

$$(\alpha_1^{(r)}x_1 + \alpha_2^{(r)}x_2)^n, \quad r = 1, 2, \ldots n,$$

are linearly independent when the factors of f are all different.

We are thus led at once to the interesting result that the necessary and sufficient condition that a form ϕ can be represented as a linear combination of the nth powers of the factors of f is that the two forms should be apolar—the condition is necessary, because as each nth power is apolar to f a linear combination of them is also, and it is sufficient, because the nth powers are linearly independent in virtue of the foregoing, supposing always that f has no multiple factors.

Hence reciprocally, if ϕ have no multiple factors, f can be expressed as a linear combination of the nth powers of the factors of ϕ.

This may be regarded as the extension of the elementary theorem that all quadratics harmonic to $ax^2 + 2bx + c$ are of the form $\lambda(x - \alpha)^2 + \mu(x - \beta)^2$, where α, β are the roots of

$$ax^2 + 2bx + c = 0.$$

II. We have still to construct the apolar system when f has multiple factors.

Suppose that f has the factor $(\alpha_1x_1 + \alpha_2x_2)^r$ replacing r different linear factors, and that ϕ is apolar to f. Since the relation is invariantive, we may change the variables so that the multiple factor is simply x_2^r, and then

$$a_0, \; a_1, \; a_2, \ldots a_{r-1} \text{ all vanish.}$$

Hence recalling the condition

$$a_0b_n - na_1b_{n-1} + \ldots + (-1)^n a_n b_0 = 0,$$

we see that it is satisfied for any values of

$$b_n, \; b_{n-1}, \ldots b_{n-r+1},$$

provided that all the other b's vanish.

Thus f is now apolar to any form for which

$$b_0, b_1, b_2, \ldots b_{n-r}$$

are all zero, that is to any form which contains the factor x_2^{n-r+1}. Among such we have the r forms

$$x_2^{n-r+1}x_1^{r-1},\ x_2^{n-r+1}x_1^{r-2}x_2, \dots\ x_2^{n-r+1}x_2^{r-1},$$

which are obviously linearly independent.

Hence, in general, when a form has a linear factor of multiplicity r it is apolar to any form containing that factor $(n-r+1)$ times, and among these apolar forms it is possible to choose r which are linearly independent. If each multiple factor be treated in the same way we obtain in all n apolar forms, viz. r from each factor of multiplicity r; those derived from the same factor are linearly independent; it remains to shew that all the n forms are so independent.

Thus for the sextic $x_1^3(x_1+x_2)^2 x_2$ we have six apolar forms in three sets

$x_1^6,\ x_1^5x_2,\ x_1^4x_2^2$ containing the factor x_1^4,

$(x_1+x_2)^5x_1, (x_1+x_2)^5x_2$ containing the factor $(x_1+x_2)^5$,

and x_2^6 derived from the single factor x_2.

The forms in each set are linearly independent, but it has not been shewn that the different sets are independent *inter se*.

The general proof that the sets so derived are independent presents no difficulty but is rather tedious owing to the complicated notation. (See Appendix II.)

Let $$f = A(x_1+\alpha_1 x_2)^{r_1}(x_1+\alpha_2 x_2)^{r_2}\dots(x_1+\alpha_p x_2)^{r_p},$$

the α's being all different, $r_1+r_2+\dots+r_p=n$, and A a constant. This assumes that in f the coefficient of x_1^n is not zero, if it be zero we can transform f by a linear substitution into one in which x_1^n actually occurs.

A form apolar to f is

$$\chi_1 = (x_1+\alpha_1 x_2)^n,$$

and $$\frac{\partial\chi_1}{\partial\alpha_1},\ \frac{\partial^2\chi_1}{\partial\alpha_1^2},\ \dots\ \frac{\partial^{r_1-1}\chi_1}{\partial\alpha_1^{r_1-1}}$$

are all apolar forms since each involves the factor $(x_1+\alpha_1x_2)^{n-r_1+1}$. We shall prove that this set and the corresponding ones derived from the other factors are linearly independent.

In fact if they are not independent the determinant

$$\begin{vmatrix} f_0(\alpha_1), & f_1(\alpha_1), & \dots & f_n(\alpha_1) \\ f_0'(\alpha_1), & f_1'(\alpha_1), & \dots & f_n'(\alpha_1) \\ \dots & \dots & \dots & \dots \\ f_0^{r_1-1}(\alpha_1), & f_1^{r_1-1}(\alpha_1), & \dots & f_n^{r_1-1}(\alpha_1) \\ \dots & \dots & \dots & \dots \\ f_0^{r_p-1}(\alpha_p), & \dots & \dots & f_n^{r_p-1}(\alpha_p) \\ f_0(q), & f_1(q), & \dots & f_n(q) \end{vmatrix},$$

where in general
$$f_s(\alpha_r) = \alpha_r^s$$
and
$$f_s^t(\alpha_r) = \frac{\partial^t \alpha_r^s}{\partial \alpha_r^t},$$
must vanish for all values of q.

Except for a non-vanishing numerical multiplier the above determinant is the limit of

$$\begin{vmatrix} f_0(\alpha_1), & f_1(\alpha_1), & \dots f_n(\alpha_1) \\ f_0(\alpha_1+t), & f_1(\alpha_1+t), & \dots f_n(\alpha_1+t) \\ \dots & \dots & \dots \\ f_0(\alpha_1+\overline{r_1-1}\,t), & f_1(\alpha_1+\overline{r_1-1}\,t), & \dots f_n(\alpha_1+\overline{r_1-1}\,t) \\ \dots & \dots & \dots \\ \dots & \dots & \dots \\ f_0(\alpha_p+\overline{r_p-1}\,t), & f_1(\alpha_p+\overline{r_p-1}\,t), & \dots f_n(\alpha_p+\overline{r_p-1}\,t) \\ f_0(q), & f_1(q), & \dots f_n(q) \end{vmatrix}$$
$$\div\, t^{(1+2+\dots+\overline{r_1-1})+(1+2+\dots+\overline{r_2-1})+\dots(1+2+\dots+\overline{r_p-1})},$$

when $t=0$.

But the determinant last written is equal to

$$\pm\, \Pi\,(\alpha_\rho+\rho' t-\alpha_\sigma-\sigma' t),$$
$$\times\, \Pi\,(\alpha_\varpi+\varpi' t-q),$$

where in the first product $\rho=1, 2, \dots p$,

$$\rho'=0, 1, 2, \dots r_\rho-1,$$
$$\sigma=1, 2, \dots p,$$
$$\sigma'=0, 1, 2, \dots r_\sigma-1,$$

except that one of the inequalities $\rho \neq \sigma$, $\rho' \neq \sigma'$ must be satisfied, and in the second product

$$\varpi=1, 2, \dots p,$$
$$\varpi'=0, 1, \dots r_\varpi-1.$$

A linear factor is a multiple of t when $\rho = \sigma$, hence t occurs as a factor of the determinant to a power equal to

$$\sum_\rho \frac{r_\rho(r_\rho - 1)}{2}, \quad \rho = 1, 2, \dots p,$$

and this is precisely the power of t in the denominator.

To find the limit we put $t = 0$ when $\rho \neq \sigma$ and take the multiplier of t in the remaining factors.

Besides numerical factors which are certainly not zero, the limit is the product of a number of factors of the types $\alpha_\rho - \alpha_\sigma$ and $\alpha_\rho - q$, and in fact it is easily seen to be

$$N \prod_{\rho \neq \sigma} (\alpha_\rho - \alpha_\sigma)^{r_\rho r_\sigma} \prod_\rho (\alpha_\rho - q)^{r_\rho},$$

where N is an integer.

Now the quantities $\alpha_1, \alpha_2, \dots \alpha_p$ are all unequal and we can choose q to be different from each of them, hence the determinant does not vanish for all values of q, and consequently the n apolar forms are linearly independent.

Ex. Evaluate the determinant

$$\begin{vmatrix} a^5, & a^4, & a^3, & a^2, & a, & 1 \\ 5a^4, & 4a^3, & 3a^2, & 2a, & 1, & 0 \\ 20a^3, & 12a^2, & 6a, & 2, & 0, & 0 \\ \beta^5, & \beta^4, & \beta^3, & \beta^2, & \beta, & 1 \\ 5\beta^4, & 4\beta^3, & 3\beta^2, & 2\beta, & 1, & 0 \\ \gamma^5, & \gamma^4, & \gamma^3, & \gamma^2, & \gamma, & 1 \end{vmatrix}.$$

179. *Forms apolar to two or more given forms.*

Suppose we have s linearly independent forms

$$f_r \equiv a_0^{(r)} x_1^n + n a_1^{(r)} x_1^{n-1} x_2 + \dots + a_n^{(r)} x_2^n, \quad r = 1, 2, \dots s,$$

then the determinants of the array formed by the coefficients cannot all vanish, because in that case values of λ which satisfy $s - 1$ of the $(n+1)$ equations

$$\lambda_1 a_p^{(1)} + \lambda_2 a_p^{(2)} + \dots + \lambda_s a_p^{(s)} = 0$$

would satisfy all these equations, and hence

$$\lambda_1 f_1 + \lambda_2 f_2 + \dots + \lambda_s f_s = 0,$$

which is contrary to hypothesis.

If
$$b_0x_1^n + nb_1x_1^{n-1}x_2 + \dots + b_nx_2^n$$
be a form apolar to each of the f's, then

$$a_0^{(r)}b_n - na_1^{(r)}b_{n-1} + \frac{n(n-1)}{2}a_2^{(r)}b_{n-2} + \dots + (-1)^n a_n^{(r)}b_0 = 0,$$

$$r = 1, 2, \dots s.$$

These s equations connecting the b's can be solved for s of the b's in terms of the others, because, as we have seen, not all the determinants of s columns formed by the coefficients are zero. Having solved the equations we obtain s of the b's expressed as linear functions of the others, and as the remaining $(n-s+1)$ b's are arbitrary, the general form apolar to all the f's involves $(n-s+1)$ constants linearly, *i.e.* there are exactly $(n-s+1)$ linearly independent forms apolar to each of s given forms.

In particular it follows that there is a unique form apolar to each of n given linearly independent forms.

Further if
$$\phi_1, \phi_2, \dots \phi_{n-s+1}$$
be $(n-s+1)$ linearly independent forms apolar to the f's, it is clear that every ϕ is apolar to every f, and that the most general form apolar to each of the ϕ's is a linear combination of the f's; thus the relation between the two sets of forms is a reciprocal one.

Some interesting results follow from this reasoning. For example, given three independent cubics, there is one form apolar to each of them, and, if its factors be all different, each of the given cubics can be expressed as a linear combination of the same three cubes, viz. the cubes of the factors of the apolar cubic. A like result applies to forms of any order and constitutes the generalisation of the problem of expressing two quadratics each as linear combinations of the same two squares.

The form apolar to the three cubics a_x^3, b_x^3, c_x^3, is
$$(bc)(ca)(ab)\, a_x b_x c_x,$$
for this is not zero if the cubics are linearly independent and it is apolar to d_x^3 if
$$(bc)(ca)(ab)(ad)(bd)(cd) = 0,$$

or, as can be readily seen, if

$$\begin{vmatrix} a_0 & a_1 & a_2 & a_3 \\ b_0 & b_1 & b_2 & b_3 \\ c_0 & c_1 & c_2 & c_3 \\ d_0 & d_1 & d_2 & d_3 \end{vmatrix} = 0,$$

which is certainly true if d_x^3 is the same as any of the three original cubics.

The reader may verify that the equation of the apolar form may be written

$$\begin{vmatrix} a_0x_1 + a_1x_2, & a_1x_1 + a_2x_2, & a_2x_1 + a_3x_2 \\ b_0x_1 + b_1x_2, & b_1x_1 + b_2x_2, & b_2x_1 + b_3x_2 \\ c_0x_1 + c_1x_2, & c_1x_1 + c_2x_2, & c_2x_1 + c_3x_2 \end{vmatrix} = 0,$$

which can be easily done by expressing this determinant symbolically.

The extension of these results to n forms of order n will present no difficulty.

180. We may illustrate some of the foregoing results and anticipate some of the developments to come by reference to the geometry of the rational plane cubic curve.

Let ξ, η, ζ be homogeneous coordinates, then for all points on the curve they are rational integral functions of one parameter t and of the third order. To apply our results more directly we shall replace t by two variables which occur homogeneously, so that we have

$$\left.\begin{aligned} \rho\xi &= a_0x_1^3 + 3a_1x_1^2x_2 + 3a_2x_1x_2^2 + a_3x_2^3 \equiv f_1 \\ \rho\eta &= b_0x_1^3 + 3b_1x_1^2x_2 + 3b_2x_1x_2^2 + b_3x_2^3 \equiv f_2 \\ \rho\zeta &= c_0x_1^3 + 3c_1x_1^2x_2 + 3c_2x_1x_2^2 + c_3x_2^3 \equiv f_3 \end{aligned}\right\},$$

and to find the point equation of the curve we must eliminate x_1 and x_2 so as to obtain a result homogeneous in ξ, η, ζ. But the properties of the curve are naturally more easily obtained by using the parametric expressions.

We remark in the first place that the cubics f_1, f_2, f_3 must be linearly independent, otherwise all such points ξ, η, ζ lie on a straight line; next the points in which the line

$$l\xi + m\eta + n\zeta = 0$$

meets the curve are given by the cubic

$$lf_1 + mf_2 + nf_3 = 0,$$

which determines their parameters; hence any straight line meets the curve in three points and the curve is therefore of the third order.

Now there is a unique cubic ϕ apolar to f_1, f_2, f_3, and ϕ is apolar to any cubic giving the parameters of three collinear points. Conversely if a cubic be apolar to ϕ the three points whose parameters are determined by it are collinear because it is of the form $lf_1 + mf_2 + nf_3$.

Thus the three points whose parameters are (x_1, x_2), (y_1, y_2), (z_1, z_2) are in a straight line if

$$\phi_x \phi_y \phi_z = 0.$$

181. *Points of Inflexion.* At a point of inflexion three successive points on the curve are in a straight line, and hence the parameters of the points of inflexion are determined by the perfect cubes apolar to ϕ, that is by

$$\phi = 0.$$

Hence there are at most three points of inflexion, and, since a cubic is apolar to itself, when there are three they are collinear.

But there are other singularities for which three consecutive points on the curve are collinear, *e.g.* cusps, and accordingly we shall examine the equation $\phi = 0$ a little more closely.

If $\alpha_x, \beta_x, \gamma_x$ be the linear factors of ϕ we have

$$\xi = \lambda \alpha_x^3 + \mu \beta_x^3 + \nu \gamma_x^3,$$

with similar expressions for η and ζ; hence on replacing ξ, η, ζ by suitable linear combinations—which is tantamount to changing the triangle of reference—we shall have

$$\xi = \alpha_x^3, \quad \eta = \beta_x^3, \quad \zeta = \gamma_x^3,$$

from which it is readily seen that the straight lines $\xi = 0, \eta = 0, \zeta = 0$ are inflexional tangents, and therefore in this case there are three distinct points of inflexion.

If $\phi_x^3 = 0$ has a double factor β_x we have

$$\xi = \lambda \alpha_x^3 + \mu \beta_x^3 + \nu \beta_x^2 x_1 \text{ etc.},$$

and hence by a similar transformation we can reduce ξ, η, ζ to the forms

$$\xi = \alpha_x^3, \quad \eta = \beta_x^3, \quad \zeta = \beta_x^2 x_1.$$

In the neighbourhood of the point $\eta = 0$, $\zeta = 0$, whose parameter is given by $\beta_x = 0$, we see that $\eta^2 \propto \zeta^3$, and hence this point is a cusp, while the line $\xi = 0$ is an inflexional tangent. The case in which $\phi = 0$ has a treble factor may be rejected because under these conditions f_1, f_2, f_3 have a common factor and, as will be readily seen, the curve breaks up into a straight line and a conic.

We shall confine the further discussion to the case in which the factors of ϕ are distinct.

182. *Double point.* To each value of the ratio $x_1 : x_2$ corresponds a point on the curve. The same ratio cannot give rise to two different points, but the same point may be obtained from two different values of the ratio and then it will be a double point on the curve, because every straight line through it meets the curve in only one other point. We might find the double points directly by developing this idea, but the search is best conducted in a different manner.

If x, y, z be the parameters of three collinear points we have

$$\phi_x \phi_y \phi_z = 0,$$

and in general this determines z uniquely when x and y are given. When x and y give rise to the same point, and only then, the above condition does not determine z but is satisfied for all values of z.

Now the equation $\qquad \phi_x \phi_y \phi_z = 0$

indicates that the quadratic whose vanishing points are x, y is apolar to

$$\phi_x^2 \phi_z,$$

the first polar of z with respect to ϕ_x^3.

Hence if x, y be the parameters of a double point the quadratic giving them must be apolar to all first polars of ϕ_x^3.

Two such polars are

$$\phi_x^2 \phi_\alpha, \quad \phi'^2_x \phi'_\beta$$

and the quadratic apolar to each of these is their Jacobian,

i.e. $$(\phi\phi')\,\phi_x\phi'_x\,\phi_\alpha\phi'_\beta = -(\phi\phi')\,\phi_x\phi'_x\,\phi'_\alpha\phi_\beta,$$

ϕ and ϕ' being equivalent symbols.

This is

$$\tfrac{1}{2}(\phi\phi')\,\phi_x\phi'_x\,(\phi_\alpha\phi'_\beta - \phi'_\alpha\phi_\beta) = \tfrac{1}{2}(\phi\phi')\,\phi_x\phi'_x\,(\phi\phi')(\alpha\beta)$$

and the quadratic required is therefore

$$(\phi\phi')^2\,\phi_x\phi'_x = 0,$$

namely the Hessian of ϕ.

Thus a rational cubic curve has always one double point, the parameters of which are given by the Hessian of the cubic giving the parameters of the points of inflexion.

It will be noticed how readily properties of points on the curve are expressed by means of the form ϕ, and this is natural since ϕ being given we can write down three forms apolar to it for f_1, f_2, f_3 and thence find the ordinary equation of the curve.

As a further example, we remark that the points of contact of the tangents drawn from the point z on the curve to the curve are given by

$$\phi_x^2\phi_z = 0,$$

that is to say there are two such tangents, and the parameters of the points of contact are given by the first polar of z. Hence the quadratic giving them is apolar to the quadratic giving the parameters of the double points.

Ex. (i). Prove that the cross ratio of the pencil formed by joining the double point to four points on the curve is equal to the cross ratio of the parameters of those four points.

Ex. (ii). If the parameters of the points of inflexion be given by $a_x = 0$, $\beta_x = 0$, $\gamma_x = 0$, the point of contact of the remaining tangent to the curve from the first is given by

$$(a\beta)\,\gamma_x + (a\gamma)\,\beta_x = 0\,;$$

hence if $f = 0$ give the points of inflexion, the cubic covariant gives the parameters of the three points L, M, N, here indicated.

Ex. (iii). Prove that the six points in which any conic meets the cubic are given by a sextic apolar to a given sextic ψ. (ψ is apolar to the squares and products f_1, f_2, f_3.) Thence shew that $\psi = 0$ gives the parameters of the points where a conic can be drawn having six-point contact with the curve, and that inasmuch as these points are the points of inflexion and L, M, N, ψ is the product of f and its cubic covariant.

183. *Apolarity of forms of different orders.* In the foregoing discussion we have seen that if x and y be the vanishing points of the Hessian of a cubic f, then

$$a_x a_y a_z = 0$$

for all values of z, so that the Hessian although only a quadratic satisfies, in a manner, the condition of apolarity to the cubic.

If α_x, β_x be the factors of the Hessian, we have

$$(a\alpha)(a\beta)\, a_x = 0,$$

that is its second transvectant with f vanishes identically (cf. § 91).

Generalising this we shall call two forms

$$f = a_x^m, \quad \phi = b_x^n, \quad m > n$$

apolar when the nth transvectant

$$(ab)^n\, a_x^{m-n}$$

vanishes identically.

Two important facts follow at once from this definition.

(i) *The form f is apolar to any form of order $n' \leqslant m$, having ϕ for factor.*

For if the new form be $\phi\psi$ we have

$$(f, \phi\psi)^{n'} = ((f, \phi)^n\, \psi)^{n'-n},$$

since ψ is of order $n' - n$ and the right-hand side vanishes by hypothesis.

A special result is that f is apolar to any m-ic containing the factor ϕ.

(ii) *ϕ is apolar to any polar form of f whose order is n.*

For let $$a_x^n\, a_y^{m-n}$$

be the apolar form, then

$$\{a_x^n a_y^{m-n},\ b_x^n\}^n = (ab)^n\, a_y^{m-n},$$

and this is zero since $$(ab)^n\, a_x^{m-n}$$

vanishes identically.

COR. ϕ is apolar to any polar form of f whose order is $\geqslant n$ but $\leqslant m$.

The proof is as above.

184. The conditions of apolarity of

$$f = (a_0, a_1, \dots a_m \between x_1, x_2)^m = a_x^m$$

and
$$\phi = (b_0, b_1, \dots b_n \between x_1, x_2)^n = b_x^n$$

are equivalent to $m - n + 1$ linear homogeneous relations among the a's.

In fact, equating to zero the several coefficients in

$$(ab)^n a_x^{m-n}$$

we have the following equations:

$$(ab)^n a_1^{m-n} = 0, \ (ab)^n a_1^{m-n-1} a_2 = 0, \ \dots \ (ab)^n a_2^{m-n} = 0.$$

On being expressed in terms of the actual coefficients the first of these relations involves

$$a_0, a_1, \dots a_n,$$

the second
$$a_1, a_2, \dots a_{n+1},$$

and so on, the last containing

$$a_{m-n}, a_{m-n+1}, \dots a_n,$$

hence if, as without loss of generality we may do, we assume that none of the coefficients b are zero, these

$$m - n + 1$$

relations among the a's are obviously linearly independent.

It follows that by means of them we can express

$$(m - n + 1)$$

of the coefficients a linearly in terms of the remaining coefficients and thence that there are

$$(m + 1) - (m - n + 1) = n$$

linearly independent forms of order m apolar to any given form of order n less than m.

185. *Construction of a linearly independent set of forms of order m ($> n$) apolar to a given form ϕ of order n.*

I. Let the factors of the given form be

$$\beta_1^{(r)} x_1 + \beta_2^{(r)} x_2 = \beta_x^{(r)}, \quad r = 1, 2, \dots n,$$

and all different.

Then the typical form β_x^m is apolar to b_x^n, for

$$(\beta_x^m, b_x^n)^n = (\beta b)^n \beta_x^{m-n}$$

and since

$$(\beta b)^n = 0,$$

when β_x is a factor of b_x^n, the result follows at once.

Next the system of forms

$$\beta_x^{(r)^m}, \quad r = 1, 2, \dots n$$

is linearly independent, for if there were a relation of the type

$$\Sigma\lambda_r \beta_x^{(r)^m} = 0,$$

on polarizing $(m-n)$ times with respect to y, we should have

$$\Sigma\lambda_r \beta_y^{(r)^{m-n}} \beta_x^{(r)^n} = 0,$$

which is contrary to the established fact that the system of forms

$$\beta_x^{(r)^n}, \quad r = 1, 2, \dots n$$

is linearly independent.

II. Suppose next that the factors are not all different but that

$$\phi = \beta_x^{(1)\mu_1} \beta_x^{(2)\mu_2} \dots \beta_x^{(s)\mu_s}$$

Then since the factor $\beta_x^{(1)}$ for example occurs μ_1 times in ϕ we know that ϕ is apolar to the n-ic

$$\beta_x^{(1)^{n-\mu_1+1}} C,$$

where C is any form of order $\mu_1 - 1$.

In like manner ϕ is apolar to the m-ic

$$\beta_x^{(1)^{m-\mu_1+1}} \Gamma,$$

Γ being any form of order $\mu_1 - 1$, for

$$(\beta_x^{(1)^{m-\mu_1+1}} \Gamma, \phi)^n$$

$$= (\beta_x^{(1)^{m-\mu_1+1}} \Gamma, b_x^n)^n$$

and this latter is an aggregate of forms each involving the factor

$$(b\beta^{(1)})^{n-\mu_1+1}.$$

But since the factor $\beta_x^{(1)}$ occurs μ_1 times in ϕ any form involving the factor

$$(b\beta^{(1)})^{n-\mu_1+1}$$

vanishes identically.

Hence choosing any s forms

$$\Gamma_1, \Gamma_2, \ldots \Gamma_s$$

of orders $\mu_1 - 1,\ \mu_2 - 1,\ \ldots\ \mu_s - 1$

respectively, forms $\beta_x^{(t)^{m-\mu_t+1}}\Gamma_t$

are all apolar to ϕ.

Next there cannot be a linear relation between them because if there were, on polarizing it $m-n$ times with respect to y we should obtain a linear relation of the type

$$\Sigma\lambda_r\beta_x^{(t)^{n-\mu_t+1}}\Gamma_t' = 0,$$

Γ_t' being of order $\mu_t - 1$. This is contrary to what was proved in constructing the apolar set of order n.

Now the form $\beta_x^{(t)^{m-\mu_t+1}}\Gamma_t$

contains μ_t arbitrary constants and so we have a form apolar to ϕ involving

$$\mu_1 + \mu_2 + \ldots + \mu_s = n$$

arbitrary constants.

The coefficients of the various constants are each apolar to ϕ and they are n in number.

The discovery of forms of order n apolar to a form of order m ($> n$) is a problem quite distinct from the foregoing.

Suppose f is the given form of order m, then a form of order n which is apolar to

$$a_x^n a_y^{m-n}$$

for all values of y will be apolar to f.

This condition has been shewn to be necessary and it is clearly sufficient because if

$$\phi = b_x^n,$$

then

$$(ab)^n a_y^{m-n}$$

vanishes for all values of y.

Hence the form ϕ sought is apolar to the $m-n+1$ forms of order n

$$a_x^n a_1^{m-n},\ a_x^n a_1^{m-n-1}a_2,\ \ldots\ a_x^n a_2^{m-n},$$

and the problem is reduced to one in apolar forms of the same order.

If the forms just written down be linearly independent there are

$$n+1-(m-n+1)$$

linearly independent forms apolar to each of them, and thus there are at least $(2n-m)$ linearly independent forms of order n apolar to a given form of order m. There may be more owing to the subsidiary system of forms not being linearly independent and we shall discuss this question more fully in the sequel.

It is clear that if $n > \frac{m}{2}$ there is at least one form of order n apolar to f.

186. The latter theory finds its natural illustration in the problem of representing one or more given forms of order n as the sum of a number of perfect nth powers. We shall discuss the case of a single quintic at length as an example.

The general cases present no difficulties—they arise for forms of special character.

Suppose the quintic is

$$f = a_0 x_1^5 + 5a_1 x_1^4 x_2 + 10a_2 x_1^3 x_2^2 + 10a_3 x_1^2 x_2^3 + 5a_4 x_1 x_2^4 + a_5 x_2^5$$
$$\equiv a_x^5 = b_x^5 = \ldots.$$

The second polars are linear combinations of

$$\frac{\partial^2 f}{\partial x_1^2}, \quad \frac{\partial^2 f}{\partial x_1 \partial x_2}, \quad \frac{\partial^2 f}{\partial x_2^2}.$$

If these are linearly independent there is a unique cubic apolar to them, and being apolar to all second polars it is apolar to the quintic itself.

On referring to § 179 we may write this cubic in the form

$$\begin{vmatrix} \frac{\partial^4 f}{\partial x_1^4}, & \frac{\partial^4 f}{\partial x_1^3 \partial x_2}, & \frac{\partial^4 f}{\partial x_1^2 \partial x_2^2} \\ \frac{\partial^4 f}{\partial x_1^3 \partial x_2}, & \frac{\partial^4 f}{\partial x_1^2 \partial x_2^2}, & \frac{\partial^4 f}{\partial x_1 \partial x_2^3} \\ \frac{\partial^4 f}{\partial x_1^2 \partial x_2^2}, & \frac{\partial^4 f}{\partial x_1 \partial x_2^3}, & \frac{\partial^4 f}{\partial x_2^4} \end{vmatrix}$$

or

$$(bc)^2 (ca)^2 (ab)^2 a_x b_x c_x,$$

so that it is the covariant denoted by j.

Now suppose

$$j = \alpha_x \beta_x \gamma_x,$$

then
$$f = \lambda \alpha_x^5 + \mu \beta_x^5 + \nu \gamma_x^5,$$

$\alpha_x \beta_x \gamma_x$ being all different and λ, μ, ν numerical.

If $j = \alpha_x^2 \beta_x$ we have for an apolar quintic

$$f = \lambda \alpha_x^5 + \mu \alpha_x^4 x_1 + \nu \beta_x^5, \text{ or } \alpha_x^4 (p x_1 + q x_2) + r \beta_x^5.$$

If $j = \alpha_x^3$ we have

$$f = \lambda \alpha_x^5 + \mu \alpha_x^4 x_1 + \nu \alpha_x^3 x_1^2, \text{ or } \alpha_x^3 (p x_1^2 + 2q x_1 x_2 + r x_2^2).$$

This exhausts the cases in which j is not identically zero.

If j be identically zero the forms $\frac{\partial^2 f}{\partial x_1^2}, \frac{\partial^2 f}{\partial x_1 \partial x_2}, \frac{\partial^2 f}{\partial x_2^2}$ are not linearly independent, because if they were they would determine a unique non-zero apolar form.

Let
$$p \frac{\partial^2 f}{\partial x_1^2} + q \frac{\partial^2 f}{\partial x_1 \partial x_2} + r \frac{\partial^2 f}{\partial x_2^2} = 0$$
be the relation.

Then we have

$$p \frac{\partial^3 f}{\partial x_1^3} + q \frac{\partial^3 f}{\partial x_1^2 \partial x_2} + r \frac{\partial^3 f}{\partial x_1 \partial x_2^2} = 0,$$

and
$$p \frac{\partial^3 f}{\partial x_1^2 \partial x_2} + q \frac{\partial^3 f}{\partial x_1 \partial x_2^2} + r \frac{\partial^3 f}{\partial x_2^3} = 0,$$

and all third polars can be expressed as linear combinations of

$$\frac{\partial^3 f}{\partial x_1^2 \partial x_2}, \frac{\partial^3 f}{\partial x_1 \partial x_2^2}.$$

If these are linearly independent they determine a unique quadratic apolar to both and being apolar to all third polars it is apolar to the quintic.

Suppose this quadratic to be

(i) $\alpha_x \beta_x$, then $f = \lambda \alpha_x^5 + \mu \beta_x^5$;

(ii) α_x^2, then $f = \alpha_x^4 (p x_1 + q x_2)$;

(iii) identically zero, then

$$\frac{\partial^3 f}{\partial x_1^2 \partial x_2}, \frac{\partial^3 f}{\partial x_1 \partial x_2^2}$$

must be identical, hence in this case all third polars are identical.

Now all fourth polars are linear combinations of

$$\frac{\partial^4 f}{\partial x_1^3 \partial x_2}, \quad \frac{\partial^4 f}{\partial x_1^2 \partial x_2^2} \quad \text{.......................} \quad \text{(A)}.$$

But since $\frac{\partial^3 f}{\partial x_1^2 \partial x_2}$ and $\frac{\partial^3 f}{\partial x_1 \partial x_2^2}$ are identical all the fourth polars are identical with either of the forms (A).

Hence the fourth polar is apolar to the quintic, for being of odd order it is apolar to itself.

In this case the quintic is a perfect fifth power unless the linear apolar form is zero identically, in which case the quintic is also identically zero.

This completes the discussion and leaves us with six canonical forms for a quintic, viz.

(i) $f = \lambda \alpha_x^5 + \mu \beta_x^5 + \nu \gamma_x^5$,

(ii) $f = \lambda \alpha_x^5 + \mu \alpha_x^4 x_1 + \nu \beta_x^5$,

(iii) $f = \lambda \alpha_x^5 + \mu \alpha_x^4 x_1 + \nu \alpha_x^3 x_1^2$,

(iv) $f = \lambda \alpha_x^5 + \mu \beta_x^5$,

(v) $f = \lambda \alpha_x^5 + \mu \alpha_x^4 x_1$,

(vi) $f = \lambda \alpha_x^5$.

The discussion for any other single form can be conducted in an exactly similar manner.

187. The reader will have no difficulty in applying the method explained for the quintic to any binary form; in particular it will be easily seen that, whereas a form of odd order $(2n+1)$ always has at least one apolar form of order $(n+1)$, a form of even order $2n$ has not an apolar form of order n unless the determinant formed by the coefficients of its nth derivatives with respect to x_1 and x_2 be zero.

Thus a form of order $(2n+1)$ can in general be expressed in one way as the sum of $(n+1)$ $(2n+1)$th powers, but a form of order $2n$ cannot be expressed as the sum of n $2n$th powers unless a certain function of the coefficients—manifestly an invariant—be zero. For example, in the sextic

$$f \equiv a_0 x_1^6 + 6a_1 x_1^5 x_2 + \ldots + a_6 x_2^6 \equiv a_x^6 = b_x^6 = c_x^6 = d_x^6,$$

there is an apolar cubic only when

$$\begin{vmatrix} a_0 & a_1 & a_2 & a_3 \\ a_1 & a_2 & a_3 & a_4 \\ a_2 & a_3 & a_4 & a_5 \\ a_3 & a_4 & a_5 & a_6 \end{vmatrix} = 0.$$

The expression of this as a symbolical form is an instructive exercise. There must be a linear relation between

$$\frac{\partial^3 f}{\partial x_1^3}, \quad \frac{\partial^3 f}{\partial x_1^2 \partial x_2}, \quad \frac{\partial^3 f}{\partial x_1 \partial x_2^2}, \quad \frac{\partial^3 f}{\partial x_2^3},$$

i.e. between $\quad a_x^3 a_1^3, \quad b_x^3 b_1^2 b_2, \quad c_x^3 c_1 c_2^2, \quad d_x^3 d_2^3,$

and hence referring to § 179 we must have

$$I = (bc)(ca)(ab)(ad)(bd)(cd)\, a_1^3 b_1^2 b_2 c_1 c_2^2 d_2^3 = 0.$$

Interchanging the letters in every possible way we find that

$$I = \tfrac{1}{24}(bc)(ca)(ab)(ad)(bd)(cd) \begin{vmatrix} a_1^3, & a_1^2 a_2, & a_1 a_2^2, & a_2^3 \\ b_1^3, & b_1^2 b_2, & b_1 b_2^2, & b_2^3 \\ c_1^3, & c_1^2 c_2, & c_1 c_2^2, & c_2^3 \\ d_1^3, & d_1^2 d_2, & d_1 d_2^2, & d_2^3 \end{vmatrix}.$$

And hence the condition is

$$(bc)^2 (ca)^2 (ab)^2 (ad)^2 (bd)^2 (cd)^2 = 0.$$

The invariant I is called the catalecticant and it will be easily seen that a similar symbolical expression holds for the catalecticant of any form of even order.

188. It has been shewn that when

$$j = 0$$

identically the quintic can in general be expressed as the sum of two fifth powers, and in the course of the work we found the conditions under which it can be expressed as the sum of a smaller number of fifth powers. A similar process would of course apply to any form, but we shall now give a direct answer to the question as to what is the smallest number of nth powers in terms of which a given binary n-ic can be expressed*.

* See Gundelfinger, *Crelle*, Bd. c. 413—424.

189. If a binary n-ic can be expressed as the sum of r nth powers it must have an apolar r-ic whose factors are all different, so in the first place we proceed to find the necessary and sufficient conditions that the form should possess an apolar r-ic.

If $r > \frac{n}{2}$ there is always at least one apolar r-ic.

Suppose, then, that $r \not> \frac{n}{2}$ and that there is an apolar r-ic, namely ϕ. Then ϕ is apolar to all derivatives of f whose order is equal to or greater than r, and since there are $(r+1)$ derivatives of order $n-r$, viz.

$$\frac{\partial^r f}{\partial x_1^r}, \quad \frac{\partial^r f}{\partial x_1^{r-1}\partial x_2}, \quad \dots \quad \frac{\partial^r f}{\partial x_2^r},$$

these cannot be linearly independent.

Hence there is an identical relation of the form

$$\lambda_0 \frac{\partial^r f}{\partial x_1^r} + \lambda_1 \frac{\partial^r f}{\partial x_1^{r-1}\partial x_2} + \dots + \lambda_r \frac{\partial^r f}{\partial x_2^r} = 0;$$

on differentiating this r times with respect to x_1 and x_2 in the $(r+1)$ different ways possible and eliminating the λ's, we have

$$\begin{vmatrix} \frac{\partial^{2r} f}{\partial x_1^{2r}}, & \frac{\partial^{2r} f}{\partial x_1^{2r-1}\partial x_2}, & \dots, & \frac{\partial^{2r} f}{\partial x_1^r \partial x_2^r} \\ \frac{\partial^{2r} f}{\partial x_1^{2r-1}\partial x_2}, & \frac{\partial^{2r} f}{\partial x_1^{2r-2}\partial x_2^2}, & \dots, & \frac{\partial^{2r} f}{\partial x_1^{r-1}\partial x_2^{r+1}} \\ \dots & \dots & \dots & \dots \\ \frac{\partial^{2r} f}{\partial x_1^r \partial x_2^r}, & \frac{\partial^{2r} f}{\partial x_1^{r-1}\partial x_2^{r+1}}, & \dots, & \frac{\partial^{2r} f}{\partial x_2^{2r}} \end{vmatrix} = 0,$$

or say $$G_r = 0,$$

and it is easy to see that G_r is a covariant of f.

Conversely when $$G_r = 0$$

by the well-known theorem of Wronski* there is a linear relation between

$$\frac{\partial^r f}{\partial x_1^r}, \quad \frac{\partial^r f}{\partial x_1^{r-1}\partial x_2}, \quad \dots \quad \frac{\partial^r f}{\partial x_2^r}.$$

By differentiating this we obtain two independent relations between the $(r+1)$th derivatives, three between the $(r+2)$th

* See Appendix II.

derivatives and in general $(p+1)$ between the $(r+p)$th derivatives.

Now there are $(r+p+1)$ derivatives of the $(r+p)$th class and hence of these only r are linearly independent.

Hence in particular there are only r linearly independent $(n-r)$th derivatives and as these are of order r there is one form of order r apolar to them and therefore apolar to the form f.

Hence when $G_r = 0$ there is an apolar form of order r.

Thus forming the successive covariants

$$G_0,\ G_1,\ G_2,\ \ldots,$$

the necessary and sufficient condition for an apolar r-ic is $G_r = 0$.

If $G_{r-1} \neq 0$ there is no apolar form of order less than r, for if there were any such apolar forms there would be at least one of order $(r-1)$ and G_{r-1} would vanish.

Hence if G_r be the first of the covariants G which vanishes the lowest order of an apolar form is r.

Finally if $G_{r-1} \neq 0$, $G_r = 0$ there is only one apolar form of order r.

Suppose in fact there are two apolar forms of order r and that for simplicity their factors are all different in both cases.

Let $$(x_1 + \alpha_s x_2), \quad s = 1, 2, \ldots, r$$

be the factors of the first, and

$$(x_1 + \beta_s x_2), \quad s = 1, 2, \ldots, r$$

be the factors of the second, then we have

$$f \equiv \sum_1^r \lambda_s (x_1 + \alpha_s x_2)^n \equiv \sum_1^r \mu_s (x_1 + \beta_s x_2)^n;$$

therefore there is an identical relation of the type

$$\sum_1^r \lambda_s (x_1 + \alpha_s x_2)^n - \sum_1^r \mu_s (x_1 + \beta_s x_2)^n = 0.$$

Now $2r$ is less than n, hence by § 178 such a relation is only possible when the coefficients of the various nth powers severally vanish; thus since the α's are all different and the β's are all different it follows that either every λ and every μ is zero or else for a certain number t of values of s

$$\alpha_s = \beta_s ; \ \lambda_s = \mu_s,$$

while for other values of s

$$\alpha_s \neq \beta_s, \ \lambda_s = 0, \ \mu_s = 0.$$

Consequently f can be expressed as the sum of t nth powers where $t < r$, hence there is an apolar form of order t. But in this case we must have $G_{r-1} = 0$ contrary to hypothesis, hence there is only one apolar form of order r.

The reader will easily establish the fact that if there are only two apolar r-ics they must have $(r-1)$ common factors and that these factors multiplied together give the apolar $(r-1)$-ic. Further the extension of the above to the case in which two or more α's are equal will present no difficulty.

Cor. Since G_1 is the Hessian of f the necessary and sufficient condition that f should be a perfect nth power is that its Hessian should vanish identically.

Ex. (i). If f is apolar to ϕ then f is apolar to every form having ϕ for factor.

Ex. (ii). If a binary form of order n have an apolar r-ic ($r < n$), then it has at least $(s+1)$ independent apolar forms of order $r+s$.

Ex. (iii). Shew that the argument of § 188 can be extended to any number of binary forms, and construct a table of canonical forms of a simultaneous system consisting of a cubic and a quartic.

Ex. (iv). Shew that in general two forms of orders n_1 and n_2 can be expressed as linear combinations of powers of p linear forms if

$$3p - 2 = n_1 + n_2 ;$$

find the p-ic giving these linear forms and extend the results to any number of forms.

Ex. (v). From the symbolical form of a catalecticant deduce the symbolical forms of the covariants G.

190. We shall conclude this chapter with a few geometrical illustrations of the foregoing theory.

Binary Quadratics in connection with the Geometry of a Conic.

If we have the equations

$$\xi = a_0x_1^2 + 2a_1x_1x_2 + a_2x_2^2 \equiv a_x^2 = f,$$
$$\eta = b_x^2 = \phi,$$
$$\zeta = c_x^2 = \psi,$$

then the point ξ, η, ζ lies on a fixed conic.

The equation of the conic is easy to find. For the line

$$\lambda\xi + \mu\eta + \nu\zeta = 0$$

touches the curve when

$$\text{Disct.}\ (\lambda f + \mu\phi + \nu\psi) = 0,$$

i.e. if $\lambda^2 i_{11} + \mu^2 i_{22} + \nu^2 i_{33} + 2\mu\nu i_{23} + 2\nu\lambda i_{31} + 2\lambda\mu i_{12} = 0,$

where $i_{12} = (f, \phi)^2$ etc.

This being the tangential equation the point equation is

$$\begin{vmatrix} i_{11} & i_{12} & i_{13} & \xi \\ i_{12} & i_{22} & i_{23} & \eta \\ i_{13} & i_{23} & i_{33} & \zeta \\ \xi & \eta & \zeta & 0 \end{vmatrix} = 0,$$

or $I_{11}\xi^2 + I_{22}\eta^2 + I_{33}\zeta^2 + 2I_{23}\eta\zeta + 2I_{31}\zeta\xi + 2I_{12}\xi\eta = 0,$

where I_{rs} is the minor of i_{rs} in the determinant

$$\begin{vmatrix} i_{11} & i_{12} & i_{13} \\ i_{12} & i_{22} & i_{23} \\ i_{13} & i_{23} & i_{33} \end{vmatrix}.$$

In particular the equation of the conic gives an identical relation between three quadratic forms and their invariants.

191. An immediate inference from the parametric expressions for ξ, η, ζ is that the cross-ratio of the pencil joining a variable point P on the conic to four fixed points x, y, z, ω on the curve is equal to the cross-ratio of the parameters of those four points, *i.e.*

$$\frac{(xy)(z\omega)}{(x\omega)(zy)}.$$

For let the equations of two lines through P be $X = 0$, $Y = 0$ and let $X + tY = 0$ be the equation of the line joining P to the point x on the curve. Then there is an algebraic relation connecting t with $\frac{x_1}{x_2}$ and since to one value of t corresponds one value of $\frac{x_1}{x_2}$ and *vice versa* we must have

$$t = \frac{Ax_1 + Bx_2}{Cx_1 + Dx_2},$$

where A, B, C, D are constants.

Hence the cross-ratio of four values of t is equal to the cross-ratio of the corresponding four values of the parameter $\frac{x_1}{x_2}$.

192. In connection with this conic there is a simple correspondence between binary quadratics and straight lines in the plane, for a binary quadratic χ equated to zero gives two points P, Q on the conic, so that if we make χ correspond to the line PQ when either is given the other is uniquely determined.

The quadratic χ can be written in one way in the form

$$\lambda f + \mu\phi + \nu\psi,$$

and then $\lambda\xi + \mu\eta + \nu\zeta = 0$ is the equation of the corresponding straight line.

Accordingly if the line pass through a fixed point in the plane, say ξ_0, η_0, ζ_0, we have

$$\begin{aligned}\chi &= \lambda f + \mu\phi + \nu\psi \\ &= \lambda f + \mu\phi - \frac{\lambda\xi_0 + \mu\eta_0}{\zeta_0}\psi \\ &= \lambda\left(f - \frac{\xi_0}{\zeta_0}\psi\right) + \mu\left(\phi - \frac{\eta_0}{\zeta_0}\psi\right),\end{aligned}$$

and χ is apolar to (*i.e.* harmonic to) the fixed quadratic apolar to

$$f - \frac{\xi_0}{\zeta_0}\psi \text{ and } \phi - \frac{\eta_0}{\zeta_0}\psi.$$

Hence if the line PQ passes through a fixed point T, the corresponding quadratic χ is harmonic to a fixed quadratic τ.

Now the perfect squares apolar to τ correspond to lines touching the conic, since in this case the points P and Q coincide, hence these perfect squares determine the points of contact of the tangents drawn from T to the conic.

If these points are R, S, then the quadratic giving R and S is apolar to the square of the linear forms giving R and S respectively, and since J is apolar to these latter squares, it follows that J corresponds to RS, as can be seen in many ways.

Thus if PQ pass through T, and RS be the polar of T, the quadratics corresponding to PQ, RS are harmonic, or in other words, when two quadratics are harmonic, the corresponding lines are conjugate with respect to the conic.

This can be easily verified by using the tangential equation of the conic.

Consider a binary quartic representing four points A, B, C, D on the conic, and let BC, AD meet in E, CA, BD in F, and AB, CD in G.

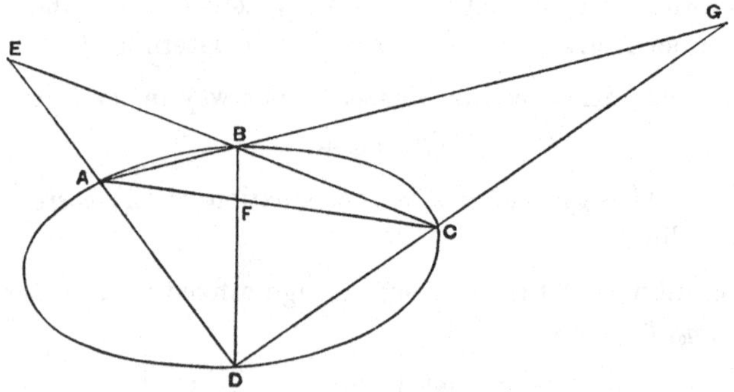

Then the quadratic corresponding to the polar of E is harmonic to the two quadratics corresponding to AD and BC, *i.e.* this polar meets the conic in two points which are the double points of the involution having B, C and A, D for conjugate elements.

Thus the polars of E, F, G meet the conic in the double points of the three involutions determined by the four points A, B, C, D; but EFG being a self-conjugate triangle of the conic, the polar of E is FG, and the lines EF, FG for example are conjugate lines with respect to the triangle, hence the pairs of double points of any two of the three involutions are harmonically conjugate.

[This corresponds to the fact that the sextic covariant of a quartic can be written as the product of three quadratics which are mutually harmonic.]

193. As another example, let us prove that a triangle and its polar triangle with respect to the conic are in perspective.

Suppose the sides of the triangle $A_1A_2A_3$ correspond to the quadratics f_1, f_2, f_3 respectively, and that the sides of the polar triangle $B_1B_2B_3$ correspond to ϕ_1, ϕ_2, ϕ_3 respectively. Then since B_1B_2 is conjugate to A_1A_3 and A_2A_3, it follows that ϕ_3 is harmonic to f_2 and f_1, so that ϕ_3 is the Jacobian of f_1 and f_2.

Hence the sides of $B_1B_2B_3$ correspond to

$$J_{23},\ J_{31},\ \text{and } J_{12} \text{ where } J_{12} = (f_1 f_2).$$

Now let B_2B_3 meet A_1A_3 in P_1, then the polar of P_1 is conjugate to both these lines, and therefore corresponds to

$$(\phi_1,\ f_1) \text{ or to } (f_1,\ J_{23}).$$

The polars of the analogous points P_2, P_3 correspond to (f_2, J_{31}), (f_3, J_{12}) respectively, and P_1, P_2, P_3 will be collinear if their polars are concurrent, *i.e.* if the quadratics

$$(f_1,\ J_{23}),\ (f_2,\ J_{31}),\ (f_3,\ J_{12})$$

are harmonic to the same quadratic.

To prove that this is so, let us calculate the quadratic harmonic to the first two.

Representing

$$f_1,\ f_2,\ f_3 \text{ by } a_x^2,\ b_x^2,\ c_x^2,$$

we have

$$J_{23} = (bc)\, b_x c_x, \text{ etc.}$$

$$\begin{aligned}
\therefore\ (f_1,\ J_{23})' &= \{a_x^2,\ (bc)\, b_x c_x\} = \tfrac{1}{2}(ac)(bc)\, a_x b_x + \tfrac{1}{2}(ab)(bc)\, a_x c_x \\
&= \tfrac{1}{4}\{(ac)^2 b_x^2 + (bc)^2 a_x^2 - (ab)^2 c_x^2\} - \tfrac{1}{4}\{(ab)^2 c_x^2 + (bc)^2 a_x^2 \\
&\qquad\qquad - (ac)^2 b_x^2\} \\
&= \tfrac{1}{2}(ac)^2 b_x^2 - \tfrac{1}{2}(ab)^2 c_x^2.
\end{aligned}$$

Similarly

$$(f_2,\ J_{31}) = \tfrac{1}{2}(ab)^2 c_x^2 - \tfrac{1}{2}(bc)^2 a_x^2,$$

and the quadratic harmonic to these two is

$$\{(ac)^2 b_x^2 - (ab)^2 c_x^2,\ (ab)^2 c_x^2 - (bc)^2 a_x^2\},$$

or

$$(ab)^2 (ac)^2 J_{23} - (ac)^2 (bc)^2 J_{21} + (ab)^2 (bc)^2 J_{31},$$

or

$$(ab)^2 (ac)^2 J_{23} + (ba)^2 (bc)^2 J_{31} + (ca)^2 (cb)^2 J_{12},$$

and by symmetry this is harmonic to (f_3, J_{12}) also.

Thus the polars of $P_1P_2P_3$ meet in a point whose polar corresponds to

$$\frac{J_{23}}{i_{23}} + \frac{J_{31}}{i_{31}} + \frac{J_{12}}{i_{12}} = 0,$$

where $i_{23} = (f_2, f_3)^2 = (bc)^2$.

And hence $P_1P_2P_3$ lie on the straight line corresponding to the quadratic

$$\frac{J_{23}}{i_{23}} + \frac{J_{31}}{i_{31}} + \frac{J_{12}}{i_{12}} = 0.$$

The reader will find it interesting to shew that the above quadratic may be written

$$\frac{f_1}{I_{23}} + \frac{f_2}{I_{31}} + \frac{f_3}{I_{12}} = 0,$$

where the I's are derived from the i's as before.

Ex. (i). *ABC* is a triangle inscribed in a conic, and the tangents at *A*, *B*, *C* meet *BC*, *CA*, *AB* in *L*, *M*, *N*. Determine the parameters of the points in which the polar of *L* meets the conic in terms of those of *A*, *B*, *C*. Hence shew that *L*, *M*, *N* are collinear, and that the line *LMN* meets the conic in the Hessian points of *ABC*.

Ex. (ii). Deduce Pascal's Theorem from the parametric representation of a conic.

Ex. (iii). Prove that six conics can be drawn through four fixed points to touch a given conic, and that their points of contact are given by the Jacobian of the quartics determined by any two conics through the four points.

194. Twisted Cubic. The relation of the rational cubic space curve to the binary cubic is exactly the same as that of the conic to the binary quadratic.

The line of argument is similar to that used in the previous case.

We have

$$\xi = a_0x_1^3 + 3a_1x_1^2x_2 + 3a_2x_1x_2^2 + a_3x_2^3 = a_x^3 = f,$$

$$\eta = b_x^3 = \phi,$$

$$\zeta = c_x^3 = \psi,$$

$$\varpi = d_x^3 = \chi,$$

and the curve represented is a cubic because the plane

$$\lambda\xi + \mu\eta + \nu\zeta + \rho\varpi = 0$$

meets it in three points.

We have now an exact correspondence between planes and binary cubics. If a plane pass through a fixed point T the corresponding cubic is apolar to a fixed cubic which determines P, Q, R the points of contact of the osculating planes of the cubic through T. Cf. § 192.

Hence there are three such points, and since a cubic is apolar to itself, the plane joining them passes through T. We shall call T the pole of the plane PQR; and we observe that when two cubics are apolar, each of the corresponding planes passes through the pole of the other.

If TLM be a chord of the cubic through T, then any plane through this line is apolar to the cubic giving PQR, so that L, M are determined by a quadratic apolar to the cubic, *i.e.* by the Hessian of the cubic.

195. We shall illustrate the above remarks by the discussion of an algebraical problem intimately connected with the cubic curve.

The planes passing through a fixed line l correspond to cubics of the form

$$\lambda_1 f_1 + \lambda_2 f_2,$$

where λ_1 and λ_2 are variable numbers. We call such a linear system of cubics a *pencil*, and then we see that the Jacobian of two members

$$\lambda_1 f_1 + \lambda_2 f_2, \quad \mu_1 f_1 + \mu_2 f_2$$

of the pencil is

$$J = (\lambda_1 \mu_2 - \lambda_2 \mu_1)(f_1 f_2),$$

so that except for a numerical factor it is the same for every pair of cubics in the pencil. Hence we may call this quartic the Jacobian of the pencil and the question arises—*Given a binary quartic J, can a pencil of cubics be found of which it is the Jacobian?*

To answer this question consider the geometrical meaning of the Jacobian with reference to the line l.

If the cubic

$$\lambda_1 f_1 + \lambda_2 f_2$$

have a double factor this factor occurs in the Jacobian, because it occurs in both

$$\lambda_1 \frac{\partial f_1}{\partial x_1} + \lambda_2 \frac{\partial f_2}{\partial x_1} \text{ and } \lambda_1 \frac{\partial f_1}{\partial x_2} + \lambda_2 \frac{\partial f_2}{\partial x_2}.$$

And further, as the discriminant of a cubic is of degree four there are only four members of the pencil which have a double factor; therefore the Jacobian of the forms contains these four double factors and no others.

But if a cubic $$\lambda_1 f_1 + \lambda_2 f_2$$

have a double factor the corresponding plane touches the curve, and hence through a given line four planes can be drawn touching the curve, and their points of contact are given by $J=0$; in other words there are four tangent lines to the curve which intersect a given line l, and they are tangents at the points given by

$$J=0.$$

Now our problem is equivalent to the following: *Being given the four points of contact, to construct the line l.* But the four tangents being given there are two lines meeting them, and, as each of these corresponds to a pencil of cubics having J for Jacobian, it follows that the question is always soluble and that there are two solutions which may coincide in particular cases.

Ex. (i). Prove that three osculating planes and one chord of a cubic can be drawn through any point. The chord determines the Hessian of the cubic giving the points of contact of the planes.

Ex. (ii). The plane joining the points of contact of the three osculating planes from a point O passes through O and is called the polar plane of O. If the polar plane of O passes through O' the polar plane of O' passes through O and the corresponding cubics are apolar.

Ex. (iii). By using line coordinates prove that the tangents to the cubic belong to a linear complex and that any two planes corresponding to apolar cubics meet in a line belonging to this complex.

Ex. (iv). If a line l belong to the above complex the points of contact of the tangents to the cubic which meet l are given by a quartic for which the invariant i vanishes. In this case the two pencils of cubics having the quartic for Jacobian coincide.

Ex. (v). If two pencils of cubics have the same Jacobian, then any member of the first pencil is apolar to any member of the second pencil.

196. The twisted quartic. The rational space curve of the fourth degree furnishes the most convenient geometrical representation of a binary quartic and its concomitants. Suppose that

$$\xi = p_x^4, \quad \eta = q_x^4, \quad \zeta = r_x^4, \quad \varpi = t_x^4$$

is the parametric representation of such a curve. Then since ξ, η, ζ, ϖ are not connected by a linear relation the four binary quartics are linearly independent and are therefore all apolar to a unique quartic f which we shall denote by

$$a_x^4 \equiv b_x^4 \equiv c_x^4 \equiv d_x^4.$$

It is evident that four points on the curve are coplanar when and only when the quartic determining their parameters is apolar to f—in fact such quartics are linear combinations of ξ, η, ζ, ϖ. Hence the four points $\alpha, \beta, \gamma, \delta$ are coplanar when

$$a_\alpha a_\beta a_\gamma a_\delta = 0.$$

Thus $a_x^4 = 0$ gives the four points of superosculation, *i.e.* the four points at which the osculating plane contains four consecutive points of the curve.

Through a point δ on the curve can be drawn three osculating planes other than that at δ and their points of contact are given by

$$a_x^3 a_\delta = 0,$$

i.e. by the first polar of δ.

By varying δ we obtain a pencil of cubics each of which determines three points on the curve such that the osculating planes at them meet in a point on the curve.

Now if a member of this pencil have a double factor we have two consecutive osculating planes whose line of intersection meets the curve, and hence the double factor gives a point the tangent at which meets the curve again. But the double factors of members of the pencil are precisely the factors of the Jacobian of the pencil, which in this case is the Hessian of f. Hence $H = 0$ gives four points on the curve the tangents at which meet the curve again.

197. We have still to interpret the sextic covariant which gives six points on the curve.

For this purpose let us seek for points P, Q on the curve such that the osculating plane at P passes through Q and the osculating plane at Q passes through P.

If λ, μ are the parameters of P and Q we have

$$a_\lambda^3 a_\mu = 0,$$

$$b_\lambda b_\mu^3 = 0.$$

To eliminate λ we note that from the second equation

$$\lambda_2 = -b_1 b_\mu^3,\ \lambda_1 = b_2 b_\mu^3,$$

and as λ occurs to degree three in the first equation we must use three different equations of this type.

Hence we have

$$(ab)(ac)(ad)\, a_\mu b_\mu^3 c_\mu^3 d_\mu^3 = 0$$

as the equation for μ.

Now

$$\begin{aligned}(ab)(ac)(ad)\, a_x b_x^3 c_x^3 d_x^3 \\ &= \tfrac{1}{2}(ab)\, a_x b_x^3 c_x^2 d_x^2 \{(ac)^2 d_x^2 + (ad)^2 c_x^2 - (cd)^2 a_x^2\} \\ &= \tfrac{1}{2}(ab)(ac)^2 a_x b_x^3 c_x^2 . d_x^4 + \tfrac{1}{2}(ab)(ad)^2 a_x b_x^3 d_x^2 . c_x^4 \\ &\qquad - \tfrac{1}{2}(ab)\, a_x^3 b_x^3 . (cd)^2 c_x^2 d_x^2 \\ &= tf,\end{aligned}$$

where t is the sextic covariant.

For $f = 0$ the points P and Q coincide, hence there are three pairs of points such that each lies in the osculating plane of the other and they are given by the sextic covariant.

198. We shall now sketch some further investigations connected with the curve.

The triple secants of the curve are given by the pencil of cubics apolar to f, so that if one cubic give three points on a triple secant and another three points whose osculating planes meet on the curve the two cubics are apolar. The two pencils have the same Jacobian, as follows from geometry and analysis.

The four points in which the tangents at the Hessian points meet the curve again are given by

$$(ab)^2 (ac)(bd)(cd)^2 a_x b_x c_x d_x = 0,$$

for the discriminant of their first polars must vanish.

This reduces to $\qquad iH - jf = 0.$

If $i = 0$ the four points of superosculation are coplanar. If $j = 0$ there is an actual double point on the curve, because in that case f has an apolar quadratic—the quartic is now the complete intersection of two quadrics which touch, whereas in general it is the partial intersection of a quadric and a cubic.

There are two other tangents meeting any given tangent to the curve and when $j = 0$ there exist pairs of chords such that the tangents at the extremities of either intersect those at the extremities of the other.

Between the parameters of the points of contact of two intersecting tangents there is a symmetrical (2, 2) relation, hence by comparison with Poncelet's porism we infer that if one twisted polygon with n sides can be circumscribed to the curve there is an infinite number of such circumscribing polygons.

By forming the expressions for the six coordinates of the tangent at any point, which are the Jacobians of the fundamental quartics taken in pairs, we see that the condition that six tangents should belong to the same linear complex is that the sextic determining their point of contact should be apolar to a fixed sextic. This fixed sextic can be shewn to be the sextic covariant either geometrically or analytically. We can in fact establish the following theorem: If two quartics are apolar to a third then their Jacobian is apolar to the sextic covariant of the latter. Ex. (ii), p. 52.

The sextic covariant therefore bears the same relation to the line geometry of the curve as the fundamental quartic does to the point geometry.

By discussion of the quartics apolar to this sextic the reader will easily prove that the tangents at four points given by a quartic covariant (of the type $\lambda iH + \mu jf$) are generators of a quadric.

References to various memoirs of Reye, Rosanes and others on the subjects discussed in this chapter will be found in Meyer's *Berichte*. See also the same author's book *Apolarität und rationale Curve*. There is an interesting paper on the twisted cubic by Sturm in *Crelle's Journal*, Vol. LXXXVI., and a very full list of references for the twisted quartic in a paper by Richmond, *Camb. Phil. Trans.* 1900.

CHAPTER XII.

TERNARY FORMS.

199. In this chapter we shall extend the symbolical notation, which has been used throughout for binary forms, to the case of forms with a higher number of variables. It may be well to remark at the outset that the methods developed up to the present are nowhere else so effective as in the case of binary forms; accordingly, while we shall explain at some length the notation and methods for ternary forms, we shall content ourselves with a very brief indication of the extension to forms with four or more variables.

200. A ternary form in one set of variables x_1, x_2, x_3 will be written in the form

$$f = \Sigma \frac{n!}{p!\,q!\,r!} a_{pqr}\, x_1^p x_2^q x_3^r\,;\ p + q + r = n,$$

and the summation is of course extended to all possible values of the integers p, q, r satisfying this condition.

In agreement with the symbols previously introduced, we shall represent f by the umbral expression

$$(\alpha_1 x_1 + \alpha_2 x_2 + \alpha_3 x_3)^n = \alpha_x^n,$$

so that

$$\alpha_1^p \alpha_2^q \alpha_3^r = a_{pqr}.$$

Just as before an expression in the α's has no actual (as opposed to symbolical) significance unless the total degree in the α's is exactly n. To represent forms whose degree in the a's is greater than unity, we have to introduce other equivalent systems of symbols β and γ, so that

$$f = \alpha_x^n = \beta_x^n = \gamma_x^n \ldots$$

201. Let us at once point out how concisely polar forms are represented in this notation; the rth polar of $y_1y_2y_3$ with respect to f is

$$\frac{(n-r)!}{n!}\left(y_1\frac{\partial}{\partial x_1}+y_2\frac{\partial}{\partial x_2}+y_3\frac{\partial}{\partial x_3}\right)^r f,$$

the numerical factor being introduced for arithmetical convenience.

This expression is

$$\frac{(n-r)!}{n!}\left(y_1\frac{\partial}{\partial x_1}+y_2\frac{\partial}{\partial x_2}+y_3\frac{\partial}{\partial x_3}\right)^r (\alpha_1x_1+\alpha_2x_2+\alpha_3x_3)^n$$

$$=\frac{(n-r)!}{n!}\,\frac{n!}{(n-r)!}\,\alpha_x^{n-r}\alpha_y^{r},$$

i.e. the rth polar is represented by

$$\alpha_x^{n-r}\alpha_y^{r}.$$

Thus for example the equation

$$\alpha_x^{n}=0$$

represents a curve of order n, and if y be a point on it the equation of the tangent at that point is

$$\alpha_x\alpha_y^{n-1}=0.$$

Again, the points of contact of the tangents drawn from a point y to the curve lie on the first polar

$$\alpha_x^{n-1}\alpha_y=0.$$

Next let us consider the effect of a linear substitution on a form such as f.

We shall write such a transformation in the form

$$\left.\begin{aligned} x_1&=\xi_1X_1+\eta_1X_2+\zeta_1X_3\\ x_2&=\xi_2X_1+\eta_2X_2+\zeta_2X_3\\ x_3&=\xi_3X_1+\eta_3X_2+\zeta_3X_3\end{aligned}\right\}.$$

The effect of this change of variables on α_x is to change it into

$$\alpha_1(\xi_1X_1+\eta_1X_2+\zeta_1X_3)+\alpha_2(\xi_2X_1+\eta_2X_2+\zeta_2X_3)+\alpha_3(\xi_3X_1+\eta_3X_2+\zeta_3X_3),$$

or
$$\alpha_\xi X_1+\alpha_\eta X_2+\alpha_\zeta X_3.$$

Hence f which is $\alpha_x{}^n$ becomes

$$(\alpha_\xi X_1 + \alpha_\eta X_2 + \alpha_\zeta X_3)^n,$$

or $\qquad \Sigma \dfrac{n!}{p!\,q!\,r!}\,\alpha_\xi{}^p\alpha_\eta{}^q\alpha_\zeta{}^r X_1{}^p X_2{}^q X_3{}^r,$ where $p + q + r = n.$

Thus the coefficient of $X_1{}^n$ is $\alpha_\xi{}^n$ or the result of putting the ξ's for the x's in f, and subsequent coefficients may be deduced from this by means of the polarizing operators

$$\left(\eta_1 \frac{\partial}{\partial \xi_1} + \eta_2 \frac{\partial}{\partial \xi_2} + \eta_3 \frac{\partial}{\partial \xi_3}\right) \text{ and } \left(\zeta_1 \frac{\partial}{\partial \xi_1} + \zeta_2 \frac{\partial}{\partial \xi_2} + \zeta_3 \frac{\partial}{\partial \xi_3}\right)$$

operating on the first.

202. The above will shew how very convenient the symbolical notation is for dealing with constantly occurring functions like polars, but, as in the case of binary forms, its great value lies in the application to the theory of invariants and covariants.

Let us recall the significance of these terms especially for ternary forms and illustrate them by some examples.

Suppose that after the application of a linear transformation

$$x_r = \xi_r X_1 + \eta_r X_2 + \zeta_r X_3, \qquad r = 1, 2, 3,$$

a ternary quantic

$$f \equiv \Sigma a_{pqr} \frac{n!}{p!\,q!\,r!} x_1{}^p x_2{}^q x_3{}^r$$

becomes $\qquad F = \Sigma A_{pqr} \dfrac{n!}{p!\,q!\,r!}\, X_1{}^p X_2{}^q X_3{}^r,$

then a rational integral function I of the coefficients is said to be an invariant when

$$I(A_{pqr}) = I(a_{pqr})\, M,$$

where M depends only on the transformation.

A covariant of f possesses a similar property, but involves the variables as well as the coefficients, and similar definitions apply to invariants and covariants of two or more forms.

After what has been said on the subject in dealing with binary forms, we need not stop to prove that it suffices to consider invariants and covariants which are homogeneous in the coefficients of each form involved, and covariants which are homogeneous in the variables—further it may be easily proved that the factor M referred to in the definition must be a power of the determinant of the transformation, and we shall assume henceforth that it is so. The reader may regard this as a simplification of the definition, or supply the necessary proof on the lines of that given for binary forms. See § 24.

As examples, we may notice that the discriminant

$$abc + 2fgh - af^2 - bg^2 - ch^2$$

is an invariant of the ternary quadric

$$ax_1^2 + bx_2^2 + cx_3^2 + 2fx_2x_3 + 2gx_3x_1 + 2hx_1x_2.$$

In this case the power of the determinant which occurs as multiplying factor is two.

Again the Hessian

$$\begin{vmatrix} \frac{\partial^2 f}{\partial x_1^2} & \frac{\partial^2 f}{\partial x_1 \partial x_2} & \frac{\partial^2 f}{\partial x_1 \partial x_3} \\ \frac{\partial^2 f}{\partial x_2 \partial x_1} & \frac{\partial^2 f}{\partial x_2^2} & \frac{\partial^2 f}{\partial x_2 \partial x_3} \\ \frac{\partial^2 f}{\partial x_3 \partial x_1} & \frac{\partial^2 f}{\partial x_3 \partial x_2} & \frac{\partial^2 f}{\partial x_3^2} \end{vmatrix}$$

is a covariant of any ternary form f.

As a further example, the Jacobian of three forms is a covariant of the three forms.

We shall not stop to verify that these expressions actually are invariants or covariants as the case may be. They are well known in the theory of curves and we mention them here to remind the reader that such invariant functions as defined above do actually exist. By the use of the symbolical notation we shall be able to verify our assertions much more easily, and also to construct any number of invariants and covariants.

203. Just as in the theory of binary forms we denoted the determinant

$$\begin{vmatrix} \alpha_1 & \alpha_2 \\ \beta_1 & \beta_2 \end{vmatrix} \text{ by } (\alpha\beta),$$

so now we shall denote

$$\begin{vmatrix} \alpha_1 & \alpha_2 & \alpha_3 \\ \beta_1 & \beta_2 & \beta_3 \\ \gamma_1 & \gamma_2 & \gamma_3 \end{vmatrix} \text{ by } (\alpha\beta\gamma).$$

We are now in a position to enunciate and prove the first theorem connecting invariants and symbolical notation, namely: *Every expression represented symbolically by factors of the type* $(\alpha\beta\gamma)$ *is an invariant, and every expression represented by factors of the types* $(\alpha\beta\gamma)$ *and* α_x *is a covariant.*

For by a linear substitution α_x becomes

$$\alpha_\xi X_1 + \alpha_\eta X_2 + \alpha_\zeta X_3,$$

and hence $(\alpha\beta\gamma)$ becomes

$$\begin{vmatrix} \alpha_\xi & \alpha_\eta & \alpha_\zeta \\ \beta_\xi & \beta_\eta & \beta_\zeta \\ \gamma_\xi & \gamma_\eta & \gamma_\zeta \end{vmatrix} \quad \text{i.e. } (\alpha\beta\gamma)(\xi\eta\zeta).$$

So that the expression formed from the new coefficients is equal to that formed from the old coefficients multiplied by a power of $(\xi\eta\zeta)$; and the power of $(\xi\eta\zeta)$ that occurs is equal to the number of symbolical factors of the type $(\alpha\beta\gamma)$ that occur in the expression.

204. The preceding proof, depending only on the way in which the symbolical letters occur in the expression under consideration, applies equally well to an invariant or covariant of any number of ternary forms. The only further condition that such a symbolical expression should actually represent a covariant is that it should have a meaning when expressed in terms of the original coefficients, *i.e.* each symbolical letter must occur to the requisite degree in every term in which it appears. It is easy to verify that the three examples already given are actually invariants or covariants as the case may be.

Thus if

$$a_{200}x_1^2 + a_{020}x_2^2 + a_{002}x_3^2 + 2a_{110}x_1x_2 + 2a_{101}x_3x_1 + 2a_{011}x_2x_3$$
$$= \alpha_x^2 = \beta_x^2 = \gamma_x^2,$$

we have by direct multiplication

$$(\alpha\beta\gamma)^2 = (\Sigma \pm \alpha_1\beta_2\gamma_3)^2 = 6 \begin{vmatrix} a_{200}, & a_{110}, & a_{101} \\ a_{110}, & a_{020}, & a_{011} \\ a_{101}, & a_{011}, & a_{002} \end{vmatrix},$$

since $\qquad \alpha_1^2 = \beta_1^2 = \gamma_1^2 = a_{200}$, etc.

Thus the discriminant of the ternary quadric is an invariant.

Again, if we have three forms

$$f_1 = \alpha_x^{n_1}, \quad f_2 = \beta_x^{n_2}, \quad f_3 = \gamma_x^{n_3},$$

then

$$(\alpha\beta\gamma)\alpha_x^{n_1-1}\beta_x^{n_2-1}\gamma_x^{n_3-1} = \begin{vmatrix} \alpha_1\alpha_x^{n_1-1}, & \alpha_2\alpha_x^{n_1-1}, & \alpha_3\alpha_x^{n_1-1} \\ \beta_1\beta_x^{n_2-1}, & \beta_2\beta_x^{n_2-1}, & \beta_3\beta_x^{n_2-1} \\ \gamma_1\gamma_x^{n_3-1}, & \gamma_2\gamma_x^{n_3-1}, & \gamma_3\gamma_x^{n_3-1} \end{vmatrix}$$

$$= \frac{1}{n_1n_2n_3} \begin{vmatrix} \frac{\partial f_1}{\partial x_1} & \frac{\partial f_1}{\partial x_2} & \frac{\partial f_1}{\partial x_3} \\ \frac{\partial f_2}{\partial x_1} & \frac{\partial f_2}{\partial x_2} & \frac{\partial f_2}{\partial x_3} \\ \frac{\partial f_3}{\partial x_1} & \frac{\partial f_3}{\partial x_2} & \frac{\partial f_3}{\partial x_3} \end{vmatrix} = \frac{1}{n_1n_2n_3} \frac{\partial (f_1f_2f_3)}{\partial (x_1x_2x_3)},$$

which shews at once that the Jacobian is a covariant.

The fact that $(\alpha\beta\gamma)^2 \alpha_x^{n-2}\beta_x^{n-2}\gamma_x^{n-2}$ represents the Hessian of the form $f = \alpha_x^n = \beta_x^n = \gamma_x^n$ may be easily verified in the same way. The deduction of the equation of the Hessian from the property that the polar conic of any point on it degenerates into two straight lines affords an instructive example of the symbolical calculus.

The polar conic of the point y_1, y_2, y_3 with respect to the curve

$$f = \alpha_x^n = \beta_x^n = \gamma_x^n = 0$$

is represented by the equation

$$\alpha_y^{n-2}\alpha_x^2 \equiv \beta_y^{n-2}\beta_x^2 \equiv \gamma_y^{n-2}\gamma_x^2 = 0,$$

and the discriminant of this is

$$(\alpha\beta\gamma)^2 \alpha_y^{n-2}\beta_y^{n-2}\gamma_y^{n-2}.$$

Hence changing y into x we get the Hessian as the locus of points possessing the property in question.

205. Geometrical Meaning of a Linear Transformation. To obtain a geometrical representation of ternary forms we naturally suppose the variables x_1, x_2, x_3 to be the homogeneous coordinates of a point in a plane referred to a certain triangle. The equation to zero of a ternary form will then represent a curve in the plane.

A linear transformation

$$x_r = \xi_r X_1 + \eta_r X_2 + \zeta_r X_3, \qquad r = 1, 2, 3,$$

may be regarded from two points of view, for if P be the point

x_1, x_2, x_3 we may either suppose P to be unaltered and the triangle of reference changed or we may suppose P to be changed and the triangle of reference unaltered, in other words X_1, X_2, X_3 may be the coordinates of the original point referred to a new fundamental triangle or they may be the coordinates of a new point (into which P is changed) referred to the original triangle of reference.

The general linear transformation as written above can be, in fact, represented as a change in the triangle of reference together with a multiplication of each coordinate by a suitable quantity—the equations of the sides of the old triangle of reference referred to the new triangle are $x_r = 0$, that is

$$\xi_r X_1 + \eta_r X_2 + \zeta_r X_3 = 0, \qquad r = 1, 2, 3,$$

and on solving for X_1, X_2, X_3 in terms of x_1, x_2, x_3 we can easily find the equations of the sides of the new triangle of reference in the old coordinates.

From this point of view, if I be an invariant of a ternary form f, when I vanishes for one triangle of reference it vanishes for any other triangle of reference; that is to say, $I = 0$ expresses a property of the curve itself and not a relation between the curve and some particular triangle.

In like manner the curve represented by a covariant is connected with the original curve by relations which are quite independent of the triangle of reference.

From the second point of view the point P is changed into a point P' by a homographic transformation, *i.e.* a transformation possessing the following properties:

(i) any point P is changed into a point P';

(ii) any straight line p is changed into a straight line p';

(iii) if P lies on p then P' lies on p';

(iv) the cross-ratio of four collinear points is equal to the cross-ratio of the four corresponding points; and the cross-ratio of four concurrent straight lines is equal to the cross-ratio of the four corresponding lines.

Of these properties (iv) is the only one we need prove.

Let $p_1 = 0$ and $p_2 = 0$ be the equations of two straight lines and $p_1' = 0$, $p_2' = 0$ the equations of the corresponding lines, then

$p_1' + \lambda p_2' = 0$ corresponds to $p_1 + \lambda p_2 = 0$ for all values of λ, and inasmuch as the cross-ratio of the four lines

$$p_1 + \lambda_r p_2 = 0, \quad r = 1, 2, 3, 4$$

is $-\dfrac{(\lambda_2 - \lambda_3)(\lambda_1 - \lambda_4)}{(\lambda_3 - \lambda_1)(\lambda_2 - \lambda_4)}$ the result follows at once.

Again, suppose that a figure in one plane is projected into another plane and that P' is the projection of P, then if x_1, x_2, x_3 be the coordinates of P referred to a fixed triangle in the first plane and X_1, X_2, X_3 be the coordinates of P' referred to a triangle in the second plane, it is very easy to see that x_1, x_2, x_3 are linear functions of X_1, X_2, X_3 and *vice versâ*. On the connection between projection and linear transformation see § 157.

Thus projection affords an example of linear transformation and in this case the four properties enunciated above are self-evident.

Hence if I be an invariant of a form f, then $I = 0$ expresses a property of the curve $f = 0$ that is unaltered by any homographic transformation and in particular by projection. A like remark applies to the connection between a covariant curve and the original curve—it is undisturbed by projection.

206. The connection between these two different points of view is interesting.

The linear transformation

$$x_r = \xi_r X_1 + \eta_r X_2 + \zeta_r X_3$$

leaves three points of the plane unaltered in general. In fact to find the points whose position is unchanged we have merely to put $X_r = \rho x_r$ in the equations above and then elimination of x_1, x_2, x_3 gives a cubic for $\dfrac{1}{\rho}$, viz.

$$\begin{vmatrix} \xi_1 - \dfrac{1}{\rho}, & \eta_1, & \zeta_1 \\ \xi_2, & \eta_2 - \dfrac{1}{\rho}, & \zeta_2 \\ \xi_3, & \eta_3, & \zeta_3 - \dfrac{1}{\rho} \end{vmatrix} = 0.$$

In general this equation has three unequal roots and to each root corresponds a single determination of the ratios $x_1 : x_2 : x_3$. If we take the triangle formed by these three points for triangle of reference the linear transformation takes the simple form

$$x_1 = k_1 X_1, \ x_2 = k_2 X_2, \ x_3 = k_3 X_3.$$

And hence in general by a suitable choice of the triangle of reference a linear transformation can be reduced to this form in which the coordinates are only multiplied by constants.

Ex. (i). Prove that by suitably choosing the coordinate system any four points may be reduced to the form

$$x_2^{(1)}=0, \quad x_3^{(1)}=0, \quad x_1^{(2)}=0, \quad x_3^{(2)}=0, \quad x_1^{(3)}=0, \quad x_2^{(3)}=0,$$
$$x_1^{(4)}=x_2^{(4)}=x_3^{(4)}.$$

And hence prove that a linear transformation can be found which changes any four points into four given points.

Ex. (ii). Given four points and their four corresponding points, shew that the point corresponding to any fifth point can be constructed by means of the ruler only. (Use the cross-ratio property.)

Ex. (iii). Hence or otherwise shew that any linear transformation is equivalent to a ruler construction.

Ex. (iv). If the linear transformation leave an isolated point and every point on a fixed line unaltered, shew that the equation of § 206 has a double root which is a root of every first minor. Give an equivalent geometrical construction in this case.

207. Starting with the definitions of invariants and covariants already given we might proceed to prove that every invariant and covariant can be represented symbolically in the form indicated in a previous theorem, and then go on to develop a theory of invariants and covariants as far as possible on the lines of binary forms. However it is essential to remember that the real and primary importance of such a theory lies in its application to geometry, and as in geometry line coordinates and tangential equations occur quite at the beginning of the analytical exposition, we are led here to introduce line coordinates as well as point coordinates from the first. Indeed it is not difficult to shew on purely analytical grounds that line coordinates are essential for the proper treatment of the algebraical questions that arise; such an explanation together with a comparison with the theory of binary forms will be given later.

The equation of a straight line being

$$u_1x_1 + u_2x_2 + u_3x_3 = 0,$$

u_1, u_2, u_3 are as usual called the coordinates of the line. Following Clebsch it is preferable to regard the equation just written as neither the point equation of the line nor the line equation of the point, but as the condition that the line (u_1, u_2, u_3) and the point (x_1, x_2, x_3) should bear a certain relation to one another.

The first fact to notice about (u_1, u_2, u_3) is that they are contragredient to x_1, x_2, x_3, for if u_1, u_2, u_3 become U_1, U_2, U_3 and x_1, x_2, x_3 become X_1, X_2, X_3, then

$$u_1x_1 + u_2x_2 + u_3x_3 \text{ must become } U_1X_1 + U_2X_2 + U_3X_3,$$

which is precisely the condition for contragredience (§ 39).

Further, the u's being contragredient to the x's are cogredient with the α's, and in fact by the linear transformation of § 201 we see at once that u_1 becomes u_ξ, u_2 becomes u_η and u_3 becomes u_ζ.

It is now evident that by a linear transformation the factor $(\alpha\beta u)$ merely becomes itself multiplied by the determinant of transformation, so that every form whose symbolical expression consists exclusively of factors of the types

$$(\alpha\beta\gamma),\ (\alpha\beta u),\ \alpha_x$$

possesses the property of invariance.

There are four classes of such invariant functions.

(i) Containing neither x's nor u's. These as already mentioned are called invariants.

(ii) Containing x's but not u's. These are covariants.

(iii) Containing u's but not x's. These are new introductions and are called contravariants.

(iv) Containing both u's and x's. These are called mixed concomitants.

Further there is the identical invariant form u_x which does not contain the coefficients of the original forms at all.

The degree to which the coefficients occur in a given invariant form is called simply its *degree*, the degree in the x's is called its *order* and the degree in the u's is called its *class*.

Examples of invariants and covariants have already been given.

As an example of a contravariant we may mention

$$\begin{vmatrix} a_{200} & a_{110} & a_{101} & u_1 \\ a_{110} & a_{020} & a_{011} & u_2 \\ a_{101} & a_{011} & a_{002} & u_3 \\ u_1 & u_2 & u_3 & 0 \end{vmatrix}$$

as a contravariant of the conic. In fact equated to zero it represents the line equation, and the reader may verify by direct multiplication that it is equivalent to

$$(\alpha\beta u)^2,$$

and therefore actually is an invariant form.

As an example of calculation similar to that given for the Hessian consider the locus of a point whose polar conic with respect to the cubic

$$f \equiv \alpha_x^3 \equiv \beta_x^3 = 0$$

touches a given straight line u.

The polar conic of the point y is

$$\alpha_x^2\alpha_y \equiv \beta_x^2\beta_y = 0,$$

and the condition that this should touch the given line is

$$(\alpha\beta u)^2 \alpha_y\beta_y = 0.$$

Thus the equation of the required locus is

$$(\alpha\beta u)^2 \alpha_x\beta_x = 0.$$

This is a mixed concomitant of the cubic. Equated to zero it represents the point equation of a curve when the u's are taken as constants, viz. the point equation of the locus mentioned. When the x's are taken to be constant it represents the line equation of a curve, viz. the line equation of the polar conic of the point x.

Similar remarks apply to mixed concomitants in general.

208. Principle of Duality. The fact that the condition that a straight line whose equation is $\alpha_x = 0$ should pass through a given point is symmetrical in the coordinates of the line and the point leads to an important remark. For if the proof of any theorem relating to a figure containing m straight lines and n

points is thrown into an analytical form, then by interchanging point and line coordinates throughout the investigation we obtain a correlative theorem for n straight lines and m points—if in the first figure one of the straight lines pass through one of the points, then in the second figure the corresponding point lies on the corresponding straight line. It must be clearly understood that such theorems as are here contemplated depend only on lines passing through points and points lying on lines, so that the process is exactly analogous to the reciprocation of descriptive properties. Naturally we can pass to point loci and line envelopes by the introduction of an infinite number of points and lines into the figures.

A fuller explanation of this principle of duality will be found in works on geometry, but the above will suffice as an indication of some general ideas which will underlie much of our subsequent work, especially on the theory of two conics.

Ex. (i). Shew that in every case the theorem dual to a given one may be obtained by reciprocating the given one (supposed descriptive) with respect to a conic.

Ex. (ii). If D, E, F be three points in the sides BC, CA, AB of a triangle, and AD, BE, CF be concurrent, then the lines EF, FD, DE meet BC, CA, AB respectively in three collinear points. What is the dual theorem?

209. Methods for transforming symbolical expressions. As in the case of binary forms there are two different methods by which a symbolical expression may be transformed:

(i) the interchange of equivalent symbols,

(ii) the use of identities among elementary symbolical expressions.

As an illustration of (i) we may notice that for the ternary quadratic $\alpha_x^2 = \beta_x^2 =$ etc. the mixed concomitant $(\alpha\beta u)\,\alpha_x\beta_x$ vanishes identically.

In fact, we have

$$(\alpha\beta u)\,\alpha_x\beta_x = (\beta\alpha u)\,\alpha_x\beta_x$$

since α, β are equivalent and

$$(\alpha\beta u)\,\alpha_x\beta_x = -(\beta\alpha u)\,\alpha_x\beta_x$$

by the properties of determinants, hence $(\alpha\beta u)\,\alpha_x\beta_x = 0$.

The fundamental identity for ternary forms is

$$(\beta\gamma\delta)\,\alpha_x - (\gamma\delta\alpha)\,\beta_x + (\delta\alpha\beta)\,\gamma_x - (\alpha\beta\gamma)\,\delta_x = 0,$$

deduced from

$$\begin{vmatrix} \alpha_1 & \beta_1 & \gamma_1 & \delta_1 \\ \alpha_2 & \beta_2 & \gamma_2 & \delta_2 \\ \alpha_3 & \beta_3 & \gamma_3 & \delta_3 \\ \alpha_x & \beta_x & \gamma_x & \delta_x \end{vmatrix} = 0.$$

On replacing δ by u

$$(\beta\gamma u)\,\alpha_x - (\gamma u\alpha)\,\beta_x + (u\alpha\beta)\,\gamma_x = (\alpha\beta\gamma)\,u_x$$

or

$$(\beta\gamma u)\,\alpha_x + (\gamma\alpha u)\,\beta_x + (\alpha\beta u)\,\gamma_x = (\alpha\beta\gamma)\,u_x,$$

an easily remembered result.

Again, if in the first identity on replacing x by $(\epsilon\zeta)$, *i.e.*

$$x_1,\ x_2,\ x_3 \text{ by } \epsilon_2\zeta_3 - \epsilon_3\zeta_2,\ \epsilon_3\zeta_1 - \epsilon_1\zeta_3,\ \epsilon_1\zeta_2 - \epsilon_2\zeta_1$$

respectively, we deduce

$$(\alpha\beta\gamma)(\delta\epsilon\zeta) - (\delta\alpha\beta)(\gamma\epsilon\zeta) + (\gamma\delta\alpha)(\beta\epsilon\zeta) - (\beta\gamma\delta)(\alpha\epsilon\zeta) = 0.$$

Another important identity is

$$\alpha_x\beta_y - \alpha_y\beta_x = (\alpha\beta u),$$

where u_1, u_2, u_3 are the coordinates of the line joining the points x and y so that

$$u_1 = x_2y_3 - x_3y_2,\quad u_2 = x_3y_1 - x_1y_3,\quad u_3 = x_1y_2 - x_2y_1.$$

In exactly the same way we have

$$v_\alpha w_\beta - v_\beta w_\alpha = (\alpha\beta x),$$

where x is the point of intersection of the lines v and w.

Ex. For the ternary quadratic $a_x^2 = \beta_x^2 =$ etc. shew that

$$(a\beta\gamma)(a\beta\delta)\,\gamma_x\delta_x = \tfrac{1}{3}(a\beta\gamma)^2 . \delta_x^2,$$

$$(a\beta u)(a\gamma u)\,\beta_x\gamma_x = \tfrac{1}{2}(a\beta u)^2 . \gamma_x^2 - \tfrac{1}{6}(a\beta\gamma)^2 . u_x^2.$$

210. We now come to the fundamental theorem in the present calculus, namely, that every invariant form can be represented symbolically by three types of factors, viz.

$$(\alpha\beta\gamma),\ (\alpha\beta u) \text{ and } \alpha_x$$

together with the identical invariant form u_x.

Some preliminary observations will enable us to simplify the proof.

In fact an invariant form containing the u's may be regarded as a pure covariant of the original form or forms together with the linear form

$$u_1x_1 + u_2x_2 + u_3x_3,$$

and hence if the theorem be proved for any number of ground forms for concomitants in which the u's are absent the general case follows at once by adjoining a linear form to the original system.

Let us then proceed to prove the theorem for invariants and covariants.

In the proof we need two lemmas regarding the operator Ω denoted by

$$\begin{vmatrix} \frac{\partial}{\partial \xi_1} & \frac{\partial}{\partial \xi_2} & \frac{\partial}{\partial \xi_3} \\ \frac{\partial}{\partial \eta_1} & \frac{\partial}{\partial \eta_2} & \frac{\partial}{\partial \eta_3} \\ \frac{\partial}{\partial \zeta_1} & \frac{\partial}{\partial \zeta_2} & \frac{\partial}{\partial \zeta_3} \end{vmatrix}$$

which is the natural extension to three variables of the operator so important in the binary theory.

211. I. *If D denote the determinant*

$$\begin{vmatrix} \xi_1 & \xi_2 & \xi_3 \\ \eta_1 & \eta_2 & \eta_3 \\ \zeta_1 & \zeta_2 & \zeta_3 \end{vmatrix}$$

then $\Omega^r D^n$ is a numerical multiple of D^{n-r}, where n, r are positive integers and $r \not> n$.

This result can be proved for $r = 1$ by straightforward differentiation and then the general result follows. We may shorten the proof by using properties of the minors of the determinant.

For a moment denote the minor of ξ_1 by X_1, that of η_1 by Y_1, of ζ_1 by Z_1, etc.

Then

$$\frac{\partial D^n}{\partial \eta_2} = nD^{n-1}Y_2$$

$$\frac{\partial^2 D^n}{\partial \eta_2 \partial \zeta_3} = n(n-1)D^{n-2}Y_2Z_3 + nD^{n-1}\xi_1$$

and

$$\frac{\partial^2 D^n}{\partial \eta_3 \partial \zeta_2} = n(n-1)D^{n-2}Y_3Z_2 - nD^{n-1}\xi_1.$$

Hence

$$\left(\frac{\partial^2}{\partial \eta_2 \partial \zeta_3} - \frac{\partial^2}{\partial \eta_3 \partial \zeta_2}\right) D^n = n(n-1)D^{n-2}(Y_2Z_3 - Y_3Z_2) + 2nD^{n-1}\xi_1$$

and

$$Y_2Z_3 - Y_3Z_2 = \xi_1 D$$

by the properties of minors, hence

$$\left(\frac{\partial^2}{\partial \eta_2 \partial \zeta_3} - \frac{\partial^2}{\partial \eta_3 \partial \zeta_2}\right) D^n = n(n+1)D^{n-1}\xi_1.$$

Consequently

$$\frac{\partial}{\partial \xi_1}\left(\frac{\partial^2}{\partial \eta_2 \partial \zeta_3} - \frac{\partial^2}{\partial \eta_3 \partial \zeta_2}\right) D^n = n(n+1)(n-1)D^{n-2}\xi_1 X_1 + n(n+1)D^{n-1},$$

and adding to this two like terms we have

$$\begin{aligned}\Omega D^n &= n(n+1)(n-1)D^{n-2}(\xi_1X_1 + \xi_2X_2 + \xi_3X_3) + 3n(n+1)D^{n-1}\\ &= n(n+1)(n+2)D^{n-1},\end{aligned}$$

since

$$D = \xi_1X_1 + \xi_2X_2 + \xi_3X_3.$$

Operating with Ω again we have

$$\Omega^2 D^2 = (n-1)n^2(n+1)^2(n+2)D^{n-2},$$

and proceeding in this way we see that $\Omega^r D^n$ is a numerical multiple of D^{n-r} and, in particular, $\Omega^n D^n$ is a numerical constant.

The reader will find it instructive to extend this property of the operator Ω to four or more variables.

212. II. *If S be a product of m factors of the type α_ξ, n factors of the type β_η, and p factors of the type γ_ζ, then $\Omega^r S$ is the sum of a number of terms each containing r factors of the type $(\alpha\beta\gamma)$, $m-r$ factors of the type α_ξ, $n-r$ of the type β_η and $p-r$ of the type γ_ζ.*

The proof of this result is exactly similar to that of the corresponding theorem in binary forms. In fact let $S = PQR$, where

$$P = \alpha_\xi^{(1)} \alpha_\xi^{(2)} \dots \alpha_\xi^{(m)},$$

$$Q = \beta_\eta^{(1)} \beta_\eta^{(2)} \dots \beta_\eta^{(n)},$$

$$R = \gamma_\zeta^{(1)} \gamma_\zeta^{(2)} \dots \gamma_\zeta^{(p)},$$

then
$$\frac{\partial^3 S}{\partial \xi_1 \partial \eta_2 \partial \zeta_3} = \sum_{\mu, \nu, \pi} \alpha_1^{(\mu)} \beta_2^{(\nu)} \gamma_3^{(\pi)} \frac{P}{\alpha_\xi^{(\mu)}} \frac{Q}{\beta_\eta^{(\nu)}} \frac{R}{\gamma_\zeta^{(\pi)}},$$

the sum being taken for

$$\mu = 1, 2, \dots m;$$

$$\nu = 1, 2, \dots n;$$

$$\pi = 1, 2, \dots p.$$

Writing down all six such terms and adding we have

$$\Omega . S = \sum_{\mu, \nu, \pi} (\alpha^{(\mu)} \beta^{(\nu)} \gamma^{(\pi)}) \frac{P}{\alpha_\xi^{(\mu)}} \frac{Q}{\beta_\eta^{(\nu)}} \frac{R}{\gamma_\zeta^{(\pi)}},$$

which proves the theorem for $r = 1$, since $\dfrac{P}{\alpha_\xi^{(\mu)}}$ contains $m - 1$ factors, and so on.

Now Ω has no effect on a factor $(\alpha\beta\gamma)$, hence applying the same result to each term in $\Omega . S$ we obtain the result for $r = 2$ and so on for any value of r. In particular if $r = m = n = p$ the result is expressed exclusively in terms of factors of the type $(\alpha\beta\gamma)$.

213. Our proof that any invariant or covariant can be represented by factors of the two types

$$(\alpha\beta\gamma), \quad \alpha_x$$

follows exactly the lines of the second proof of the corresponding theorem for binary forms. As the proof is exactly the same for any number of forms, we give it explicitly for one only.

First suppose that $I(a)$ is an *invariant* of the form

$$\Sigma a_{pqr} \frac{n!}{p!q!r!} x_1^p x_2^q x_3^r,$$

then if $\Sigma A_{pqr} \dfrac{n!}{p!q!r!} X_1^p X_2^q X_3^r$ be the transformed expression, we have

$$I(A) = I(a) . (\xi\eta\zeta)^w \text{ identically.}$$

Now express the left-hand side of this equation in symbols. We have

$$a_{pqr} = \alpha_1^p \alpha_2^q \alpha_3^r, \text{ and hence } A_{pqr} = \alpha_\xi^p \alpha_\eta^q \alpha_\zeta^r.$$

Thus on the left we have the sum of a number of terms. Each contains w factors of the type α_ξ, w of the type α_η, and w of the type α_ζ.

Hence operating on both sides of the equation with Ω^w, the left-hand side becomes an aggregate of terms each made up entirely of factors of the type $(\alpha\beta\gamma)$ and the right-hand side is a numerical multiple of $I(a)$, so that we have expressed $I(a)$ in the required manner.

The proof for a covariant is similar in form to the above, but as in the case of binary forms, a little care is necessary. Of course, here as always, we confine ourselves to covariants that are homogeneous in the variables x_1, x_2, x_3.

Suppose that $C(a, x)$ is a covariant of order m, then by definition

$$C(A, X) = (\xi\eta\zeta)^w \times C(a, x).$$

Now on solving the equations of transformation

$$x_1 = \xi_1 X_1 + \eta_1 X_2 + \zeta_1 X_3, \text{ etc.}$$

we have

$$X_1 = \frac{x_1(\eta_2\zeta_3 - \eta_3\zeta_2) + x_2(\eta_3\zeta_1 - \eta_1\zeta_3) + x_3(\eta_1\zeta_2 - \eta_2\zeta_1)}{(\xi\eta\zeta)}, \text{ etc.}$$

On replacing x_1 by $(v_2w_3 - v_3w_2)$, x_2 by $(v_3w_1 - v_1w_3)$ and x_3 by $(v_1w_2 - v_2w_1)$—a set of equations which may be written $x = (vw)$—we have

$$X_1 = \frac{v_\eta w_\zeta - v_\zeta w_\eta}{(\xi\eta\zeta)},$$

$$X_2 = \frac{v_\zeta w_\xi - v_\xi w_\zeta}{(\xi\eta\zeta)},$$

$$X_3 = \frac{v_\xi w_\eta - v_\eta w_\xi}{(\xi\eta\zeta)}.$$

Again, $A_{pqr} = \alpha_\xi^p \alpha_\eta^q \alpha_\zeta^r$, and hence on substituting for the A's and x's their values as given above in $C(A, X)$, and multiplying across by $(\xi\eta\zeta)^m$, the left-hand side becomes an aggregate of products of factors having the suffixes ξ, η, ζ, and the right-hand side becomes $(\xi\eta\zeta)^{w+m} \times C(a, x)$. Moreover each term on the left-hand side must contain $(w+m)$ factors with each suffix.

Now if we operate on the left-hand side once with Ω, we obtain an aggregate of terms each containing one determinantal factor and $(w+m-1)$ factors with each suffix.

The determinantal factors are of three types:

$$\text{(i)}\quad (\alpha\beta\gamma),$$
$$\text{(ii)}\quad (\alpha v w) = \alpha_x,$$
$$\text{(iii)}\quad (\alpha\beta v),$$

where the first two are of the form we require in the result, but the third is not.

Now a term involving a factor $(\alpha\beta v)$ must have arisen from operating with Ω on a term containing a factor such as $\alpha_\xi\beta_\eta v_\zeta$ and this term must therefore contain a further factor w_η (or w_ξ). Accordingly let the term be in the original expression

$$G\alpha_\xi\beta_\eta v_\zeta w_\eta,$$

then from the mode in which v and w occur in the expression on the left-hand side there must also be present a term

$$-G\alpha_\xi\beta_\eta v_\eta w_\zeta,$$

and on operating with Ω this gives

$$-G(\alpha\beta w)\,v_\eta,$$

whereas the former term is

$$+G(\alpha\beta v)\,w_\eta.$$

But by the fundamental identity

$$(\alpha\beta v)\,w_\eta + (\beta w v)\,\alpha_\eta + (w\alpha v)\,\beta_\eta = (w\alpha\beta)\,v_\eta,$$

or $$(\alpha\beta v)\,w_\eta - (\alpha\beta w)\,v_\eta = (\beta v w)\,\alpha_\eta - (\alpha v w)\,\beta_\eta = \beta_x\alpha_\eta - \alpha_x\beta_\eta,$$

and the sum of the two terms mentioned is

$$G(\beta_x\alpha_\eta - \alpha_x\beta_\eta).$$

Thus although factors like $(\alpha\beta v)$ do appear explicitly after operating with Ω, the terms containing them may be paired in such a way that the aggregate may be expressed entirely in terms of factors of the types $(\alpha\beta\gamma)$ and α_x together with factors having the suffixes ξ, η, or ζ.

Hence if we operate with Ω, $w+m$ times on the left and transform at each step as indicated above, we finally have an aggregate of terms each containing $(w+m)$ factors of the types $(\alpha\beta\gamma)$ and α_x only; also after performing the same operations on the right-hand side, we are left with a numerical multiple of $C(a, x)$. We have therefore expressed the covariant as an aggregate of terms each of which is the product of a number of factors of the types $(\alpha\beta\gamma)$ and α_x, and since the order must be m, it follows that in each term there are w factors of the type $(\alpha\beta\gamma)$ and m of the type α_x.

214. Leading Coefficients of Concomitants. In § 33 it was shewn that a covariant of a binary form could be deduced from its leading coefficient; we shall in what follows consider the extension of this idea to ternary forms, but as the results are easily obtained and not necessary for present purposes our remarks will be somewhat brief; the reader who desires further information is referred to a memoir by Forsyth, *Am. Journal of Mathematics*, 1889.

If a mixed concomitant be of class m' and order n', the coefficient of $x_1^{n'}u_1^{m'}$ is called the leading coefficient.

Given the concomitant, the leading coefficient is unique of course, but the reverse is not true, because if we multiply a concomitant by any power of u_x we obtain another concomitant with the same leading coefficient.

However, save as to a power of u_x, a concomitant can be found when its leading coefficient is given, as we proceed to shew.

The leading coefficient is an aggregate of products of factors of the types

$$(a\beta\gamma), \quad (a_2\beta_3 - a_3\beta_2), \quad a_1;$$

and on replacing

$$a_1 \text{ by } a_x,$$
$$a_2 \text{ by } a_y,$$
$$a_3 \text{ by } a_z,$$

and so on, the above factors become

$$(a\beta\gamma)\,u_x, \quad (a\beta u), \quad a_x$$

respectively, where $u=(yz)$. Cf. § 209.

The symbolical substitution is equivalent to replacing the coefficient a_{pqr} by

$$\frac{p\,!}{n\,!}\left(y\frac{\partial}{\partial x}\right)^q\left(z\frac{\partial}{\partial x}\right)^r a_x^n,$$

so that if in a leading coefficient we replace a_{pqr} by the value just given we obtain the corresponding concomitant multiplied by a power of u_x—the exponent of u_x is in fact equal to the number of factors of the type $(a\beta\gamma)$ that occur in the symbolical expression of the concomitant.

Hence, as in binary forms, it follows that "If there be an identical relation between a number of leading coefficients, the same relation exists among the corresponding concomitants."

We leave the reader to establish this on the lines of § **33**; the concomitants corresponding to the various leading coefficients must be chosen so as to make the whole expression of uniform order and class.

On multiplication by a sufficiently high power of u_x a concomitant can be completely expressed by means of factors of the types $(\alpha\beta u)$ and a_x.

The leading coefficient can therefore be expressed in terms of factors of the types

$$(\alpha_2\beta_3-\alpha_3\beta_2) \text{ and } \alpha_1.$$

It follows immediately that the leading coefficient is an invariant of the binary forms

$$(a_2x_2+a_3x_3)^n, \quad (a_2x_2+a_3x_3)^{n-1}a_1, \quad \ldots\ldots \quad (a_2x_2+a_3x_3)\,a_1^{n-1},$$

which are n in number.

These binary forms are the coefficients of the various powers of x_1 in the original ternary form, hence they can be obtained at once from f.

This is the fundamental result of Forsyth on algebraically complete sets of ternariants and from it readily flow all the results in the memoir cited. Similar methods may be applied to obtain the results of another paper of Forsyth's*.

215. We shall now explain an important principle due to Clebsch which establishes a connection between the invariants of binary forms and the contravariants of ternary forms. To facilitate the discussion we shall commence with a special case.

Suppose we require to find the condition that the line $u_x=0$ should touch the conic $a_x^2=b_x^2=0$. Let the points on the line be $y_1y_2y_3$ and $z_1z_2z_3$, so that

$$u_1 : u_2 : u_3 = y_2z_3-y_3z_2 : y_3z_1-y_1z_3 : y_1z_2-y_2z_1.$$

The coordinates x_1, x_2, x_3 of any point on this line are of the form

$$\xi_1y_1+\xi_2z_1, \quad \xi_1y_2+\xi_2z_2, \quad \xi_1y_3+\xi_2z_3,$$

and ξ_1, ξ_2 may clearly be regarded as the coordinates of a variable point on the line referred to the points y and z as base points.

We have

$$a_x=a_1x_1+a_2x_2+a_3x_3=\xi_1a_y+\xi_2a_z,$$

* See *London Math. Soc. Proc.*, 1898.

and the points in which the line meets the conic are given by

$$(a_y \xi_1 + a_z \xi_2)^2 = (b_y \xi_1 + b_z \xi_2)^2 = \ldots = 0.$$

As the line touches the conic this expression regarded as a quadratic $\alpha_\xi{}^2 = \beta_\xi{}^2 = \ldots$ in ξ must have its invariant $(\alpha\beta)^2$ zero.

Hence $$(a_y b_z - a_z b_y)^2 = 0,$$

and by an identity already given

$$(a_y b_z - a_z b_y) = (abu);$$

hence we have $$(abu)^2 = 0$$

as the required condition.

Ex. If the line $u_x = 0$ cuts the conics $a_x{}^2 = 0$, $b_x{}^2 = 0$ in harmonic points, prove that $(abu)^2 = 0$.

Thus when the points on the line satisfy the invariant relation $(\alpha\beta)^2 = 0$, the line satisfies the relation $(abu)^2 = 0$.

This principle is true in general, that is to say, in order to pass from the invariant relation satisfied by the points in which the line meets any number of curves to the condition satisfied by the coordinates of the line, we have merely to change every factor of the type $(\alpha\beta)$, in the expression of the binary invariant, into a factor (abu).

The proof is very simple—suppose the equations of the curves are

$$a_x{}^m = 0, \quad b_x{}^n = 0, \text{ etc.},$$

and that y, z are two points on the line, then the points in which the line meets the curves are given by the binary forms

$$(a_y \xi_1 + a_z \xi_2)^m = 0, \quad (b_y \xi_1 + b_z \xi_2)^n = 0, \text{ etc.}$$

But an invariant of these forms is expressible in terms of factors of the type

$$(a_y b_z - a_z b_y)$$

entirely, and inasmuch as

$$a_y b_z - a_z b_y = (abu)$$

the result follows at once.

Thus, knowing the discriminant of a binary form of order n, we can find the tangential equation of a curve of the nth degree, for the discriminant being equated to zero gives the condition that two points of ntersection should coincide, and in this case the line touches the curve; *e.g.* the discriminant of a binary cubic being

$$(ab)^2 (ac) (bd) (cd)^2,$$

the tangential equation of the curve

$$a_x^3 = b_x^3 = c_x^3 = d_x^3 = 0$$

is

$$(abu)^2 (acu) (bdu) (cdu)^2 = 0.$$

216. The principle can be extended to covariants and mixed concomitants, as we shall explain by an example.

A line $u_x = 0$ cuts two conics $a_x^2 = 0$, $b_x^2 = 0$ in pairs of points PQ, RS; to find the pair of points harmonic to both P, Q and R, S.

We know that the pair of points harmonic to those given by

$$\alpha_\xi^2 = 0, \quad \beta_\xi^2 = 0,$$

are given by equating the Jacobian $(\alpha\beta)\,\alpha_\xi\beta_\xi$ to zero.

Now if u be the join of the points y and z, then ξ_1, ξ_2 are given by

$$(a_y\xi_1 + a_z\xi_2)^2 = 0, \quad (b_y\xi_1 + b_z\xi_2)^2 = 0,$$

hence the common harmonic pair are given by

$$(a_y b_z - a_z b_y)(a_y\xi_1 + a_z\xi_2)(b_y\xi_1 + b_z\xi_2) = 0,$$

or they satisfy

$$(abu)\, a_x b_x = 0,$$

and are therefore the points in which this conic meets the given line.

The extension to covariants in general will now be sufficiently obvious.

217. There are also dual methods which enable us to determine, for example, the locus of a point such that the pencils of tangents drawn from it to a number of fixed curves possess a given projective property.

After the study of a simple example it will be easy for the reader to enunciate and prove the necessary theorems.

Let us find the locus of a point such that the pairs of tangents drawn from it to two conics

$$u_\alpha^2 = 0, \quad u_\beta^2 = 0,$$

are harmonically conjugate.

If any two lines through the point are $v_x = 0$, $w_x = 0$, then the coordinates of a variable line through it are

$$\xi_1 v_1 + \xi_2 w_1, \quad \xi_1 v_2 + \xi_2 w_2, \quad \xi_1 v_3 + \xi_2 w_3,$$

where ξ_1, ξ_2 may be regarded as the coordinates of a variable line in the pencil.

The tangents to the two conics are given by the binary quadratics

$$(u_1 \alpha_1 + u_2 \alpha_2 + u_3 \alpha_3)^2, \text{ or } (v_\alpha \xi_1 + w_\alpha \xi_2)^2 = 0,$$

and
$$(v_\beta \xi_1 + w_\beta \xi_2)^2 = 0.$$

And since these are harmonic

$$(v_\alpha w_\beta - v_\beta w_\alpha)^2 = 0,$$

but
$$v_\alpha w_\beta - v_\beta w_\alpha = (\alpha\beta x),$$

where x is the point of intersection of the lines v and w.

Hence the required locus is

$$(\alpha\beta x)^2 = 0.$$

If α and β are equivalent symbols, this gives the point equation of the conic whose tangential equation is

$$u_\alpha^2 = u_\beta^2 = 0.$$

Ex. (i). Prove that if three binary quadratics a_x^2, b_x^2, c_x^2 are in involution, then $(ab)(bc)(ca) = 0$; deduce the locus of a point such that the tangents from it to three conics form a pencil in involution, and state the correlative theorem.

Ex. (ii). Interpret the envelope loci $(abu)^4 = 0$, $(abu)^2(bcu)^2(cau)^2 = 0$ in connection with the quartic curve $a_x^4 = b_x^4 = c_x^4 = 0$; shew that the inflexional tangents touch each of the above curves, and are therefore in general completely determined as the common tangents.

Ex. (iii). A straight line p meets a cubic curve in P, Q, R, shew that

(i) there are two points H, H' on the line such that their polar conics touch the line;

(ii) the polar conic of H touches the line in H' and that of H' touches the line in H;

(iii) H and H' are the Hessian points of P, Q, R;

(iv) if p pass through a fixed point y the locus of H and H' is the quartic obtained by putting $u = (xy)$ in $(abu)^2 a_x b_x = 0$;

(v) the equation of the quartic is $fP - C^2 = 0$, where $f = 0$ is the cubic, C the first and P the second polar of y.

218. Quaternary Forms. The general quaternary form of order n is

$$\Sigma \frac{n!}{p!\,q!\,r!\,s!} x_1^p x_2^q x_3^r x_4^s,$$

where $$p + q + r + s = n.$$

In accordance with the methods used in this work, we write this in the umbral form

$$(\alpha_1 x_1 + \alpha_2 x_2 + \alpha_3 x_3 + \alpha_4 x_4)^n,$$

and this in turn we denote by α_x^n.

To represent expressions of degree higher than unity in the coefficients, we have to introduce equivalent symbols β, γ, etc., so that

$$f = \alpha_x^n = \beta_x^n = \gamma_x^n = \text{etc.}$$

Denoting the determinant

$$\begin{vmatrix} \alpha_1 & \alpha_2 & \alpha_3 & \alpha_4 \\ \beta_1 & \beta_2 & \beta_3 & \beta_4 \\ \gamma_1 & \gamma_2 & \gamma_3 & \gamma_4 \\ \delta_1 & \delta_2 & \delta_3 & \delta_4 \end{vmatrix}$$

by $(\alpha\beta\gamma\delta)$ and extending the definitions of invariants and covariants, we can shew that every expression completely represented by factors of the type $(\alpha\beta\gamma\delta)$ is an invariant, and every expression represented by factors of the types $(\alpha\beta\gamma\delta)$ and α_x is a covariant; of course in a symbolical expression each symbol must occur to degree n.

Then using plane coordinates u, such that

$$u_1 x_1 + u_2 x_2 + u_3 x_3 + u_4 x_4 = 0$$

is the equation of the plane u, we have mixed concomitants containing the three types of factors

$$\alpha_x, \ (\alpha\beta\gamma\delta), \ (\alpha\beta\gamma u).$$

Finally, introducing a second plane v, we have a more comprehensive type of mixed concomitant expressed by factors of the four types

$$\alpha_x, \ (\alpha\beta\gamma v), \ (\alpha\beta\gamma u), \ (\alpha\beta uv).$$

There is no need to introduce a third set of plane coordinates w, because

$$(\alpha uvw)$$

is really of the form α_x, where x is the point of intersection of the planes u, v, w.

The reader will not have more than purely algebraical difficulties in extending the methods of §§ 211, 212 to four variables.

219. In illustration of the foregoing we may mention that if

$$f = \alpha_x^2 = \beta_x^2 = \gamma_x^2 = \delta_x^2$$

be a quaternary quadratic, then

(i) $(\alpha\beta\gamma\delta)^2$ is its discriminant;

(ii) $(\alpha\beta\gamma u)^2 = 0$ is the condition that the plane u should touch the quadric $f = 0$;

(iii) $(\alpha\beta uv)^2 = 0$ is the condition that the line of intersection of the planes u and v should touch the quadric;

(iv) $(\alpha uvw)^2 = 0$ is the condition that the point of intersection of the three planes u, v, and w should be on the quadric.

220. It is clear that when any invariant form expresses a relation of a projective nature between a straight line in space and a surface, factors of the type

$$(\alpha\beta uv)$$

must occur in its symbolical form, because the straight line is given in the first place as the intersection of two planes, which are u and v in this case.

It is easy to see that

$$(\alpha\beta uv) = \Sigma (\alpha_1\beta_2 - \alpha_2\beta_1)(u_3v_4 - u_4v_3);$$

the six quantities

$$u_2v_3 - u_3v_2,\ u_3v_1 - u_1v_3,\ u_1v_2 - u_2v_1,\ u_1v_4 - u_4v_1,\ u_2v_4 - u_4v_2,\ u_3v_4 - u_4v_3$$

are called the six coordinates of the line of intersection of the planes u and v. It can be verified that they are altered in the same ratio, when instead of the planes u, v we take any other two planes through the line.

Further, if x, y be any two points on the line, the quantities

$$x_1y_4 - x_4y_1,\ x_2y_4 - x_4y_2,\ x_3y_4 - x_4y_3,\ x_2y_3 - x_3y_2,\ x_3y_1 - x_1y_3,\ x_1y_2 - x_2y_1$$

are altered all in the same ratio, when instead of x, y we use two other points on the same line.

Finally, from the equations

$$u_x = 0, \quad u_y = 0, \quad v_x = 0, \quad v_y = 0,$$

it follows that

$$x_1 y_4 - x_4 y_1 = \rho (u_2 v_3 - u_3 v_2) = p_{14},$$
$$x_2 y_4 - x_4 y_2 = \rho (u_3 v_1 - u_1 v_3) = p_{24},$$
$$x_3 y_4 - x_4 y_3 = \rho (u_1 v_2 - u_2 v_1) = p_{34},$$
$$x_2 y_3 - x_3 y_2 = \rho (u_1 v_4 - u_4 v_1) = p_{23},$$
$$x_3 y_1 - x_1 y_3 = \rho (u_2 v_4 - u_4 v_2) = p_{31},$$
$$x_1 y_2 - x_2 y_1 = \rho (u_3 v_4 - u_4 v_3) = p_{12},$$

in other words, that the six coordinates of a line are either the six determinants of the array

$$\begin{vmatrix} x_1 & x_2 & x_3 & x_4 \\ y_1 & y_2 & y_3 & y_4 \end{vmatrix}$$

or those of the array

$$\begin{vmatrix} u_1 & u_2 & u_3 & u_4 \\ v_1 & v_2 & v_3 & v_4 \end{vmatrix}.$$

The expression $(\alpha\beta uv)$ is written $(\alpha\beta p)$ for convenience, but it must not be confused with the similar expression for a three-rowed determinant.

The discussion of concomitants involving line coordinates is complicated by the existence of the relation

$$p_{23} p_{14} + p_{31} p_{24} + p_{12} p_{34} = 0$$

between the six coordinates of a line.

221. As a simple illustration of the use of the above methods, we shall take the extension of the principle of Clebsch explained in § 215.

To explain this extension we take a particular problem, viz. to find the condition that the plane $u_x = 0$ should touch the quadric surface represented by

$$\alpha_x^2 = \beta_x^2 = \gamma_x^2 = 0.$$

Let x, y, z be three points in the plane, then any other point (X_1, X_2, X_3, X_4) in the plane is given by

$$X_r = \xi_1 x_r + \xi_2 y_r + \xi_3 z_r, \quad r = 1, 2, 3, 4,$$

and ξ_1, ξ_2, ξ_3 may be regarded as the coordinates of (X_1, X_2, X_3, X_4) in the plane referred to x, y, z as fundamental triangle.

Now
$$\sum_1^4 \alpha_r X_r = \alpha_x \xi_1 + \alpha_y \xi_2 + \alpha_z \xi_3,$$

hence the equation of the conic in which the plane meets the quadric may be written

$$(\alpha_x \xi_1 + \alpha_y \xi_2 + \alpha_z \xi_3)^2 = 0.$$

If the plane touch the quadric, this conic must be a pair of straight lines; the condition for this is

$$\begin{vmatrix} \alpha_x & \alpha_y & \alpha_z \\ \beta_x & \beta_y & \beta_z \\ \gamma_x & \gamma_y & \gamma_z \end{vmatrix}^2 = 0,$$

and inasmuch as the determinant here written is equal to

$$(\alpha\beta\gamma u),$$

the condition required is

$$(\alpha\beta\gamma u)^2 = 0.$$

The reader will have no difficulty in applying the same method to find the envelope of a plane cutting a surface in a curve which has a definite projective property, of which the invariant equivalent is given.

222. We can in like manner solve problems leading to line coordinates, *e.g.* to find the condition that a line should cut the quadrics

$$\alpha_x^2 = 0, \quad \beta_x^2 = 0,$$

in pairs of points harmonically conjugate.

Let x, y be two points on the line, then any other point X is given by

$$X_r = \xi_1 x_r + \xi_2 y_r, \quad r = 1, 2, 3, 4,$$

and
$$\alpha_X = \alpha_x \xi_1 + \alpha_y \xi_2,$$

so that the quadratics giving the ratio in which the line is divided by the quadrics are

$$(\alpha_x \xi_1 + \alpha_y \xi_2)^2 = 0, \quad (\beta_x \xi_1 + \beta_y \xi_2)^2 = 0.$$

The condition that the pairs should be harmonically conjugate is

$$(\alpha_x \beta_y - \alpha_y \beta_x)^2 = 0,$$

or

$$(\alpha\beta p)^2 = 0,$$

where p is a typical coordinate of the line.

The construction of other examples and the discovery of the dual principles will be easy for the reader who has grasped the corresponding results relating to ternary forms.

CHAPTER XIII.

TERNARY FORMS (*continued*).

223. In this chapter we shall give some further theorems relating to ternary forms; mainly such as arise in obtaining the irreducible systems in the simpler cases. In connection with Hilbert's proof of Gordan's theorem it has already been remarked that this proof can be extended without much difficulty to the case of ternary forms; we are therefore certain that for any number of such forms the irreducible system is finite, *i.e.* there exists a finite number of invariant forms in terms of which all others can be expressed as rational integral algebraic functions. But although the existence of the finite system is thus established, no clue is given as to the method of discovering such systems and, as a matter of fact, very little is known in this branch of the subject. The more important complete systems known up to the present are those for one, two* and three† ternary quadratics, that for a single ternary cubic‡ form, and quite recently Gordan has given the complete system for two quaternary quadratics§.

224. In his paper on the ternary cubic Gordan gave a systematic method of searching for all irreducible forms by proceeding from those of degree $r-1$ to those of degree r in the manner of Chapter V.; in any but the simplest cases the application of this method requires uncommon skill and patience. The reader will find an introductory sketch of the method at the end of this chapter; we shall content ourselves with obtaining the complete systems for one and two quadratics by an elegant and ingenious

* Clebsch, *Geometrie*, p. 174.

† Ciamberlini, *Battaglini*, Vol. XXIV.

‡ Gordan, *Math. Ann.*, Band I.

§ *Math. Ann.*, Band LVI.

process, also due to Gordan, which has been applied with success to three, and would probably be equally successful in dealing with any number of, quadratics. The complete systems are in all cases, except for linear forms or one quadratic form, so large that any method for their discovery will involve a great deal of labour.

Ex. (i). For a single linear form a_x the complete system is a_x.

Ex. (ii). For two linear forms a_x and b_x the complete system is

$$a_x, \quad b_x, \quad (abu).$$

Ex. (iii). For any number of linear forms $a_x, b_x, c_x, \ldots$ the complete system consists of

(α) Forms of the type a_x,

(β) Contravariants of the type (abu),

(γ) Invariants of the type (abc).

These results all follow immediately from the symbolical expressions for invariant forms. Cf. § 84.

225. One Quadratic Form. Let the form be

$$f = a_x^2 = b_x^2 = c_x^2 = \text{etc.},$$

then the typical invariant factors are

$$(abc), \quad (abu), \quad a_x,$$

and there are three invariant forms

$$(abc)^2, \quad (abu)^2, \quad a_x^2;$$

we proceed to shew that these constitute the complete irreducible system.

I. *Any invariant form P containing the factor (abc) can be transformed so as to contain the factor $(abc)^2$.*

First suppose that no two of the letters a, b, c occur again in the same factor, then

$$P = (abc)\, a_p b_q c_r M,$$

where M does not contain a, b or c, and

$$p = (de), \ (du) \text{ or } x \text{ etc.}$$

Hence by interchanging a, b, c in all possible ways

$$\begin{aligned} 6P &= (abc)\, M\, (a_p b_q c_r - a_p b_r c_q - a_q b_p c_r + a_q b_r c_p - a_r b_q c_p + a_r b_p c_q) \\ &= (abc)\, M\, (abc)\, (pqr), \\ &= (abc)^2\, M\, (pqr). \end{aligned}$$

Thus P involves the factor $(abc)^2$ which is an invariant, hence the other factor must be an invariant form* and P is therefore reducible.

Secondly, suppose that

$$P = (abc)(abp)\, c_q N,$$

where

$$p = d \text{ or } u,$$

$$q = (de), \quad (du) \text{ or } x.$$

Then

$$\begin{aligned} 3P &= (abc)\, N \{(abp)\, c_q + (bcp)\, a_q + (cap)\, b_q\} \\ &= (abc)\, N\, (abc)\, p_q \\ &= (abc)^2\, N p_q, \end{aligned}$$

and as before it follows that P is reducible.

Thus any form containing the factor (abc) is reducible, and we may neglect such forms in future.

II. *A form Q containing the factor (abu) can be expressed in terms of forms containing the factor $(abu)^2$ and reducible forms.*

Rejecting forms containing factors of the type (cde), we have the three following possibilities:

(i) $Q = (abu)(acu)(bdu)\, M$,

(ii) $Q = (abu)(acu)\, b_x M$,

(iii) $Q = (abu)\, a_x b_x M$.

As regards (i), on interchanging a and b,

$$\begin{aligned} 2Q &= (abu)\, M \{(acu)(bdu) + (cbu)(adu)\} \\ &= (abu)\, M \{-(bau)(cdu)\} \\ &= (abu)^2 (cdu)\, M, \end{aligned}$$

and Q is reducible.

(ii) Here

$$\begin{aligned} 2Q &= (abu)\, M \{(acu)\, b_x - (bcu)\, a_x\} \\ &= (abu)\, M \{(abu)\, c_x - (abc)\, u_x\} \\ &= (abu)^2 M c_x - (abc)\, u_x (abu)\, M, \end{aligned}$$

and the latter part is reducible because it contains the factor (abc). Thus again Q is reducible.

* We shall return to this point later, see § 226.

(iii) In this case Q vanishes, as we see by interchanging a and b.

Now every symbolical expression representing an invariant form contains a factor of one of the types (abc) or (abu) unless it be a_x^2; it follows that all such are reducible except

$$(abc)^2, \quad (abu)^2, \quad a_x^2,$$

and these form the complete irreducible system.

Of course, to be strictly accurate we should add to this and every complete system the identical concomitant u_x.

Ex. (i). Express $(abu)(acu)\,b_x c_x$ in terms of the irreducible system.

Ex. (ii). Prove the symbolical identity

$$\{ab\,(pq)\} = a_p b_q - a_q b_p.$$

226. The foregoing discussion leads to some general remarks that will be useful in the sequel.

The artifice of replacing (abc) by a_p, or, what is the same thing, writing

$$(bc) = p,$$

and more explicitly

$$(b_2c_3 - b_3c_2) = p_1; \quad (b_3c_1 - b_1c_3) = p_2; \quad (b_1c_2 - b_2c_1) = p_3,$$

will be frequently used. (It is clear that if b and c be two straight lines then p is their point of intersection.)

In particular the contravariant $(abu)^2$ will be written u_α^2, thus

$$\alpha = (ab),$$

and the invariant of the quadratic is c_α^2.

The α's may be regarded as expressible in terms of the original symbols, but we shall more often regard them as independent symbols; the contravariant above will thus be

$$u_\alpha^2 = u_\beta^2 = u_\gamma^2 = \dots,$$

the original form is

$$f = a_x^2 = b_x^2 = c_x^2 = \dots,$$

and the invariant is

$$a_\alpha^2 = a_\beta^2 = b_\alpha^2 = \text{etc.}$$

Of course in replacing an α by symbols belonging to f, we must use no symbol already occurring in the expression because then that symbol would occur more than twice in the transformed expression.

The α's being formed from the a's and b's, as line coordinates are formed from point coordinates, it follows that the α's are contragredient to a and u but cogredient with x.

We shall now return to a point that arose in the discussion of a single quadratic, where we deduced from first principles that if a_p, a_q, a_r be all invariant factors then (pqr) is an invariant factor —it will be shewn here how to express it in terms of factors of the types (abc), (abu) and a_x.

The possible forms for p, q, r are (de), (du) and x; if each of them were x, (pqr) would vanish, so we may suppose that p for example is not x.

Let $p=(dv)$, where v is either e or u, then

$$(pqr)=\{(dv)\,qr\}=d_q v_r - d_r v_q, \qquad (\S\ 209)$$

but since b_q is an invariant factor, d_q and u_q are both invariant factors; for if in such a factor we replace a letter of the type a by another of the same type the factor is still invariant, while if we replace it by u the resulting factor either vanishes or is an invariant factor.

Hence (pqr) can be expressed entirely in terms of factors of the types (abc), (abu) and a_x; it is therefore an invariant factor.

227. To make the matter clearer we shall illustrate it by some examples.

Suppose a quadratic is

$$a_x^2 = b_x^2 = c_x^2 = \ldots,$$

and its contravariant is

$$u_\alpha^2 = u_\beta^2 = u_\gamma^2 = \ldots,$$

then a_α, a_β, b_α, b_β etc. are invariant factors.

Hence $(\alpha\beta x)$ must be an invariant factor and $(\alpha\beta x)^2$ an invariant form of the quadratic. We proceed to express it in terms of the members of the irreducible system as follows:

$$\begin{aligned}(\alpha\beta x)^2 &= (\overline{ab}\ \beta x)^2 = (a_\beta b_x - a_x b_\beta)^2\\ &= a_\beta^2 \,.\, b_x^2 + a_x^2 \,.\, b_\beta^2 - 2a_x b_x a_\beta b_\beta,\end{aligned}$$

and we have now to deal with the term $a_x b_x a_\beta b_\beta$.

As a guide to the further reduction we remark that since a_β is a factor the form must involve the factor a_β^2 after suitable transformation.

Thus

$$a_x b_x a_\beta b_\beta = a_x b_x (acd)(bcd) = (bcd)\, a_x (acd)\, b_x,$$

and $$3a_x b_x a_\beta b_\beta = (bcd)\, a_x \{(acd)\, b_x - (abd)\, c_x - (acb)\, d_x\},$$

or since $$(acd)\, b_x - (abd)\, c_x - (acb)\, d_x = a_x (bcd),$$

$$3a_x b_x a_\beta b_\beta = (bcd)^2\, a_x^2.$$

Consequently

$$(\alpha\beta x)^2 = 2a_\alpha^2 \,.\, b_x^2 - \tfrac{2}{3} a_\alpha^2 \,.\, b_x^2 = \tfrac{4}{3} a_\alpha^2 \,.\, b_x^2,$$

and except for a numerical factor the invariant form is equal to the product of the original form and its invariant.

The following example will shew the advantage of introducing symbols like α, β, γ above in geometrical investigations.

Let $a_x = 0$, $b_x = 0$, $c_x = 0$ be the equations of three straight lines, then if $\alpha = (bc)$, $\beta = (ca)$, $\gamma = (ab)$,

$$(\beta\gamma) = (abc)\, a.$$

This is not surprising, because α, β, γ are the points of intersection of b, c; c, a; a, b respectively and $(\beta\gamma)$ must therefore give the line joining β and γ, that is a*.

The equation of the straight line joining the point y to the point of intersection of $a_x = 0$ and $b_x = 0$ is

$$(\overline{ab}, xy) = 0 \text{ or } a_x b_y - a_y b_x = 0,$$

which is the well-known form.

The line joining the point (a, b) to the point (c, d) is

$$(\overline{ab}, \overline{cd}, x) = 0, \text{ or } (acd)\, b_x - (bcd)\, a_x = 0.$$

This may also be written $(abc)\, d_x - (abd)\, c_x = 0$, and the two forms are equivalent in virtue of the fundamental identity from which both can, in fact, be at once inferred.

Now let a', b', c' be the sides of a second triangle and α', β', γ' its angular points.

* The similarity of these ideas to the methods of Grassmann's *Ausdehnungslehre* will not escape the notice of the reader who is acquainted with the Calculus.

The lines joining corresponding vertices of the two triangles are $(\alpha\alpha')$, $(\beta\beta')$, $(\gamma\gamma')$, and these are concurrent if

$$(\overline{\alpha\alpha'},\ \overline{\beta\beta'},\ \overline{\gamma\gamma'}) = 0,$$

i.e. if $$(\alpha\beta\beta')(\alpha'\gamma\gamma') - (\alpha'\beta\beta')(\alpha\gamma\gamma') = 0.$$

But since $(\alpha\beta) = (abc)\,c$ etc. this condition is

$$(abc)(a'b'c')\{c_{\beta'}b'_{\gamma} - c'_{\beta}b_{\gamma'}\} = 0,$$

or $$(abc)(a'b'c')\{(a'c'c)(abb') - (acc')(a'b'b)\} = 0,$$

or $$(abc)(a'b'c')\{(abb')(a'cc') - (acc')(a'bb')\} = 0,$$

or $$(abc)(a'b'c')(\overline{aa'},\ \overline{bb'},\ \overline{cc'}) = 0.$$

Now $(\overline{aa'},\ \overline{bb'},\ \overline{cc'}) = 0$ is the condition that the points of intersection of corresponding sides should be collinear.

Hence when the joins of corresponding vertices are concurrent the intersections of corresponding sides are collinear—a well-known theorem. A direct analytical proof by the ordinary methods without using a particular triangle of reference is by no means easy.

Ex. (i). Prove that the cross ratio of the range in which $u_x = 0$ is met by $a_x = 0$, $b_x = 0$, $c_x = 0$, $d_x = 0$ is $-\dfrac{(bcu)(adu)}{(abu)(cdu)}$; hence the general equation of a conic touching the four lines is

$$(abu)(cdu) + \lambda(bcu)(adu) = 0.$$

If the six lines a, b, c, d, e, f touch a conic, then

$$(abe)(cde)(bcf)(adf) = (bce)(ade)(abf)(cdf).$$

Ex. (ii). State and prove the results dual to those of (i).

Ex. (iii). If the points $\alpha, \beta, \gamma, \delta, \epsilon, \zeta$ be on a conic and

$$(\alpha\beta) = p,\quad (\beta\gamma) = q,\quad (\gamma\delta) = r,\quad (\delta\epsilon) = p',\quad (\epsilon\zeta) = q',\quad (\zeta\alpha) = r',$$

then $$(\overline{pp'},\ \overline{qq'},\ \overline{rr'}) = 0.$$

Deduce Pascal's theorem and prove Brianchon's theorem in the same way.

228. Two Quadratics. Suppose the quadratics are

$$f = a_x^2 = b_x^2 = c_x^2 = \text{etc.},$$
$$f' = a'^2_x = b'^2_x = c'^2_x = \text{etc.}$$

and that their contravariants are

$$\phi = u_\alpha^2 = u_\beta^2 = u_\gamma^2 = \text{etc.}$$
$$\phi' = u_{\alpha'}^2 = u_{\beta'}^2 = u_{\gamma'}^2 = \text{etc.}$$

respectively, then their invariants are a_α^2 and $a'^2_{\alpha'}$ respectively.

The types of invariant factors are

$$(abc) = a_\alpha, \quad (aba') = a'_\alpha, \quad (ab'c') = a_{\alpha'},$$

$$(a'b'c') = a'_{\alpha'}, \quad (abu) = u_\alpha, \quad (a'b'u) = u_{\alpha'},$$

$$(aa'u), \quad a_x, \quad a'_x,$$

and in the course of the work we shall use the additional types

$$(\alpha\beta x), \ (\alpha'\beta' x), \ (\alpha\alpha' x)$$

which, as has been pointed out, certainly are invariant factors.

There are four types of symbols occurring in any expression, viz.

$$a, \ a', \ \alpha, \ \alpha',$$

and the leading idea of the investigation is to reduce the number of symbols in the expression for any invariant form to a minimum.

Thus $(abu)^2$ involves two symbols a, b, but it can be written u_α^2 and then only involves one.

Again $$2a_{\alpha'}b_{\alpha'}a_x b_x$$

is equivalent to $$(a_{\alpha'}b_x - a_x b_{\alpha'})^2$$

except for reducible terms; then since

$$a_{\alpha'}b_x - a_x b_{\alpha'} = (\alpha\alpha' x)$$

we see that $$a_{\alpha'}b_{\alpha'}a_x b_x$$

can be expressed in terms of reducible forms and forms containing a smaller number of symbols.

This example will make clear what we mean in the sequel by reducing the number of symbols.

I. *If a symbolical product contain a factor of the type a_α or $a'_{\alpha'}$ it is reducible.*

For let $$P = (abc)\, a_p b_q c_r M,$$

then, as in § 225,

$$6P = (abc)^2 . (pqr)\, M,$$

and (pqr) is an invariant factor; hence P is reducible.

This shews incidentally how factors of the type $(\alpha\beta x)$ arise naturally in the course of the work, for we might have

$$p = \alpha, \ q = \beta, \ r = x.$$

The case in which a and b occur again in the same factor, *e.g.* (abu), is treated as in § 225.

II. *If a symbolical product contain a factor of the type* (abv), *where* v *is of the type* a' *or is* u, *then it can be so transformed as to contain an additional factor* (abw).

Suppose in fact that

$$P = (abv)\, a_p b_q M,$$

then

$$2P = (abv)(a_p b_q - a_q b_p) M$$
$$= (abv)(\overline{ab},\, pq) M$$
$$= v_\alpha (\alpha pq) M,$$

and (αpq) can be expressed in terms of α and the symbols occurring in p and q.

Hence in the transformed expression there are two factors involving the symbol α.

A like theorem applies to α', and here we see a great advantage in introducing these symbols, because just as when one a occurs in a product there must be another, so when α occurs the expression can be transformed so as to be of degree two in α.

III. *If an expression contain a factor of the type* $(\alpha\beta y)$ *or* $(\alpha'\beta' y)$ *it is reducible.*

For let

$P = (\alpha\beta y) M$, where y must be of the type α, α' or x.

Then since

$$(\alpha\beta y) = (\overline{ab},\, \beta y) = a_\beta b_y - b_\beta a_y,$$
$$P = a_\beta b_y M - b_\beta a_y M$$

and by I. the form P contains the invariant a_α^2 as a factor.

Summing up these results we infer that any concomitant other than a_x^2, a'^2_x, a_α^2, $a'^2_{\alpha'}$ must be composed of factors of the types

$$a_x,\ a'_x,\ a_{\alpha'},\ a'_\alpha,\ u_\alpha,\ u_{\alpha'},\ (\alpha\alpha' x),\ (aa'u),$$

and if one α occurs there must be another present after a suitable transformation.

IV. *Any expression containing two equivalent symbols can be expressed in terms of concomitants that are either reducible or contain a smaller number of symbols.*

There are two cases according as the equivalent symbols are of the type a or α, so that we have to consider the expressions

$$a_p a_q b_r b_s M \text{ and } \varpi_\alpha \rho_\alpha \sigma_\beta \tau_\beta M,$$

with the exactly similar ones

$$a'_p a'_q b'_r b'_s M \text{ and } \varpi_{\alpha'} \rho_{\alpha'} \sigma_{\beta'} \tau_{\beta'} M.$$

Since $$a_p b_r - a_r b_p = (\alpha pr),$$

and $$a_q b_s - a_s b_q = (\alpha qs),$$

it follows that expressions derived from

$$a_p a_q b_r b_s M,$$

by permuting the symbols p, q, r, s, differ from the original expression by forms in which the two symbols a and b are replaced by the simple symbol α.

Hence if any one of the three expressions

$$a_p a_q b_r b_s M,\ a_p a_r b_q b_s M,\ a_p a_s b_q b_r M$$

can be expressed in terms of reducible forms and forms containing fewer symbols the same is true of the other two.

The same result can be easily established for the three expressions

$$\varpi_\alpha \rho_\alpha \sigma_\beta \tau_\beta M,\ \varpi_\alpha \sigma_\alpha \rho_\beta \tau_\beta M,\ \varpi_\alpha \tau_\alpha \rho_\beta \sigma_\beta M;$$

in fact the difference between any two is reducible by III.

Consider now the expression

$$a_p a_q b_r b_s M.$$

There are only five possible types for p, q, r, s, viz. α, α', (bu), $(a'u)$ and x; of these α may be neglected because if it occur the form is reducible, and (bu) may be neglected because if it occur we can replace a and b at once by the single symbol α.

Hence there are only three remaining possibilities, viz. α', $(a'u)$, x, and of p, q, r, s some two must certainly be of the same type.

First suppose that two are identical, say p and r, then

$$a_p a_r b_q b_s M = a_p^2 . b_q b_s M;$$

thus $a_p a_r b_q b_s M$ is reducible, and hence $a_p a_q b_r b_s M$ can be expressed in terms of reducible forms and forms containing fewer symbols.

Similar reasoning applies to the expression

$$\varpi_\alpha \rho_\alpha \sigma_\beta \tau_\beta M$$

when any two of the symbols ϖ, ρ, σ, τ are identical; here, as in the other case, there are three possibilities, viz. a, a' and u, so some two must be of the same type.

Proceed now to the general case in which no two of the letters p, q, r, s are identical.

In this case some two must be equivalent without being identical.

Suppose the equivalent symbols be p and r, then there are three cases to consider.

(i) Let $p = r = x$. Here p and r are identical and the result has been established already.

(ii) Let $p = \alpha'$, $r = \beta'$, then

$$a_p a_r b_q b_s M = a_{\alpha'} a_{\beta'} b_q b_s M.$$

And since the symbols α', β' each occur once the expression can be transformed by II. so that each occurs twice, hence

$$a_p a_r b_q b_s M = a_{\alpha'} a_{\beta'} \gamma_{\alpha'} \delta_{\beta'} N.$$

But $a_{\alpha'} a_{\beta'} \gamma_{\alpha'} \delta_{\beta'} N$ falls under that class of expression

$$\varpi_{\alpha'} \rho_{\alpha'} \sigma_{\beta'} \tau_{\beta'} M$$

in which two of the symbols ϖ, ρ, σ, τ are identical, for $\varpi = \sigma = a$; hence, as we have seen, it can be expressed in terms of simpler forms.

Thus $a_p a_r b_q b_s M$ can be expressed in terms of forms that are either reducible or contain fewer symbols, and the same is true of $a_p a_q b_r b_s M$.

(iii) Let $p = (a'u)$, $r = (b'u)$, then

$$a_p a_r b_q b_s M = (aa'u)(ab'u) b_q b_s M.$$

And the latter expression can be written

$$(a'au)(b'au) a'_{q'} b'_{s'} M',$$

or

$$a'_{p'} b'_{r'} a'_{q'} b'_{s'} M',$$

where $p' = (au)$, $r' = (au)$, *i.e.* p' and r' are identical.

Thus, as we have already shewn

$$a'_{p'}a'_{q'}b'_{r'}b'_{s'}M'$$

can be expressed in terms of simpler forms; hence $a_p a_r b_q b_s M$ can be expressed in terms of forms that are either reducible or contain fewer symbols, and the same is true of $a_p a_q b_r b_s M$.

This completely establishes the Theorem IV. when the equivalent symbols are of the type a or a'.

We have still to consider the general expression

$$\varpi_\alpha \rho_\alpha \sigma_\beta \tau_\beta M,$$

where ϖ, ρ, σ, τ are each of the types a', $(\alpha' x)$ or u. The case in which some two are identical has already been discussed and reasoning exactly similar to that of (i), (ii), (iii) above enables us to prove Theorem IV. for the general expression.

The theorem we have just proved simplifies vastly the evolution of the irreducible system inasmuch as it shews that in an irreducible form there cannot be more than one symbol of any of the types

$$a, \quad a', \quad \alpha, \quad \alpha'.$$

A further limitation is imposed by the following theorem.

V. *A form containing both the factors $(aa'u)$ and $(\alpha\alpha' x)$ may be rejected in constructing the irreducible system.*

In fact by direct multiplication

$$(aa'u) \times (\alpha\alpha' x) = \begin{vmatrix} a_\alpha & a'_\alpha & u_\alpha \\ a_{\alpha'} & a'_{\alpha'} & u_{\alpha'} \\ a_x & a'_x & u_x \end{vmatrix},$$

hence if a form involve both these factors it can be transformed so as to only contain the simpler factors

$$a_\alpha, \; a'_{\alpha'}, \; a'_\alpha, \; a_{\alpha'}, \; u_\alpha, \; u_{\alpha'}, \; a_x, \; a'_x \text{ and } u_x.$$

With the aid of the above five theorems the problem of the complete system is reduced to a very simple one; in fact we have only to write down such products of factors of the types

$$a_\alpha, \; a'_{\alpha'}, \; a_{\alpha'}, \; a'_\alpha, \; u_\alpha, \; u_{\alpha'}, \; a_x, \; a'_x, \; (aa'u), \; (\alpha\alpha' x),$$

as satisfy the conditions implied in the theorems.

Since no two equivalent symbols can occur in the same product we need introduce no more symbols; further a_α, $a'_{\alpha'}$ can only occur in the invariants a_α^2 and $a'^2_{\alpha'}$ respectively; therefore it only remains to write down all products of the factors

$$a_{\alpha'},\ a'_\alpha,\ u_\alpha,\ u_{\alpha'},\ a_x,\ a'_x,\ (aa'u),\ (\alpha\alpha'x),$$

in which every letter except u and x which appears at all appears twice and no more; besides the two last factors must not appear in the same product.

Hence we have the following forms:

(A) $a_\alpha^2,\ a'^2_{\alpha'},\ a_{\alpha'}^2,\ a'^2_\alpha,\ u_\alpha^2,\ u_{\alpha'}^2,\ a_x^2,\ a'^2_x,\ (aa'u)^2,\ (\alpha\alpha'x)^2,$

obtained by squaring each factor.

(B) $a_{\alpha'}a_x u_{\alpha'},\ a_{\alpha'}a_x(\alpha\alpha'x)u_\alpha,\ a_{\alpha'}a_x(\alpha\alpha'x)a'_\alpha a'_x,$

$a_{\alpha'}(aa'u)a'_x,\ a_{\alpha'}(aa'u)a'_\alpha u_\alpha u_{\alpha'},$

containing the factor $a_{\alpha'}$.

(C) $a'_\alpha a'_x u_\alpha,\ a'_\alpha a'_x(\alpha\alpha'x)u_{\alpha'},\ a'_\alpha(aa'u)a_x,$

containing a'_α but not $a_{\alpha'}$.

(D) $u_\alpha(\alpha\alpha'x)u_{\alpha'},$

containing u_α but neither a'_α nor $a_{\alpha'}$.

(E) $a_x(aa'u)a'_x,$

containing a_x but neither a'_α nor $a_{\alpha'}$.

There are twenty forms in all, viz.:

Four Invariants,

$$a_\alpha^2 = A_{111},\ a'^2_{\alpha'} = A_{222},\ a_{\alpha'}^2 = A_{122},\ a'^2_\alpha = A_{112}.$$

Four Covariants,

$$a_x^2 = S_1,\ a'^2_x = S_2,\ (\alpha\alpha'x)^2 = F,\ (\alpha\alpha'x)a_{\alpha'}a'_\alpha a_x a'_x.$$

Four Contravariants,

$$u_\alpha^2 = \Sigma_1,\ u_{\alpha'}^2 = \Sigma_2,\ (aa'u)^2 = \Phi,\ (aa'u)a_{\alpha'}a'_\alpha u_\alpha u_{\alpha'}.$$

Eight Mixed Concomitants,

$$a_x a_{\alpha'} u_{\alpha'},\ a'_x a'_\alpha u_\alpha,\ (aa'u)a_x a'_x,\ (aa'u)a_{\alpha'}a'_x u_{\alpha'},$$

$$(aa'u)a'_\alpha a_x u_\alpha,\ (\alpha\alpha'x)u_\alpha a_{\alpha'}a_x,\ (\alpha\alpha'x)u_{\alpha'}a'_\alpha a'_x,$$

$$(\alpha\alpha'x)u_\alpha u_{\alpha'}.$$

To these twenty should be added the identical concomitant u_x.

It should be noticed in conclusion that, although it follows from the five theorems of this article that every other invariant form is expressible in terms of these as a rational integral algebraic function, it has not been shewn that these twenty forms are themselves irreducible. Cf. note, p. 131.

Ex. (i). Prove that

$$(a\beta\gamma)^2 = \tfrac{4}{3}A_{111}{}^2, \quad (a\beta\gamma')^2 = \tfrac{4}{3}A_{111}A_{112}.$$

Ex. (ii). Prove that

$$a_\alpha a'_\alpha a_{\alpha'} a'_{\alpha'} = \tfrac{1}{3}A_{111}A_{222},$$

and thence that $$(aa', \overline{aa'})^2 = \tfrac{1}{3}A_{111}A_{222} + A_{112}A_{122}.$$

Ex. (iii). Prove that

$$a_x a'_x a_\alpha a'_\alpha = \tfrac{1}{3}A_{111}S_2.$$

Ex. (iv). Prove that

$$a_\alpha u_{\alpha'} a_x (aa'x) = 0.$$

229. A complete irreducible system of invariant forms may be regarded as giving rise to two inquiries.

(i) What is the geometrical meaning of each member of the irreducible system?

(ii) What is the expression in terms of the members of that system of the invariant forms which arises in the analytical treatment of a given problem?

To the first of these inquiries an answer can generally be given, provided a sufficiently complex geometrical apparatus be allowed, but it commonly happens that the significance of some members of the system is so remote as to render them of little geometrical importance.

The second inquiry is naturally unanswerable until the problem be named, and thus all we can do is to illustrate it by the discussion of some simple problems.

Before going further it may be well to add that the second inquiry is the really important one; in a manner it includes the first as a particular case, and in fact there being no direct method of prooeeding from the invariant to the geometrical meaning, the answer of the first inquiry is obtained fortuitously in pursuing the second. If it be not obtained we should console ourselves with the reflexion that the uninterpreted forms are of little geometrical interest in the present state of knowledge; besides, if we regard the algebra as being merely helpful to geometry in the analytical formulation of results, it does not follow that everything in the algebra need be taken seriously from the geometrical point of view*.

In spite of what we have said, we shall begin the geometrical theory of two conics by interpreting the members of the irreducible system; it will be seen that they are all of importance in elementary geometry.

230. Geometrical Theory of two Quadratics. The irreducible system. The forms themselves when equated to zero represent two conics, viz.

$$a_x^2 = 0, \ a'^2_x = 0,$$

which we write $S_1 = 0$, $S_2 = 0$, and call S_1 and S_2.

Invariants. The meaning of the invariant of a single conic is expressed by the fact that $a_\alpha^2 = 0$ is the condition for two straight lines.

Again, $a'^2_\alpha = 0$ is satisfied when the point equation of S_1 involves only product terms and the line equation S_2 only squared terms, or when the point equation of S_1 involves only squared terms, and the line equation of S_2 only product terms.

It is then the poristic condition† that there should be an infinite number of triangles inscribed in the first conic and self-conjugate in the second, or, what is the same thing, that there should be an infinite number of triangles self-conjugate to the first conic and inscribed in the second. We shall consider this type of invariant more fully in the next chapter.

Covariants, etc. The simple ones are well known, viz.

$a_x^2 = 0$ is the condition that (x) should be on the first conic,

$u_\alpha^2 = 0$ is the condition that (u) should touch the first conic.

* Reducibility itself is a purely algebraical idea and the reader will soon convince himself that it is generally hard to obtain any geometrical satisfaction from the fact that a covariant is reducible. See a curious remark of Clifford's, *Collected Papers*, p. 81.

† The sufficiency of the condition can easily be seen by taking a triangle inscribed in the first conic and having two pairs of vertices conjugate with respect to the second for fundamental triangle.

Of those not involving determinantal factors, there remain

$$a'_\alpha a'_x u_\alpha, \; a_{\alpha'} a_x u_{\alpha'}.$$

It will be sufficient to consider one of these. Now $a'_x a'_\alpha u_\alpha = 0$ is the polar of the point $(\alpha_1 u_\alpha, \alpha_2 u_\alpha, \alpha_3 u_\alpha)$ with respect to $a'^2_x = 0$.

This point is the pole of u with respect to the first conic

$$u_\alpha^2 = 0 \text{ or } a_x^2 = 0$$

and hence $a'_x a'_\alpha u_\alpha = 0$, when u is constant, represents the polar with respect to the second conic of the pole of u with respect to the first. When x is constant, it is the tangential equation of the pole with respect to the first conic of the polar of x with respect to the second.

Again $(aa'u)^2 = 0$ is the equation of the envelope of the lines cutting the two conics in harmonic point pairs, and in like manner $(\alpha\alpha'x)^2$ is the locus of a point such that the tangents drawn to the two conics from it are harmonic line pairs.

Consider now the forms involving one determinant factor.

(i) $(aa'u)\, a_x a'_x = 0$ is the condition that the lines

$$a_x a_y = 0, \; a'_x a'_y = 0, \; u_y = 0,$$

should be concurrent, y being the current coordinate, *i.e.* when u is constant it is the locus of points whose polars intersect on the line u, and when x is constant it is the equation of the point of intersection of the polars of x with respect to the conics.

(ii) $(\alpha\alpha'x)\, u_\alpha u_{\alpha'}$ is dual to the last; when x is constant it represents the envelope of a line such that the line joining its poles passes through x, and when u is constant it represents the equation of the line joining the poles of u with respect to the two conics.

(iii) $(aa'u)\, a'_\alpha a_x u_\alpha$ is the Jacobian with respect to y of the quantities

$$a_x a_y, \; a'_\alpha u_\alpha a'_y, \; u_y.$$

Equated to zero, these represent three straight lines, namely, the polar of x with respect to the first conic, the polar with respect to the second conic of the pole of u with respect to the first, and the line u. The vanishing of the concomitant is the condition that the three lines should be concurrent; hence, for example, when u is constant, the equation represents a straight line constructed as follows: let P_1 be the pole of u with respect to the first conic, and v the polar of P_1 with respect to the second conic, then the line represented is the polar of the point (u, v) with respect to the first conic.

(iv) $(\alpha\alpha' x)\, a_{\alpha'} a_x u_\alpha$ is the Jacobian with respect to the v's of

$$u_\alpha v_\alpha,\quad v_{\alpha'} a_{\alpha'} a_x,\quad v_x,$$

and hence equated to zero is the condition that these three points should be collinear.

Now $u_\alpha v_\alpha = 0$ represents the pole of u with respect to the first conic, and $a_x a_{\alpha'} v_{\alpha'} = 0$ represents the pole with respect to the second conic of the polar of x with respect to the first. Thus we can interpret the concomitant geometrically. See also Ex. ix. p. 294.

(v) $(aa'u)\, a_{\alpha'} a'_\alpha u_\alpha u_{\alpha'}$ is a contravariant of the third class, and moreover the only such contravariant that can be built up from the members of the system. But the angular points of the common self-conjugate triangle must be given by equating to zero a contravariant of the third class. Hence the contravariant in question represents the vertices of the common self-conjugate triangle of the conics $a_x^2 = 0$ and $a'^2_x = 0$.

We may prove this as follows*: The three conics $u_\alpha^2 = 0$, $u_{\alpha'}^2 = 0$, and $(aa'u)^2 = 0$ have a common self-conjugate triangle if there be a proper triangle self-conjugate to the first two, as we see by taking it for triangle of reference. Further, when three conics are referred to their common self-conjugate triangle, the Jacobian of their tangential equations is the tangential equation of the vertices.

Hence the Jacobian of u_α^2, $u_{\alpha'}^2$, $(aa'u)^2$ represents the vertices of the common self-conjugate triangle.

Now it is

$$(aa'u)\, u_\alpha u_{\alpha'} \left(\overline{aa'}\alpha\alpha'\right) = (aa'u)\, u_\alpha u_{\alpha'} \left(a_\alpha a'_{\alpha'} - a_{\alpha'} a'_\alpha\right).$$

But $$(aa'u)\, u_\alpha u_{\alpha'} a_\alpha a'_{\alpha'} = 0,$$

for otherwise it would involve the factors a_α^2 and $a'^2_{\alpha'}$, whereas its total degree is only six, so that the remaining factor would be a contravariant of zero degree which is impossible.

* See also Ex. vi. p. 293.

Hence the Jacobian in question equated to zero is equivalent to

$$(aa'u)\, u_a u_{a'} a_{a'} a'_a = 0.$$

(vi) By exactly similar methods we can shew that

$$(\alpha\alpha' x)\, a_{\alpha'} a'_\alpha a'_x a_x = 0$$

represents the sides of the common self-conjugate triangle.

231. Let us now consider some problems bearing on two conics with a view to illustrating the second inquiry of § 229.

(i) To find the equation of the reciprocal of the conic S_2 with respect to S_1.

If y be a point on the reciprocal conic its polar with respect to $a_x^2 = 0$ must touch $u_{\alpha'}^2 = 0$.

The polar is

$$a_x a_y \equiv b_x b_y = 0,$$

and if $u_x = 0$ touches $u_{\alpha'}^2 = 0$

$$0 = u_{\alpha'}^2 = \tfrac{1}{4}\left(\frac{\partial^2}{\partial v_1 \partial x_1} + \frac{\partial^2}{\partial v_2 \partial x_2} + \frac{\partial^2}{\partial v_3 \partial x_3}\right)^2 \{u_x^2 v_{\alpha'}^2\}.$$

Hence in this case

$$0 = \left(\frac{\partial^2}{\partial v_1 \partial x_1} + \frac{\partial^2}{\partial v_2 \partial x_2} + \frac{\partial^2}{\partial v_3 \partial x_3}\right)^2 \{a_x a_y\, b_x b_y\, v_{\alpha'}^2\}$$

$$= 4 a_{\alpha'} b_{\alpha'}\, a_y b_y,$$

so that the reciprocal is

$$a_x b_x\, a_{\alpha'} b_{\alpha'} = 0 \quad \text{or} \quad -(a_x b_{\alpha'} - a_{\alpha'} b_x)^2 + a_x^2 b_{\alpha'}^2 + b_x^2 a_{\alpha'}^2 = 0$$

or $$0 = 2 a_x^2 b_{\alpha'}^2 - (\alpha\alpha' x)^2 = 2A_{122} \,.\, S_1 - F,$$

which expresses the equation in terms of the irreducible system.

(ii) To find the point equation of the covariant conic

$$(aa'u)^2 = 0.$$

The point equation of $u_\alpha^2 = 0$ is $(\alpha\beta x)^2 = 0$, and hence in our case the point equation is

$$\{\overline{aa'}\,\overline{bb'}\,x\}^2 = 0,$$

or $$\{(abb')\, a'_x - (a'bb')\, a_x\}^2 = 0,$$

i.e. $$(abb')^2 a'^2_x + (a'bb')^2 a_x^2 - 2\,(abb')\,(a'bb')\, a_x a'_x = 0.$$

Now

$$(abb')(a'bb')a_x a'_x = -(abb')(a'ab')a'_x b_x$$
$$= \tfrac{1}{2}(abb')a'_x\{a_{\alpha'}b_x - b_{\alpha'}a_x\}, \text{ where } \alpha' = (a'b')$$
$$= \tfrac{1}{2}(\alpha\alpha' x)(abb')a'_x$$
$$= -\tfrac{1}{2}(\alpha\alpha' x)(aba')b'_x = -\tfrac{1}{4}(\alpha\alpha' x)\{a'_\alpha b'_x - a'_x b'_\alpha\} = \tfrac{1}{4}(\alpha\alpha' x)^2,$$

thus the point equation is

$$(abb')^2 a'^2_x + (a'bb')^2 a_x^2 - \tfrac{1}{2}(\alpha\alpha' x)^2 = 0,$$

or

$$A_{112} . S_2 + A_{122} S_1 - \tfrac{1}{2} F = 0.$$

(iii) To find the locus of the point of intersection of harmonic pairs of tangents to F and S.

The locus for $u_\alpha^2 = 0$ and $u_{\alpha'}^2 = 0$ is $(\alpha\alpha' x)^2 = 0$ and in our case the locus is accordingly

$$\{\overline{aa'}\alpha x\}^2 = 0$$

or $\quad (a_x a'_\alpha - a'_x a_\alpha)^2 = 0,\ i.e.\ S_1 A_{112} + S_2 A_{111} - 2a_x a'_x a_\alpha a'_\alpha = 0.$

To reduce the last term we perceive that it must contain A_{111} and we write it

$$(abc)\,a'_x(a'bc)\,a_x = \tfrac{1}{3}(abc)\,a'_x\{(a'bc)\,a_x + (caa')\,b_x - (aa'b)\,c_x\}$$
$$= \tfrac{1}{3}(abc)\,a'_x(bca)\,a'_x = \tfrac{1}{3}A_{111} . S_2.$$

The equation required is

$$S_1 A_{112} + S_2 A_{111} - \tfrac{2}{3} S_2 A_{111} = 0,$$

or

$$A_{111} S_2 + 3 S_1 A_{112} = 0.$$

The following gives an easy means of verifying the results of (i), (ii), (iii) above and of the examples which follow.

Taking the conics in the canonical forms

$$S_1 = a_1 x_1^2 + a_2 x_2^2 + a_3 x_3^2$$
$$S_2 = \quad x_1^2 + \quad x_2^2 + \quad x_3^2$$

we have

$$\Sigma_1 = 2(a_2 a_3 u_1^2 + a_3 a_1 u_2^2 + a_1 a_2 u_3^2)$$
$$\Sigma_2 = 2(u_1^2 + u_2^2 + u_3^2),$$

$$A_{111} = 6a_1 a_2 a_3, \quad A_{222} = 6, \quad A_{112} = 2(a_2 a_3 + a_3 a_1 + a_1 a_2),$$
$$A_{122} = 2(a_1 + a_2 + a_3),$$

$$\Phi = (a_2 + a_3)\,u_1^2 + (a_3 + a_1)\,u_2^2 + (a_1 + a_2)\,u_3^2$$
$$F = 4\{a_1(a_2 + a_3)\,x_1^2 + a_2(a_3 + a_1)\,x_2^2 + a_3(a_1 + a_2)\,x_3^2\}.$$

Ex. (i). The locus of points whose polars with respect to S_2 cut S_1, S_2 in pairs of points harmonically conjugate is

$$(aa'b')\,(aa'c')\,b'_x c'_x = 0,$$

or

$$A_{222} S_1 - 3A_{122} S_2 = 0.$$

Ex. (ii). Prove that

$$(abu)^2 = \tfrac{1}{8}\,\Omega^2\,(a_x^2 b_y^2 u_z^2)$$

where

$$\Omega = \begin{vmatrix} \dfrac{\partial}{\partial x_1} & \dfrac{\partial}{\partial x_2} & \dfrac{\partial}{\partial x_3} \\ \dfrac{\partial}{\partial y_1} & \dfrac{\partial}{\partial y_2} & \dfrac{\partial}{\partial y_3} \\ \dfrac{\partial}{\partial z_1} & \dfrac{\partial}{\partial z_2} & \dfrac{\partial}{\partial z_3} \end{vmatrix},$$

and hence that

$$(\overline{\alpha\beta},\ \overline{\alpha'\beta'},\ u)^2 = \tfrac{16}{9} A_{111} A_{222}\,(aa'u)^2.$$

Interpret this result geometrically.

Ex. (iii). The line equation of $F=0$ is

$$(\overline{aa'}\ \overline{\beta\beta'}\ u)^2 = 0,$$

or

$$(a\beta\beta')^2 u_{\alpha'}{}^2 + (a'\beta\beta')^2 u_{\alpha}{}^2 - \tfrac{1}{2}\,(\overline{\alpha\beta}\ \overline{\alpha'\beta'}\ u)^2.$$

Thence

$$(\overline{aa'}\ \overline{\beta\beta'}\ u)^2 = \tfrac{4}{3}\{3A_{111} A_{122}\Sigma_2 + 3A_{222} A_{112}\Sigma_1 - 2A_{111} A_{222}\Phi\}.$$

Cf. Ex. (ii).

Ex. (iv). Deduce from Ex. (iii) that the discriminant of F is

$$(\overline{aa'}\ \overline{\beta\beta'}\ \overline{\gamma\gamma'})^2 = \tfrac{8}{27}\,(9A_{112} A_{122} - A_{111} A_{222})\,A_{111} A_{222}.$$

Ex. (v). Deduce from Ex. (ii),

$$(\overline{aa'}\ \overline{bb'}\ \overline{cc'})^2 = \tfrac{1}{6}\,(9A_{112} A_{122} - A_{111} A_{222}).$$

Ex. (vi). Prove that the discriminant of $(aa'u)\,a_x a'_x$ is

$$(aa'u)\,(bb'u)\,(cc'u)\,(ab'c')\,(a'bc).$$

Hence if the conic $(aa'u)\,a_x a'_x = 0$ be two straight lines

$$(aa'u)\,a_{\alpha'} a'_{\alpha} u_{\alpha} u_{\alpha'} = 0.$$

Thence verify that this equation represents the angular points of the common self-conjugate triangle and work out the dual results.

Ex. (vii). Prove that if the point equation of a conic be

$$\lambda_1 S_1 + \lambda_2 S_2 = 0,$$

then its tangential equation is

$$\lambda_1^2 \Sigma_1 + 2\lambda_1\lambda_2 \Phi + \lambda_2^2 \Sigma_2 = 0,$$

and its discriminant is

$$\lambda_1^3 A_{111} + 3\lambda_1^2 \lambda_2 A_{112} + 3\lambda_1 \lambda_2^2 A_{122} + \lambda_2^3 A_{222}.$$

Ex. (viii). Prove that the point equation of any covariant conic is of the form

$$\lambda_1 S_1 + \lambda_2 S_2 + \lambda F = 0,$$

where λ_1, λ_2, λ are invariants, and that its line equation is

$$\lambda_1{}^2\Sigma_1 + 2\lambda_1\lambda_2\Phi + \lambda_2{}^2\Sigma_2 + \tfrac{2}{3}\lambda\lambda_1(A_{111}\Sigma_2 + 3A_{122}\Sigma_1) + \tfrac{2}{3}\lambda\lambda_2(A_{222}\Sigma_1 + 3A_{112}\Sigma_2)$$
$$+\lambda^2\tfrac{4}{9}(3A_{111}A_{122}\Sigma_2 + 3A_{222}A_{112}\Sigma_1 - 2A_{111}A_{222}\Phi) = 0.$$

Ex. (ix). The locus of points whose polars with respect to S_1 and F meet on the line u is

$$(a\,\overline{\alpha\alpha'}\,u)\,a_x\,(\alpha\alpha' x) = 0,$$

or

$$(\alpha\alpha' x)\,a_{\alpha'}u_{\alpha}a_x = 0.$$

This gives a simpler interpretation of the irreducible form than that given in § 230.

Ex. (x). Use the method of Ex. (ix) to interpret

$$(aa'u)\,a'_{\alpha}a_x u_{\alpha} = 0.$$

Examples (ix) and (x) enable us to interpret all the irreducible mixed concomitants very simply in connection with the conics S_1, S_2, F, Φ.

Ex. (xi). The locus of points whose polars with respect to F and Φ meet on u is

$$A_{112}(\alpha\alpha' x)\,a_x u_{\alpha} a_{\alpha'} + A_{122}(\alpha\alpha' x)\,a'_x u_{\alpha'} a_{\alpha'} = 0.$$

(Use the point equation of Φ given in Ex. (ii).)

Ex. (xii). The equation of the line joining the poles of u with respect to F and Φ is

$$A_{222}A_{112}(aa'u)\,u_{\alpha}a_x a'_{\alpha} + A_{111}A_{122}(aa'u)\,u_{\alpha'}a'_x a_{\alpha'} = 0.$$

(Use the result of Ex. (iii).)

Ex. (xiii). Calculate the four invariants of the conics S_1 and F.

We have

$$C_{222} = \tfrac{8}{27}(9A_{112}A_{122} - A_{111}A_{222})\,A_{111}A_{222} \text{ by Ex. (iv).}$$
$$C_{122} = (a\,\overline{\alpha\alpha'}\,\overline{\beta\beta'})^2 = \tfrac{4}{3}(3A_{111}A^2_{122} + A_{222}A_{112}A_{111}).$$
$$C_{112} = \tfrac{4}{3}A_{111}A_{122}.$$
$$C_{111} = A_{111}.$$

232. Gordan's general method. Consider a concomitant of any number of forms containing the r letters $a, b, c, \ldots h, k$. If we replace each factor of the type (aku) by a_x, each factor of the type (abk) by (abu) and delete all factors k_x the resulting expression is still an invariant form but only of degree $r-1$ because it only contains $r-1$ symbols.

Hence reversing the operation, *i.e.* replacing a proper number of factors of the type a_x by (aku), some of the type (abu) by (abk) and introducing a sufficient power of k_x we can deduce the form of degree r from one of degree $r-1$. Applying this process in all possible ways to all invariant forms of degree $r-1$ we certainly obtain all invariant forms of degree r. We pause to explain more precisely what we mean by applying the reverse process in all possible ways to an invariant ϕ. Suppose in fact that the newly introduced letter k belongs to a form of order n, then we replace any p factors of the type a_x by (aku) any q factors of the type (abu) by (abk) and multiply by k_x^r where of course we must have

$$p+q+r=n,$$

and we must take all values of p, q, r satisfying this equation subject, of course, to the condition that there exist p factors of the given type to alter and a like restriction for q.

For example, let the original form be

$$(abu)(acu)\,b_x c_x,$$

and suppose the new letter d like a, b, c belongs to a form of order 2.

We can only change 2 factors at most in this case and we have

$$p+q=1 \text{ or } 2,$$

there are five cases,

$$p=1,\ q=0;\ p=0,\ q=1;\ p=2,\ q=0;\ p=1,\ q=1;\ p=0,\ q=2;$$

and in following out the case $p=1$, $q=1$, $r=0$, for example, we deduce four forms from the given one, viz.

$$(abd)(acu)(bdu)\,c_x,$$
$$(abd)(acu)\,b_x(cdu),$$
$$(abu)(acd)(bdu)\,c_x,$$
$$(abu)(acd)\,b_x(cdu).$$

The above indicates the general method of procedure, but some introductory lemmas are necessary to render the method of any practical value—for example for all we know at present a form of degree $r-1$ which is identically zero might lead to irreducible forms of degree r, and we need hardly say that this would complicate matters enormously.

233. The reader will have observed some likeness between the above and the methods used in Chapter v. on binary forms to deduce the invariants of degree r from those of degree $r-1$. This analogy will be further exemplified in the rest of the argument.

Consider the effect of the operator Ω^p on the expression

$$\phi k_y^n u_z^p$$

ϕ being a covariant of degree $r-1$ and Ω being the operator

$$\begin{vmatrix} \dfrac{\partial}{\partial x_1} & \dfrac{\partial}{\partial x_2} & \dfrac{\partial}{\partial x_3} \\ \dfrac{\partial}{\partial y_1} & \dfrac{\partial}{\partial y_2} & \dfrac{\partial}{\partial y_3} \\ \dfrac{\partial}{\partial z_1} & \dfrac{\partial}{\partial z_2} & \dfrac{\partial}{\partial z_3} \end{vmatrix}.$$

As we saw in the last chapter, the result is the sum of a number of terms each containing the determinantal factors of ϕ together with p of the form (aku), there being in the end p fewer factors of the type a_x, $(n-p)$ factors k_y and no factors of the type u_z in each term.

Next operate on the complete result of which a typical term is ψk_y^{n-p}

$$Q \equiv \frac{\partial^2}{\partial u_1 \partial y_1} + \frac{\partial^2}{\partial u_2 \partial y_2} + \frac{\partial^2}{\partial u_3 \partial y_3}$$

q times in succession.

Since the effect of this operator on $(abu)\, k_y^{n-p}$ is to give a multiple of $(abk)\, k_y^{n-p-1}$, it is very easy to see that the effect of the operator q times is to give a number of terms each containing the same determinantal factors as ψ, with the exception that q factors of the type (abu) are replaced by (abk); the power of k_y remaining in each term is k_y^{n-p-q}, and k replaces u in q places in all possible ways.

Hence if we operate with $Q^q \Omega^p$ on the product $\phi k_y^n u_z^p$ and then put $y = x$ we obtain the sum of a number of terms each of which has the same determinantal factors as ϕ except that in any q of them u is replaced by k and p new ones of the type (aku) are introduced while p factors of the type a_x disappear and finally the factor k_x^{n-p-q} is introduced in each term.

Consequently in the resulting expression there will be contained every term derived from ϕ in the reverse process explained above with these definite values of p and q.

We may conveniently call

$$Q^q \Omega^p \{\phi,\ k_y^n u_z^p\}_{y=x}$$

the transvectant of ϕ and k_x^n whose indices are p, q, the order of the indices being essential, and we have the result that every concomitant of degree r is the sum of a number of terms each occurring as a term in a transvectant of a form of degree $r-1$ with k_x^n. Naturally when there are different forms we have to introduce in turn a symbol belonging to each.

234. We next require certain relations that exist among the terms of the same transvectant, and to establish them we shall alter our notation for a moment.

Suppose in fact that

$$\phi = a_x^{(1)} a_x^{(2)} \ldots a_x^{(p')} u_{\alpha_1} u_{\alpha_2} \ldots u_{\alpha_{q'}} M,$$

where M contains neither u nor x.

In each term of the transvectant

$$\{\phi,\ k_x^n\}^{p,q}$$

we have q of the u's replaced by k, p of the terms $a_x^{(r)}$ by (aku).

We shall call two terms N_1 and N_2 *adjacent* when $p+q-1$ of the factors affected in ϕ to obtain them are common, and two cases will arise according as the remaining factor affected is an a_x or a u_α.

In the first case we have, supposing that $a_x^{(r)}$ and $a_x^{(s)}$ are the additional altered factors in the two terms respectively,

$$\begin{aligned} N_1 - N_2 &= N\{(a^{(r)}ku)\, a_x^{(s)} - (a^{(s)}ku)\, a_x^{(r)}\} \\ &= N\{(a^{(r)}a^{(s)}k)\, u_x - (a^{(r)}a^{(s)}u)\, k_x\}. \end{aligned}$$

Now $N(a^{(r)}a^{(s)}k)$ is a term in

$$(\phi_1, k_x^n)^{p-1, q+1}$$

where ϕ_1 is deduced from ϕ by changing $a_x^{(r)}a_x^{(s)}$ into $(a^{(r)}a^{(s)}u)$ and further

$$N(a^{(r)}a^{(s)}u)\,k_x$$

is a term in

$$\{\phi_2, k_x^n\}^{p-1, q}$$

where ϕ_2 is the same as ϕ_1.

In the second case let u_{a_i} and u_{a_j} be the additional factors altered in the two terms, then

$$(N_1 - N_2) = N'\{u_{a_i}k_{a_j} - u_{a_j}k_{a_i}\}$$

$$= -N'\{a_i a_j \overline{ku}\}$$

and this latter is a term in

$$\{\phi_3, k_x^n\}^{p+1, q-1}$$

where ϕ_3 is deduced from ϕ by changing $u_{a_i}u_{a_j}$ into $(a_i a_j x)$.

Now if we call the sum of the order and class of a function its *grade* it is evident that ϕ_1, ϕ_2, ϕ_3 are each of grade less by unity than that of ϕ.

Further between any two terms of the transvectant we can insert a number of others such that any two of the whole sequence are adjacent in our sense of the word and accordingly we have the important theorem:

"*The difference between any two terms of the same transvectant can be expressed in terms of transvectants of functions of lower grade than ϕ with k_x^n.*"

Thus if we consider our function ϕ of degree $r-1$ in ascending order of grade we need only retain one term out of each transvectant that we consider—or if we please the sum of any number of terms will equally serve our purpose and in particular the transvectant itself might be used. It follows at once that if a transvectant contains a single reducible term it may be neglected entirely.

Again, if there be a linear relation among a number of the forms of degree $r-1$ there will be a linear relation among the transvectants of given index formed from them, so that we need only consider linearly independent forms of degree $r-1$. In particular, zero forms of degree $r-1$ can be entirely put out of account.

A knowledge of the irreducible system up to and including degree $r-1$ therefore gives us immediately all the forms ϕ of which transvectants need be considered, for we have only to include the irreducible forms of degree $r-1$ and such simple products of the others as are of total degree $r-1$.

We have now effected our purpose of making the method at present under discussion of real value, and we proceed to illustrate it by reference to the complete system for a single quadratic.

235. Quadratics. We have here five different sets of indices for transvectants, namely

$$p=1, q=0\,;\quad p=0, q=1\,;\quad p=2, q=0\,;\quad p=1, q=1\,;\quad p=0, q=2.$$

Consider now how far products need be taken into account, if $p+q=1$ then all products may be neglected because only one factor is modified and hence some terms of the transvectant of a product are certainly reducible. If $p=2$, $q=0$, then a transvectant of $\phi_1\phi_2$ with k_x^2 will contain reducible terms unless the orders of ϕ_1 and ϕ_2 are both unity, also for $p=0, q=2$, we need only consider in like manner products of two forms whose class is unity. If $p=1$, $q=1$, we need only consider the product of two forms when one is of zero order and the other of zero class. Throughout products of more than two forms need not be taken into account.

Further, in every case pure invariants of degree $r-1$ can give rise to no new forms.

236. Single Quadratic Form. Of the first degree we have

$$a_x^2=b_x^2=c_x^2=\ldots$$

Proceeding to the second degree we have

$$(abu)\,a_x b_x \text{ for } p=1,\ q=0$$

and

$$(abu)^2 \text{ for } p=2,\ q=0$$

of which the first is zero.

Third degree. From $(abu)^2$ we get

$$(abc)\,(abu)\,c_x\,(01)$$
$$(abc)^2\,(02)$$

and from $a_x^2 b_x^2$ we can only get reducible forms.

Now $$(abc)\,\{(abu)\,c_x\}=\tfrac{1}{3}\,(abc)\,\{(abu)\,c_x+(bcu)\,a_x+(abc)\,u_x\}$$
$$=\tfrac{1}{3}\,(abc)^2\,u_x^2$$

so that of the third degree we have only the invariant $(abc)^2$.

Fourth degree. From $(abu)^2 c_x^2$ we need only consider

$$(abd)\,(abu)\,(cdu)\,c_x$$

arising from $p=1$ and $q=1$.

This is $\tfrac{1}{4}\,(abu)\,(cdu)\,\{(abd)\,c_x+(bdc)\,a_x+(dca)\,b_x+(cab)\,d_x\}=0.$

For further forms we need only consider transvectants of products of powers of c_x^2 and $(abu)^2$ with $p+q \not> 2$.

These all contain reducible terms and hence there are no new forms so that the complete system consists of a_x^2, $(abu)^2$ and $(abc)^2$ as already indicated.

For the case of three quadratics and incidentally two see Baker, *Camb. Phil. Trans.* Vol. XV.

CHAPTER XIV.

APOLARITY (*continued*).

237. Apolar Conics. Two conics S and S' whose equations in point and line coordinates are respectively

$$S \equiv a_x^2 \equiv ax_1^2 + bx_2^2 + cx_3^2 + 2fx_2x_3 + 2gx_3x_1 + 2hx_1x_2 = 0,$$

and $\Sigma' \equiv u_{\alpha'}^2 \equiv A'u_1^2 + B'u_2^2 + C'u_3^2 + 2F'u_2u_3 + 2G'u_3u_1 + 2H'u_1u_2 = 0$

are said to be apolar when the invariant $a_{\alpha'}^2$, or what is the same thing,

$$aA' + bB' + cC' + 2fF' + 2gG' + 2hH'$$

vanishes.

This relation between the two conics is not a symmetrical one, inasmuch as it arises from the point equation of one and the line equation of the other; it is convenient to have an alternative name shewing the exact relation between the curves. For reasons to be explained later we shall say that S is harmonically inscribed in S', and that S' is harmonically circumscribed to S.

The curves are also apolar when $a'^2_\alpha = 0$, but in this case S' is harmonically inscribed in S.

As is well known from the geometry of conics, $a_{\alpha'}^2 = 0$ is the condition that there should exist an infinite number of triangles self-conjugate to S and circumscribed to S', or an infinite number of triangles inscribed in S and self-conjugate to S'—in fact the equations $a_x^2 = 0$ and $u_{\alpha'}^2 = 0$ can be so transformed that the first has no product terms and the second has no square terms, or that the first has no square terms and the second has no product terms.

The relation $a_{\alpha'}^2 = 0$ is linear in the coefficients of the equations

$$a_x^2 = 0$$

and $$u_{\alpha'}^2 = 0,$$

hence a conic apolar to the conics $S_1, S_2, \dots S_r$ is apolar to any conic

$$\lambda_1 S_1 + \lambda_2 S_2 + \dots + \lambda_r S_r = 0.$$

Further if these r conics are linearly independent there are $(6-r)$ linearly independent conics apolar to them. In particular there is a unique conic apolar to (harmonically circumscribed to) five given linearly independent conics.

The same remarks apply to ρ given conics

$$\Sigma_1' = 0, \quad \Sigma_2' = 0, \dots \Sigma_\rho' = 0,$$

and in particular there is a unique conic apolar to (harmonically inscribed in) five given conics.

238. Particular Cases. (i) If $a_x^2 = 0$ represents two straight lines and $a_{\alpha'}^2 = 0$, the two lines are conjugate with respect to $u_{\alpha'}^2 = 0$. If $a_x^2 = 0$ represents two straight lines coinciding in l then the line l touches the conic $u_{\alpha'}^2 = 0$.

(ii) If $u_{\alpha'}^2 = 0$ represents two points then these points are conjugate with respect to $a_x^2 = 0$. If $u_{\alpha'}^2 = 0$ represents two points coinciding in l then the point l lies on the conic $a_x^2 = 0$.

All these statements can be verified immediately by using the apolar condition expressed in terms of actual coefficients or symbolically, *e.g.* if $a_x^2 = v_x w_x$ the apolar condition is

$$0 = \{a_x^2 u_{\alpha'}^2\}^{0,2} = (v_x w_x,\ u_{\alpha'}^2)^{0,2} = v_{\alpha'} w_{\alpha'},$$

which is the condition that the lines $v_x = 0$, $w_x = 0$ should be conjugate with respect to $u_{\alpha'}^2 = 0$.

239. Ex. (i). *If two pairs of opposite vertices of a complete quadrilateral are conjugate with respect to a given conic so also is the third pair.*

Let the conic be $a_x^2 = 0$, and suppose the two pairs of opposite vertices are given tangentially by

$$u_p = 0,\ u_{p'} = 0\ ;\quad u_q = 0,\ u_{q'} = 0.$$

The general equation of a conic inscribed in the quadrilateral is then

$$\lambda u_p u_{p'} + \mu u_q u_{q'} = 0,$$

and since $u_p u_{p'} = 0$ and $u_q u_{q'} = 0$ are both apolar to $a_x^2 = 0$ it follows that every conic inscribed in the quadrilateral is apolar to $a_x^2 = 0$. But the third pair of opposite vertices is one such conic, hence these remaining vertices are conjugate with respect to the given conic. We shall call such a quadrilateral a quadrilateral harmonically inscribed in the conic $a_x^2 = 0$.

Ex. (ii). *Four conics have in general one common harmonic quadrilateral.*

Let the conics be S_1, S_2, S_3, S_4, then the apolar system is of the type

$$\lambda_1 \Sigma_1 + \lambda_2 \Sigma_2 = 0,$$

consequently the apolar conics in general all touch four fixed straight lines. The opposite vertices of the quadrilateral formed by these lines taken in pairs constitute conics of the apolar system, and hence pairs of opposite vertices are conjugate with respect to each of our four conics. Hence the quadrilateral is harmonically inscribed in each of the given conics.

Ex. (iii). *A triangle ABC and its polar triangle with respect to a conic are in perspective.* For if the polars of B and C meet the sides CA and AB respectively in Q, R, and the line QR meet BC in P, then the quadrilateral formed by BC, CA, AB and the line PQR has two pairs of opposite vertices, viz. (B, Q), (C, R) conjugate with respect to the conic; therefore (A, P) are conjugate with respect to the conic, or the polar of A meets BC in P. Thus the polars of A, B, C meet the opposite sides in three collinear points, and they therefore form a triangle in perspective with ABC.

Ex. (iv). If $\quad u_x^{(1)} = 0,\ u_x^{(2)} = 0,\ u_x^{(3)} = 0,\ u_x^{(4)} = 0$

be the sides of a quadrilateral harmonic with respect to the conic $S = 0$, then we have

$$S = \lambda_1 u^{(1)}{}_x{}^2 + \lambda_2 u^{(2)}{}_x{}^2 + \lambda_3 u^{(3)}{}_x{}^2 + \lambda_4 u^{(4)}{}_x{}^2.$$

For let two pairs of opposite vertices be (pp') and (qq'); then apolar to the conics (pp') and (qq') we have the five conics

$$u^{(1)}{}_x{}^2 = 0,\ u^{(2)}{}_x{}^2 = 0,\ u^{(3)}{}_x{}^2 = 0,\ u^{(4)}{}_x{}^2 = 0 \text{ and } S = 0.$$

But the first four are linearly independent and hence S is a linear combination of them.

240. Some interesting applications can also be made to the metrical geometry of conics.

In fact, suppose that the tangential equation of the circular points at infinity (I, J) is

$$\phi \equiv u_\gamma{}^2 = 0.$$

Then a conic apolar to ϕ has I, J for conjugate points and is therefore a rectangular hyperbola.

Again, the tangential equation of a circle whose centre is p is of the form

$$u_p{}^2 = \lambda\phi$$

where λ varies with the radius.

If a circle C be apolar to a conic

$$\Sigma = u_a^2 = 0$$

then the director circle of the conic cuts the circle C orthogonally.

Use rectangular Cartesian coordinates, and let the equations of the conic and circle respectively be

$$Al^2 + 2Hlm + Bm^2 + 2Gl + 2Fm + C = 0,$$

and $$x^2 + y^2 + 2gx + 2fy + c = 0,$$

so that we have

$$A + B + 2gG + 2fF + cC = 0.$$

But the equation of the director circle of the conic is

$$C(x^2 + y^2) - 2Gx - 2Fy + A + B = 0,$$

and this cuts the given circle at right angles if

$$-2g\frac{G}{C} - 2f\frac{F}{C} - \frac{A+B}{C} - c = 0,$$

i.e. if $$2gG + 2fF + cC + A + B = 0$$

which is precisely the condition of apolarity.

The director circle of a conic inscribed in a triangle cuts the self-conjugate circle orthogonally.

For since the self-conjugate circle has the triangle for a self-conjugate triangle and the conics are inscribed in the triangle, each of the conics is apolar to the circle. Hence their director circles cut the self-polar circle at right angles. Or thus,—the system apolar to the inscribed conics is of the form

$$\lambda p_x^2 + \mu q_x^2 + \nu r_x^2 = 0,$$

where $p_x = 0$, $q_x = 0$, $r_x = 0$ represent the sides of the triangle. By suitably choosing λ, μ, ν this equation may be made to represent a circle, and from the form of its equation it is the self-polar circle of the triangle.

The locus of the centre of a circle which has two fixed pairs of conjugate lines is a rectangular hyperbola.

In fact, suppose the lines are

$$\left.\begin{array}{ll} p_x = 0, & q_x = 0 \\ r_x = 0, & s_x = 0 \end{array}\right\}.$$

The system apolar to the tangential system of conics having these two pairs of conjugate lines is

$$\lambda p_x q_x + \mu r_x s_x = 0.$$

There is one value of the ratio $\lambda : \mu$ for which this represents a rectangular hyperbola.

Let $S = 0$ be the rectangular hyperbola in question and let

$$u_p^2 = \lambda\phi$$

be one of the circles.

Then S is apolar to ϕ, because $S = 0$ is a rectangular hyperbola, and as it is apolar to

$$u_p^2 = \lambda\phi$$

it is apolar to u_p^2. Hence the point p must lie on S and therefore S is the centre locus of the circles.

In general when a rectangular hyperbola S is apolar to a circle Σ the centre of the circle lies on the rectangular hyperbola.

241. Apolar Curves in general. The two curves whose equations are

$$\left.\begin{aligned} f &\equiv a_x^m = 0 \\ \text{and} \qquad \phi &\equiv u_\alpha^n = 0 \end{aligned}\right\}$$

are said to be apolar when the form $a_\alpha^n a_x^{m-n}$, which we denote by ψ, is identically zero.

Except for a numerical multiple we have

$$\begin{aligned} \psi &= \left(\frac{\partial^2}{\partial u_1 \partial x_1} + \frac{\partial^2}{\partial u_2 \partial x_2} + \frac{\partial^2}{\partial u_3 \partial x_3}\right)^n a_x^m u_\alpha^n \\ &= (a_x^m,\ u_\alpha^n)^{0,n}. \end{aligned}$$

The following are analogous to theorems on binary forms.

I. *The form ϕ is apolar to any polar of f whose order is not less than n.*

For $\quad \{a_x^{m-r} a_y^r,\ u_\alpha^n\}^{0,n} = a_\alpha^n a_x^{m-r-n} a_y^r$

and this is zero as we see by polarizing the identity

$$a_\alpha^n a_x^{m-n} = 0$$

r times with respect to y.

II. *The form f is apolar to any form which contains ϕ as a factor and whose class does not exceed m.*

For if ϕ' be any form of class n' we have

$$\{f, \phi\phi'\}^{0, n+n'} = \{(f, \phi)^{0,n}, \phi'\}^{0,n'}$$
$$= 0,$$

since $(f, \phi)^{0,n}$ vanishes identically. Hence f is apolar to $\phi\phi'$. We have supposed that $m \geqslant n$ hitherto. Exactly similar remarks apply to the case in which $n > m$.

The search for the forms of given class (n) apolar to a given form f is facilitated by the fact that the necessary and sufficient conditions for ϕ are that it should be apolar to every $(m-n)$th polar of f.

For an $(m-n)$th polar is

$$a_x{}^n a_y{}^{m-n}$$

and this is apolar to $u_\alpha{}^n$ if

$$a_\alpha{}^n a_y{}^{m-n} = 0.$$

But if this relation be true for all values of y then the form ϕ is apolar to f.

242. Ex. (i). *A ternary cubic has three linearly independent apolar conics.*

For the first polars of the cubic are linear combinations of

$$\frac{\partial f}{\partial x_1}, \frac{\partial f}{\partial x_2}, \frac{\partial f}{\partial x_3}$$

which are three linearly independent quadratic forms. Hence there are three linearly independent conics apolar to all first polars and therefore apolar to the cubic itself.

Ex. (ii). *A ternary quartic has an apolar conic only when the determinant of the coefficients of its second differential coefficients vanishes.*

For an apolar conic must be apolar to all second polars and they are linear combinations of

$$\frac{\partial^2 f}{\partial x_1{}^2}, \frac{\partial^2 f}{\partial x_2{}^2}, \frac{\partial^2 f}{\partial x_3{}^2}, \frac{\partial^2 f}{\partial x_2 \partial x_3}, \frac{\partial^2 f}{\partial x_3 \partial x_1}, \frac{\partial^2 f}{\partial x_1 \partial x_2}.$$

In general there is no conic apolar to each of these six, but there will be an apolar conic if the six be not linearly independent, *i.e.* if the determinant of six rows and six columns be zero.

243. General Theory of Curves which possess an Apolar Conic. By using suitable coordinates the analysis of ternary forms apolar to a given conic may be reduced to that of binary forms.

Suppose that the fixed conic is

$$x_1x_3 - x_2^2 = 0$$

or in line coordinates $\quad 4u_1u_3 - u_2^2 = 0.$

There is no loss of generality in taking the equations in this form, because by suitably choosing the triangle of reference, the equation of a proper conic can be always reduced to the form

$$x_1x_3 - x_2^2 = 0.$$

Thus we may take for the parametric representation of points on the conic

$$x_1 = \nu_1^2, \quad x_2 = \nu_1\nu_2, \quad x_3 = \nu_2^2,$$

and we shall call this point (x_1, x_2, x_3) the point ν.

If the line $u_x = 0$ meet the conic in the points (λ, μ) the quantities λ, μ are given by

$$u_1\nu_1^2 + u_2\nu_1\nu_2 + u_3\nu_2^2 = 0$$

so that

$$\left.\begin{aligned} u_1 &= \lambda_2\mu_2 \\ u_2 &= -(\lambda_1\mu_2 + \lambda_2\mu_1) \\ u_3 &= \lambda_1\mu_1 \end{aligned}\right\} \quad \text{......................(A)}$$

except for a constant factor.

Hence we may regard the quantities

$$\lambda_2\mu_2, \quad -(\lambda_1\mu_2 + \lambda_2\mu_1), \quad \lambda_1\mu_1,$$

as the coordinates of the line, and a homogeneous relation of order m connecting the u's becomes a homogeneous symmetrical relation of order $2m$ between the λ's and μ's. Thus any symmetrical relation between λ and μ is equivalent to the tangential equation of a certain curve whose class is one-half of the order of the given relation.

The coordinates of the tangent at the point ν are

$$\nu_2^2, \quad -2\nu_1\nu_2, \quad \nu_1^2$$

and hence the points of contact of the tangents from the point x to the curve are given by

$$x_1\nu_2^2 - 2x_2\nu_1\nu_2 + x_3\nu_1^2 = 0,$$

so that we may take

$$\left.\begin{aligned} x_1 &= \lambda_1\mu_1 \\ x_2 &= \tfrac{1}{2}(\lambda_1\mu_2 + \lambda_2\mu_1) \\ x_3 &= \lambda_2\mu_2 \end{aligned}\right\} \quad \text{....................(B).}$$

Consequently a homogeneous symmetrical relation between the λ's and μ's of order $2n$ represents a curve of order n. Being given a symmetrical relation we therefore deduce two curve equations from it, one in line coordinates and the other in point coordinates. The curves represented are reciprocal with respect to the fundamental conic because, taking λ, μ to be fixed, the line u given by (A) is the line joining them, and the point x given by (B) is the intersection of the tangents at λ and μ, *i.e.* the pole of the line u.

This method of representing a point by the parameters of the tangents drawn from it to a fixed conic and a line by the parameters of the points in which it meets the conic was practically used by Hesse and first explicitly used by Darboux.

244. By means of this system of coordinates we can readily find all curves apolar to the given conic.

I. Suppose that $$u_\gamma^n = 0$$
is a class curve apolar to the conic, then we have
$$(\gamma_2^2 - \gamma_1\gamma_3)\, u_\gamma^{n-2} \equiv 0.$$

Thus $(\gamma_2^2 - \gamma_1\gamma_3)$ multiplied by any function of the γ's is zero if it be interpretable, hence this symbolical expression must be zero, and we may write
$$\gamma_1 = a_2^2,$$
$$\gamma_2 = - a_1 a_2,$$
$$\gamma_3 = a_1^2,$$
and our equation is
$$(a_2^2 u_1 - a_1 a_2 u_2 + a_1^2 u_3)^n = 0.$$

The a's are now the only symbols used, and it is clear that as any expression of degree n in the γ's represents an actual quantity, any expression of degree $2n$ in the a's is an actual coefficient, or in other words the a's are the symbols of a binary form of order $2n$.

On introducing the λ's and μ's our equation becomes
$$\{a_2^2\lambda_2\mu_2 + a_1 a_2(\lambda_1\mu_2 + \lambda_2\mu_1) + a_1^2\lambda_1\mu_1\}^n = 0,$$
or
$$\{(a_1\lambda_1 + a_2\lambda_2)(a_1\mu_1 + a_2\mu_2)\}^n = 0,$$
that is finally
$$a_\lambda^n a_\mu^n = 0.$$

The binary $2n$-ic of which the a's are symbols has an important significance, for if we make $\lambda = \mu$, the line u_x is the tangent to the fundamental conic at the point λ; consequently the equation

$$a_\lambda^{2n} = 0$$

gives the parameters of the points of contact of the $2n$ common tangents of the apolar curve and the conic.

Conversely, when the equation

$$a_\lambda^{2n} = 0$$

is given, the equation $\quad a_\lambda^n a_\mu^n = 0$

is uniquely determined by polarizing, and hence we have the theorem that a class curve apolar to a conic is uniquely determined when its common tangents with the conic are given.

By proving this theorem from first principles, and then observing that

$$a_\lambda^n a_\mu^n = 0,$$

or its equivalent

$$(a_2^2 u_1 - a_1 a_2 u_2 + a_1^2 u_3)^n = 0,$$

certainly represents a curve apolar to a conic, we can shew that any apolar curve may be reduced to the form

$$a_\lambda^n a_\mu^n = 0,$$

without using a parametric representation of the symbolical equation

$$\gamma_1 \gamma_3 - \gamma_2^2 = 0^*.$$

II. Suppose that the curve

$$c_x^n = 0$$

of order n is apolar to the given conic

$$4u_1 u_3 - u_2^2 = 0,$$

then we must have

$$(4c_1 c_3 - c_2^2)\, c_x^{n-2} \equiv 0,$$

and reasoning as before, we have

$$4c_1 c_3 - c_2^2 = 0.$$

* See Schlesinger, *Math. Ann.* Band XXII.

We may now use the parametric representation

$$c_1 = a_1^2,$$
$$c_2 = 2a_1a_2,$$
$$c_3 = a_2^2,$$

and the equation of the curve becomes

$$(a_1^2x_1 + 2a_1a_2x_2 + a_2^2x_3)^n = 0,$$

or introducing the λ's and μ's

$$\{a_1^2\lambda_1\mu_1 + a_1a_2(\lambda_1\mu_2 + \lambda_2\mu_1) + a_2^2\lambda_2\mu_2\}^n = 0,$$

i.e. $$a_\lambda^n a_\mu^n = 0,$$

and as before the a's are the symbols of the binary $2n$-ic a_λ^{2n}, which, equated to zero, gives the points of intersection of the apolar curve and the conic.

Example. To find the conic apolar to $x_1x_3 - x_2^2 = 0$ which touches the tangents to this conic at the points given by $\nu_1^4 - \nu_2^4 = 0$.

Here $$a_\lambda^{2n} = 0 \text{ is } \lambda_1^4 - \lambda_2^4 = 0,$$

thence $$a_\lambda^n a_\mu^n = 0$$

is $$\left(\mu\frac{\partial}{\partial\lambda}\right)^2(\lambda_1^4 - \lambda_2^4),$$

or $$\lambda_1^2\mu_1^2 - \lambda_2^2\mu_2^2 = 0.$$

On using the substitutions

$$u_1 = \lambda_2\mu_2, \text{ etc.}$$

the equation becomes $$u_3^2 - u_1^2 = 0,$$

or $$(u_3 + u_1)(u_3 - u_1) = 0,$$

so that the conic consists of the two points (1, 0, 1) (1, 0, −1), and in fact it is easy to see that these points are conjugate with respect to the conic.

245. Theorems on conics apolar to the fundamental conic. The equation of a conic apolar to $x_1x_3 - x_2^2 = 0$, and touching the tangents at the points given by

$$a_\tau^4 = 0,$$

is equivalent to $$a_\lambda^2 a_\mu^2 = 0,$$

and hence to $$u_\gamma^2 = 0,$$

where $$u_\gamma \equiv a_\lambda a_\mu.$$

Now suppose that A, B, C, D are four points $(\lambda, \mu, \nu, \rho)$ on the fundamental conic, and that the lines AB, CD whose equations are

$$v_x = 0 \text{ and } w_x = 0,$$

are conjugate with respect to the apolar conic $u_\gamma^2 = 0$.

Then since the condition of conjugacy is $v_\gamma w_\gamma = 0$ and

$$v_\gamma \equiv a_\lambda a_\mu, \quad w_\gamma \equiv a_\nu a_\rho,$$

we have

$$a_\lambda a_\mu a_\nu a_\rho = 0,$$

that is the quartic giving λ, μ, ν, ρ is apolar to $a_\tau^4 = 0$.

This is one of the simplest geometrical representations of forms apolar to a given form. From the symmetry of the result, we see that each pair of opposite sides of the quadrangle $ABCD$ are conjugate with respect to the apolar conic.

Thus there is an infinite number of quadrangles inscribed in the fundamental conic and harmonic to the apolar conic; the four vertices are given by forms apolar to the form

$$a_\tau^4 = 0.$$

Now if λ, μ, ν be chosen so that ρ is arbitrary, any line through A is conjugate to BC, so that A is the pole of BC, and hence ABC is a self-conjugate triangle of the apolar conic $u_\gamma^2 = 0$. In this case the cubic giving λ, μ, ν is apolar to $a_\tau^4 = 0$, and therefore to every first polar of this form; hence there is an infinite number of triangles inscribed in the fundamental conic and self-conjugate with respect to the apolar conic, and their vertices are given by the singly infinite number of cubic forms apolar to $a_\tau^4 = 0$.

Next suppose that the linear factors of the quartic giving λ, μ, ν, ρ are l_τ, m_τ, n_τ, r_τ, then

$$a_\tau^4 = Ll_\tau^4 + Mm_\tau^4 + Nn_\tau^4 + Rr_\tau^4,$$

where L, M, N, R are independent of τ.

By polarizing we obtain the identity

$$a_\xi^2 a_\eta^2 = L(l_\xi l_\eta)^2 + M(m_\xi m_\eta)^2 + N(n_\xi n_\eta)^2 + R(r_\xi r_\eta)^2,$$

where ξ, η are any two points on the conic.

Now by means of the usual substitutions

$$u_1 = \xi_2\eta_2,$$
$$u_2 = -(\xi_1\eta_2 + \xi_2\eta_1),$$
$$u_3 = \xi_1\eta_1,$$

the left-hand side becomes u_γ^2.

Consider next the term $l_\xi l_\eta$.

Since $l_1 = \lambda_2$ and $l_2 = -\lambda_1$, this becomes

$$l_1^2 u_3 - l_1 l_2 u_2 + l_2^2 u_1$$

or

$$\lambda_1^2 u_1 + \lambda_1\lambda_2 u_2 + \lambda_2^2 u_3,$$

and λ_1^2, $\lambda_1\lambda_2$, λ_2^2, are the coordinates of A, so that

$$l_\xi l_\eta = u_A,$$

where $u_A = 0$ is the tangential equation of A.

Hence we have

$$u_\gamma^2 \equiv L u_A^2 + M u_B^2 + N u_C^2 + R u_D^2,$$

and the conic is represented as the sum of four squares.

In particular, if A, B, C be the vertices of a self-conjugate triangle, we obtain in like manner

$$u_\gamma^2 \equiv L u_A^2 + M u_B^2 + N u_C^2.$$

These results are well known and easily obtained otherwise, but the methods here used may be applied with equal success to more difficult problems as we shall presently shew.

Exactly the same reasoning applies to a conic c_x^2 apolar to

$$4u_1u_3 - u_2^2 = 0,$$

and now the triangles are circumscribed to the fundamental conic and self-conjugate to the apolar conic.

246. Condition of apolarity of two conics apolar to the standard conic.

Suppose the two conics are

$$c_x^2 = 0, \quad u_\gamma^2 = 0,$$

that the first meets the standard conic in the points $a_\lambda^4 = 0$, and the second touches the tangents to the standard conic in the points $b_\lambda^4 = 0$.

Then we have

$$c_x \equiv a_\lambda a_\mu \text{ and } u_\gamma \equiv b_\lambda b_\mu,$$

and in particular

$$\left.\begin{aligned} c_1 &= a_1^2 \\ c_2 &= 2a_1a_2 \\ c_3 &= a_2^2 \end{aligned}\right\}, \quad \left.\begin{aligned} \gamma_1 &= b_2^2 \\ \gamma_2 &= -b_1b_2 \\ \gamma_3 &= b_1^2 \end{aligned}\right\},$$

hence

$$c_\gamma = a_1^2 b_2^2 + a_2^2 b_1^2 - 2a_1a_2b_1b_2 = (ab)^2,$$

and the conics are apolar when $c_\gamma^2 = 0$, that is when

$$(ab)^4 = 0,$$

or when the two binary quartics a_λ^4 and b_λ^4 are apolar. This gives another simple geometrical representation of apolar quartics.

247. To many of the theorems developed for conics there are analogues for all curves possessing an apolar conic. For brevity we introduce a definition.

If the equation of a curve of order m be

$$f(x_1, x_2, x_3) = 0$$

and f can be written as a linear combination of the forms

$$u_x^{(r)m}, \qquad r = 1, 2, \dots n,$$

then the n lines $\qquad u_x^{(r)} = 0$

are said to form a *conjugate n-line* with respect to the curve.

In like manner if the tangential equation of a curve of class m be

$$\phi(u_1, u_2, u_3) = 0$$

and ϕ can be written as a linear combination of the forms

$$u_{x_{(r)}}^m, \qquad r = 1, 2, \dots n,$$

then the n points $\qquad u_{x_{(r)}} = 0$

are said to form a *conjugate n-point* with respect to the curve.

Suppose then that the curve

$$c_x^n = 0$$

is apolar to the fundamental conic

$$x_1 x_3 - x_2^2 = 0,$$

the equation may be written

$$a_\lambda^n a_\mu^n = 0$$

where $\qquad a_\lambda^{2n} = 0$

gives the $2n$ points in which the curve meets the conic.

If a_λ^{2n} be apolar to the form

$$b_\lambda^r = 0$$

whose linear factors are

$$p_\lambda^{(1)}, \ p_\lambda^{(2)}, \ \dots \ p_\lambda^{(r)},$$

then a_λ^{2n} can be expressed as a linear combination of

$$p_\lambda^{(1)^{2n}},\ p_\lambda^{(2)^{2n}},\ \dots\ p_\lambda^{(r)^{2n}},$$

so that $a_\xi^n a_\eta^n$ is a linear combination of

$$p_\xi^{(1)^n} p_\eta^{(1)^n},\ \text{etc.}$$

and hence just as in the case of the conic it follows that the tangents to the conic at the points given by

$$p_\lambda^{(1)} = 0,\ p_\lambda^{(2)} = 0,\ \dots\ p_\lambda^{(r)} = 0,$$

i.e. by $$b_\lambda^r = 0,$$

form a conjugate r-line with respect to the curve whose equation can accordingly be expressed as a sum of r nth powers.

The above will suffice to indicate the general principles which we shall now apply to the ternary cubic and quartic.

248. Ternary Cubic. A cubic curve, as we have seen, § 242, always possesses an infinite number of apolar conics. Take the fundamental conic for one of these and let

$$c_x^3 = 0$$

be the equation of the cubic.

This may be written

$$a_\lambda^3 a_\mu^3 = 0$$

and meets the conic in the points given by

$$a_\lambda^6 = 0.$$

This binary sextic has three linearly independent second polars, and therefore a singly infinite number of apolar binary quartics but not in general an apolar cubic.

Hence a ternary cubic may be written in an infinite number of ways as the sum of four cubes for each apolar conic it possesses, but not in general as the sum of three cubes for an arbitrary apolar conic.

The condition that the binary sextic

$$a_\lambda^6 \equiv a'^6_\lambda \equiv a''^6_\lambda \equiv a'''^6_\lambda$$

may have an apolar cubic is that the determinant formed by the coefficients of its third differential coefficients may be zero, *i.e.* that any four third polars may be linearly dependent.

This condition is

$$(aa')^2 (a'a'')^2 (a''a)^2 (aa''')^2 (a'a''')^2 (a''a''')^2 = 0. \qquad (\S\ 187)$$

Now

$$a_1^2 = c_1,$$
$$2a_1 a_2 = c_2,$$
$$a_2^2 = c_3,$$

hence

$$(aa')(a'a'')(a''a)$$
$$= \begin{vmatrix} a_1^2 & a_1 a_2 & a_2^2 \\ a'^2_1 & a_1' a_2' & a'^2_2 \\ a''^2_1 & a_1'' a_2'' & a''^2_2 \end{vmatrix} = \tfrac{1}{2}(cc'c'')$$

and the condition reduces to

$$(cc'c'')(c'c''c''')(c''c'''c)(c'''cc') = 0.$$

This is an invariant of the cubic

$$c_x^3 = c'^3_x = c''^3_x = c'''^3_x$$

and hence however we choose the apolar conic we cannot reduce the cubic to the sum of three cubes unless a certain invariant of degree four vanishes.

249. Ternary Quartic. Here there is no apolar conic unless the six second differential coefficients are linearly dependent, *i.e.* unless a certain invariant called the *Catalecticant* vanishes.

Now if the quartic can be written as the sum of five fourth powers it must have an apolar conic, because a conic can be chosen apolar to any five fourth powers—in fact we have only to describe a conic touching the five straight lines represented by the linear forms.

Hence in general a ternary quartic cannot be expressed as the sum of five fourth powers.

But if the catalecticant be zero there is an apolar conic and, taking it for a fundamental conic, the equation of the quartic may be written

$$a_\lambda^4 a_\mu^4 = 0,$$

where $a_\lambda^8 = 0$ gives the points of intersection with the conic.

Now a singly infinite number of quintics can be found apolar to a binary octavic, hence in this case the quartic curve has a singly infinite number of conjugate five-lines, and all such lines touch the apolar conic.

250. We shall conclude this chapter with a brief account of the class of invariant forms known as combinants, confining ourselves to binary forms.

An invariant or covariant of any number of binary forms

$$f_1, f_2, \dots f_r$$

of the same order is said to be a combinant if it be unaltered, except as regards a factor independent of the forms, when each form f is replaced by a linear combination of the type

$$l_1 f_1 + l_2 f_2 + \dots + l_r f_r,$$

in which the l's are constants.

For example in the case of two binary forms we have

$$(l_1 f_1 + l_2 f_2,\ m_1 f_1 + m_2 f_2)$$
$$= (lm)(f_1 f_2),$$

so that the Jacobian of two binary forms is a combinant.

For the sake of brevity we shall deal with the combinants of three binary forms

$$f_1 = a_0 x_1^n + n a_1 x_1^{n-1} x_2 + \dots + a_n x_2^n$$
$$f_2 = b_0 x_1^n + n b_1 x_1^{n-1} x_2 + \dots + b_n x_2^n$$
$$f_3 = c_0 x_1^n + n c_1 x_1^{n-1} x_2 + \dots + c_n x_2^n.$$

A combinant is not only unaltered by a linear substitution effected on the variables, but also by a linear substitution of the type

$$a_r' = l_1 a_r + m_1 b_r + n_1 c_r$$
$$b_r' = l_2 a_r + m_2 b_r + n_2 c_r$$
$$c_r' = l_3 a_r + m_3 b_r + n_3 c_r$$

effected on the coefficients.

Regarded as a function of the coefficients the combinant is therefore an invariant of the linear forms

$$a_r \xi_1 + b_r \xi_2 + c_r \xi_3, \qquad r = 1,\ 2 \dots\dots n,$$

because if we put

$$\xi_1 = l_1 \xi_1' + l_2 \xi_2' + l_3 \xi_3' \text{ etc.}$$

we find $$a_r' = l_1 a_r + m_1 b_r + n_1 c_r \text{ etc.}$$

which are the substitutions above.

Hence, by the fundamental theorem on the symbolical representation of invariants, a combinant, as far as the coefficients a, b, c are concerned, is a rational integral function of determinants of the type

$$\begin{vmatrix} a_p & a_q & a_r \\ b_p & b_q & b_r \\ c_p & c_q & c_r \end{vmatrix}.$$

A like result applies to any number of binary forms.

Thus for example for two quadratics

$$a_0x_1^2 + 2a_1x_1x_2 + a_2x_2^2$$
$$b_0x_1^2 + 2b_1x_1x_2 + b_2x_2^2,$$

a combinant is a function of

$$(a_0b_1 - a_1b_0),\ (a_1b_2 - a_2b_1),\ (a_0b_2 - a_2b_0),$$

as far as the coefficients are concerned.

But the Jacobian is

$$(a_0b_1 - a_1b_0)\,x_1^2 + (a_0b_2 - a_2b_0)\,x_1x_2 + (a_1b_2 - a_2b_1)\,x_2^2,$$

and hence a combinant is a rational integral function of the coefficients of the Jacobian and the variables. Hence any combinant is an invariant form of the Jacobian, and therefore the complete system of combinants in this case consists of the Jacobian and its discriminant—the latter is equivalent to the resultant of the two original forms.

It is easy to form any number of combinants of two binary forms, for

(i) An invariant or covariant of a combinant is itself a combinant, since it is manifestly an invariant form and further involves the coefficients of the original forms in the manner peculiar to combinants.

(ii) Let f_1 and f_2 be two binary forms, I_0 an invariant form of f_1 and the corresponding form for $\lambda_1 f_1 + \lambda_2 f_2$,

$$\lambda_1^m I_0 + \lambda_1^{m-1}\lambda_2 I_1 + \ldots + \lambda_2^m I_m,$$

then an invariant of this expression considered as a binary form in (λ_1, λ_2) is a combinant of f_1 and f_2.

For such an invariant is unaltered when we effect a linear substitution on the x's because each I is an invariant form; and it is unaltered when we effect a linear substitution on the coefficients because it is an invariant of the form in λ written above.

251. Combinants naturally occur in the discussion of rational curves as we shall now shew.

Suppose such a curve is parametrically represented by

$$\xi_1 = a_x^n \equiv f_1$$
$$\xi_2 = b_x^n \equiv f_2$$
$$\xi_3 = c_x^n \equiv f_3, \qquad \text{(cf. § 196)}$$

then the curve is unaltered by a linear substitution effected on the x's since its equation is found by eliminating the x's.

Now if a set of points on the curve be defined by some projective property the equation giving their parameters is derived from f_1, f_2, f_3 in a definite way, hence if by means of a linear substitution f_1, f_2, f_3 become f_1', f_2', f_3' the transformed equation for the parameters is derived from f_1', f_2', f_3' in the same way as its original form was derived from f_1, f_2, f_3.

Thus if the equation be $C=0$ it follows that C is a covariant of f_1, f_2, f_3.

Next, keeping the parameters fixed, to change the triangle of reference we replace ξ_1, ξ_2, ξ_3 by linear functions of themselves, so that f_1, f_2, f_3 are replaced by linear combinations of the form

$$l_1 f_1 + l_2 f_2 + l_3 f_3.$$

Now the equation giving the parameters of the set of points must be independent of the triangle of reference, for such points depend on the curve itself, and the parameters of every point of the curve are unchanged when we alter the triangle of reference; hence C is not only a covariant but a combinant of the forms f_1, f_2, f_3, and the rational curve is the natural geometrical representation of the system of combinants.

The curve can be equally well defined by the system of forms apolar to f_1, f_2, f_3, because these determine f_1, f_2, f_3, and the projective properties of the curve are also given by the combinants of the apolar system of forms.

We are therefore led to the theorem that the combinants of two apolar systems of forms are identical, and in fact a rigorous algebraic proof of its truth will be found in Meyer's *Apolarität*, § 11.

As an example the reader may verify that in the quartic curve

$$\xi_1 = a_x^4, \ \xi_2 = b_x^4, \ \xi_3 = c_x^4,$$

the points of inflexion are given by

$$(bc)(ca)(ab)\, a_x^2 b_x^2 c_x^2 = 0,$$

and that, if d_x^4 and e_x^4 be two forms belonging to the apolar system, they are also given by

$$(de)\, d_x^3 e_x^3 = 0.$$

The first equation follows from the ordinary methods of the differential calculus—the second from the fact that the conditions of collinearity of four points are easily expressed by means of the apolar system; if λ be a point of inflexion, and μ the point in which the inflexional tangent meets the curve again, we have $d_\lambda^3 d_\mu = 0$, $e_\mu^3 e_\mu = 0$ so that μ may be eliminated.

The full discussion of the theory of combinants would lead us too far from the methods of the present treatise, and accordingly we shall content ourselves with the explanation of a "translation-principle" connecting the combinants of binary forms with the covariants of ternary forms.

252. It will be convenient to change the notation and to suppose a rational curve given by

$$\left.\begin{aligned} \xi_1 &= a_1 x_1^n + n b_1 x_1^{n-1} x_2 + \ldots\ldots + k_1 x_2^n \\ \xi_2 &= a_2 x_1^n + n b_2 x_1^{n-1} x_2 + \ldots\ldots + k_2 x_2^n \\ \xi_3 &= a_3 x_1^n + n b_3 x_1^{n-1} x_2 + \ldots\ldots + k_3 x_2^n \end{aligned}\right\} .$$

Consider the problem of finding the locus of the point of intersection of two straight lines which meet the curve in two sets of n points given by binary forms for which a certain combinant is zero.

Let the two lines be

$$u_\xi = 0, \ v_\xi = 0,$$

and denote by ξ their point of intersection.

The two binary forms are

$$a_u x_1^n + n b_u x_1^{n-1} x_2 + \ldots + k_u x_2^n$$

and

$$a_v x_1^n + n b_v x_1^{n-1} x_2 + \ldots + k_v x_2^n,$$

and any combinant is a function of determinants of the type

$$\begin{vmatrix} a_u & b_u \\ a_v & b_v \end{vmatrix}.$$

Now $$a_u b_v - a_v b_u = (ab\xi),$$

hence the equation of the locus is found by changing (ab) in the expression of the vanishing combinant into $(ab\xi)$.

For example if two lines meet the cubic curve

$$\xi_1 = a_1 x_1^3 + 3b_1 x_1^2 x_2 + 3c_1 x_1 x_2^2 + d_1 x_2^3$$
etc.

in two apolar sets of points we have

$$(a_u d_v - a_v d_u) - 3(b_u c_v - b_v c_u) = 0$$

or $$(ad\xi) - 3(bc\xi) = 0,$$

and hence the locus of their common point ξ is a straight line.

It is evident that if A be a point of inflexion then the tangent at A and any line through A satisfy the conditions of the problem, so that all the points of inflexion of the curve lie on this straight line.

As a second example let us find the equation of the cubic. Here the two straight lines meet on the curve and the vanishing combinant is the λ eliminant of the two binary forms.

Following Bezout's method, the eliminant of

$$px_1^3 + qx_1^2 x_2 + rx_1 x_2^2 + sx_2^3$$

and $$p'x_1^3 + q'x_1^2 x_2 + r'x_1 x_2^2 + s'x_2^3$$

is $$\begin{vmatrix} (pq') & (pr') & (ps') \\ (pr') & (ps') + (qr') & (qs') \\ (ps') & (qs') & (rs') \end{vmatrix} = 0,$$

and hence making $p = a_u$, $q = 3b_u$ etc. the equation required is

$$\begin{vmatrix} 3(ab\xi) & 3(ac\xi) & (ad\xi) \\ 3(ac\xi) & (ad\xi) + 9(bc\xi) & 3(bd\xi) \\ (ad\xi) & 3(bd\xi) & 3(cd\xi) \end{vmatrix} = 0.$$

It is clear that a similar method applies to the curve of the nth degree.

CHAPTER XV.

TYPES.

253. It was proved in § 35 that the effect of operating with

$$\left(A\frac{\partial}{\partial B}\right) \equiv A_0\frac{\partial}{\partial B_0} + A_1\frac{\partial}{\partial B_1} + \dots + A_n\frac{\partial}{\partial B_n}$$

on a covariant Φ of a simultaneous system of binary forms, which includes a_x^n and b_x^n where

$$a_x^n \equiv (A_0, A_1, \dots A_n \between x_1, x_2)^n$$

$$b_x^n \equiv (B_0, B_1, \dots B_n \between x_1, x_2)^n,$$

is itself a covariant of the system.

All covariants thus obtained from Φ are said to be of *the same type* as Φ. In other words two covariants are said to be of the same type if one of them is obtainable from the other by means of operators of this kind. For example the invariants

$$(f, f)^2, \ (f, \phi)^2, \ (\phi, \phi)^2$$

of two quadratics f, ϕ are all of the same type.

It should be noticed that this connection between two covariants is not necessarily reciprocal; two covariants Φ_1, Φ_2, where Φ_2 is obtainable from Φ_1 by operators of the required kind, are of the same type, even if Φ_1 is not so obtainable from Φ_2. Thus if $F(a, a', \dots)$ is a simultaneous covariant of a system of quantics which includes $f = a_x^n = a'^n_x$, $\phi = b_x^n$, and if F is of the second degree in the coefficients of f but does not contain those of ϕ, the covariant

$$F(a, b, \dots) + F(b, a, \dots) = \left(\phi\frac{\partial}{\partial f}\right) F(a, a', \dots)$$

is of the same type as $F(a, a', \ldots)$; but here

$$F(a, a', \ldots) = \left(f\frac{\partial}{\partial\phi}\right) F(a, b, \ldots)$$

and we see that $F(a, a', \ldots)$ and $F(a, b, \ldots)$ are of the same type.

It will be seen in this way that two covariants Φ_1, Φ_2 may each be of the same type as a third covariant Φ, although neither Φ_1 nor Φ_2 is obtainable from the other by an operator of the kind considered. In view of this the further statement is necessary that covariants which are each of the same type as a third covariant are (by definition) of the same type as each other.

254. Every covariant of degree m, of one or more quantics, is of the same type as a covariant which is linear in the coefficients of each of m quantics—the number of quantics in the system being, of course, increased if necessary. Any such representative covariant is, for convenience, called a *type*; a type is then a covariant which is linear in the coefficients of each of the quantics concerned, it being understood that these are not special quantics of the system and that the word type is used in a purely formal sense.

Thus for three quadratics (§ 139 A)

$$(ab)(bc)(ca)$$

is an irreducible type, and furnishes only one invariant of the system; $(ab)^2$ is also an irreducible type and furnishes six invariants.

It should be noticed that if f_1, f_2, f_3 are the quadratics, and

$$J_{1,2} = (f_1, f_2),$$

the invariant $$(J_{1,2}, f_3)^2$$

is not of the same type as $(ab)^2$, because $J_{1,2}$ is not one of the fundamental quantics of the system.

Consider the covariants of a simultaneous system of binary forms of the same order. When the number of binary forms is indefinitely increased, the number of irreducible covariants will also be increased without limit; in fact the number of irreducible covariants belonging to any one type will be indefinitely increased. The question arises—does the number of irreducible types increase indefinitely too? This question has been answered in the negative by Peano*. Peano's theorem is the following: *Every type of a system of binary n-ics which does not furnish irreducible covariants for a system of n n-ics is reducible, with the single possible exception of the invariant type*

$$\begin{vmatrix} A_0 & A_1 & \dots & A_n \\ B_0 & B_1 & \dots & B_n \\ \dots & \dots & \dots & \dots \\ K_0 & K_1 & \dots & K_n \end{vmatrix},$$

where

$$(A_0, A_1, \dots A_n \between x_1, x_2)^n$$
$$(B_0, B_1, \dots B_n \between x_1, x_2)^n$$
$$\dots\dots\dots\dots\dots$$
$$(K_0, K_1, \dots K_n \between x_1, x_2)^n$$

are $n+1$ n-ics. But if this invariant is reducible, all types are reducible which do not furnish irreducible covariants for $n-1$ n-ics.

A proof of this theorem is given in the next chapter.

255. As was pointed out in § 21 there are two principles by means of which the reduction of a covariant has to be attempted, viz.:

(i) The fundamental identities

$$(bc)(ad) + (ca)(bd) + (ab)(cd) = 0$$

$$(bc)\, a_x + (ca)\, b_x + (ab)\, c_x = 0.$$

(ii) The fact that the interchange of two symbols which refer to the same quantic does not alter the actual value of a symbolical product.

To effect the reduction of a type the first of these two principles must alone be employed.

256. The quadratic types. The quadratics will be denoted, as usual, by

$$a_x^2,\ b_x^2,\ c_x^2,\ \dots.$$

* *Atti di Torino*, t. XVII. p. 580 (1881). See also Jordan (*Liouville*, 1876, 2 Sér. III.), who proved that the number of irreducible types belonging to any simultaneous system of forms, the order of each of which is less than some fixed number n, is finite.

For invariants the only symbolical products to be considered are

$$(ab)^2$$

$$(ab)(bc)(ca)$$

$$(ab)(bc)(cd)(da)$$

$$\dots\dots\dots\dots\dots\dots$$

The first two of these are irreducible, for the fundamental identities give us no relations by which we may reduce them.

The other invariant types are all reducible. For

$$2(ab)(bc)(cd)(da) = (ab)^2(cd)^2 + (bc)^2(da)^2 - (ac)^2(bd)^2;$$

operate on this identity with

$$e_1 \frac{\partial}{\partial a_1} + e_2 \frac{\partial}{\partial a_2},$$

then

$$(ab)(bc)(cd)(de) + (eb)(bc)(cd)(da)$$
$$= -(cd)^2(ab)(be) - (bc)^2(ad)(de) + (bd)^2(ac)(ce).$$

But

$$(ab)(bc)(cd)(de) - (eb)(bc)(cd)(da) = (bc)(cd)(db)(ae).$$

Hence

$$2(ab)(bc)(cd)(de)$$
$$= (bc)(cd)(db)(ae) - (cd)^2(ab)(be) - (bc)^2(ad)(de) + (bd)^2(ac)(ce).$$

By means of these two identities all the invariant types of degree greater than three are at once reduced. Now any covariant of a system of quantics of even order must be itself of even order (§ 20); hence any covariant type of the quadratic may always be obtained by replacing one or more letters in the symbolical expression for some invariant type by the variable. For example, from $(ab)^2$ we obtain a_x^2 on replacing b_1 by $-x_2$ and b_2 by x_1. Hence the irreducible covariant types are

$$a_x^2,\ (ab)\,a_x b_x.$$

The quadratic has then only four irreducible types (compare § 139 A),

$$(ab)^2,\ (ab)(bc)(ca),$$
$$a_x^2,\ (ab)\,a_x b_x.$$

The second is in fact the determinant type referred to above, for as has already been pointed out,

$$(ab)(bc)(ca) = - \begin{vmatrix} a_1^2 & a_1a_2 & a_2^2 \\ b_1^2 & b_1b_2 & b_2^2 \\ c_1^2 & c_1c_2 & c_2^2 \end{vmatrix}.$$

257. The cubic types. It is possible to obtain the complete system of types for binary forms of a given order by a method almost identical with that of Chapter VI. for covariants of a single binary form. The reductions in this method are generally very difficult to obtain. The cubic types, however, can be thus obtained simply. It is thought unnecessary to go through the general argument, the alterations in Chapter VI. to meet the case of types being mainly verbal. It should be noticed that the finiteness of the complete irreducible system of types could thus be demonstrated.

Let $a_x^3, b_x^3, c_x^3, \ldots$ be the cubics. The symbol F will be used to denote any one of them indifferently. The types of degree two are

$$(ab)\, a_x^2 b_x^2 = J, \;\; (ab)^2 a_x b_x = H, \;\; (ab)^3.$$

Consider first the types of grade unity. These all contain a factor (ab), and hence are terms of transvectants of J with types of grade not greater than unity.

In fact any such type is a term of a transvectant of the form

$$(\overline{J_1 J_2 \ldots J_r}, \;\; F_1 F_2 \ldots F_s)^\lambda,$$

where the bar over the left-hand member indicates any type obtained by convolution from the product there written down. It follows at once that every type of unit grade can be expressed as a sum of numerical multiples of such transvectants. Now by §§ 74, 75 any type obtained by convolution from $J_1, J_2, \ldots J_r$ is of grade two at least. Hence the only irreducible types of unit grade are expressible as transvectants of the form

$$(J_1 J_2 \ldots J_r, \;\; F_1 F_2 \ldots F_s)^\lambda.$$

If $\lambda = 1$ this is clearly reducible—for J is a Jacobian.

If $\lambda > 1$, this contains a term of grade two.

Therefore the only irreducible type of unit grade is J.

Now the irreducible types of the quadratic H are

$$H,\ (H_1, H_2) = K,\ (H_1 H_2)^2,\ (H_1 H_2)(H_2 H_3)(H_3 H_1).$$

Hence the types of grade two are expressible as transvectants of the form

$$(H_1 H_2 \dots H_\alpha K_1 K_2 \dots K_\beta,\ F_1 F_2 \dots F_\gamma J_1 J_2 \dots J_\delta)^\lambda$$

or of these multiplied by invariant types.

In the first place we notice that K and J are Jacobians and hence we may suppose that neither β nor δ exceeds unity.

We have the following types to consider:

$$(H, F),\ (H, F)^2,\ (H_1 H_2, F)^3,\ (H_1 H_2 H_3, F_1 F_2)^6$$
$$(H, J)^2,\ (H_1 H_2, J)^3,\ (H_1 H_2, J)^4$$
$$(K, F)^2,\ (HK, F)^3,\ (H_1 H_2 K, F_1 F_2)^6,\ (K, J)^2,\ (HK, J)^3,\ (HK, J)^4.$$

Of these $(H_1 H_2 H_3, F_1 F_2)^6$ contains the term

$$(a_1 b_1)^2 (a_2 b_2)^2 (a_3 b_3)^2 (a_1 c_1)(b_1 c_1)(a_2 c_2)(b_2 c_2)(a_3 c_1)(b_3 c_2)$$
$$= (a_1 b_1)(b_1 c_1)(c_1 a_1) \, . \, (a_2 b_2)(b_2 c_2)(c_2 a_2) \, . \, (a_3 b_3)^2 (a_1 b_1)(a_2 b_2)(a_3 c_1)(b_3 c_2)$$
$$= \tfrac{1}{2} \begin{vmatrix} (a_1 a_2)^2 & (a_1 b_2)^2 & (a_1 c_2)^2 \\ (b_1 a_2)^2 & (b_1 b_2)^2 & (b_1 c_2)^2 \\ (c_1 a_2)^2 & (c_1 b_2)^2 & (c_1 c_2)^2 \end{vmatrix} (a_3 b_3)^2 (a_1 b_1)(a_2 b_2)(a_3 c_1)(b_3 c_2), \dots \ (\S\ 77).$$

This is a sum of terms obtained by convolution from products of four types H, and hence is reducible. In exactly the same way the type $(H_1 H_2 K, F_1 F_2)^6$ may be reduced.

The type $(H, J)^2$ contains a term

$$(ab)^2 (bc)(ad)(cd)\, c_x d_x$$
$$= -\tfrac{1}{2} (ab) \{(ab)^2 (cd)^2 + (bc)^2 (ad)^2 - (ac)^2 (bd)^2\} c_x d_x,$$

which may be expressed in terms of the type $(H_1 H_2)$ and reducible forms.

The type $(K, J)^2$ contains the term

$$(H_1 H_2)(H_2 c)(H_1 d)(cd)\, c_x d_x$$
$$= -\tfrac{1}{2} \{(H_1 H_2)^2 (cd)^2 + (H_2 c)^2 (H_1 d)^2 - (H_2 d)^2 (H_1 c)^2\} c_x d_x,$$

and hence is reducible.

The type $(H_1 H_2, J)^3$ contains the term

$$(H_1, (H_2, J)^2) = \Sigma (H_1, (H_2, H_3)) + \text{reducible terms},$$

and hence is reducible. In the same way the types $(H_1 H_2, J)^4$, $(HK, J)^3$, $(HK, J)^4$ may be reduced.

The type

$$(K, F)^2 = (H_1H_2)(H_2F)(H_1F)F_x$$
$$= (H_1F)^2(H_2F)H_{2x} - (H_2F)^2(H_1F)H_{1x};$$

and $$2(H_1H_2, F)^3 = (H_1F)^2(H_2F)F_x + (H_2F)^2(H_1F)F_x.$$

Hence both $(K, F)^2$ and $(H_1H_2, F)^2$ can be expressed in terms of the type

$$(H_1F)^2(H_2F)F_x = L_x.$$

Lastly, $(HK, F)^3$ contains the term

$$(a_1b_1)^2(a_2b_2)^2(a_3b_3)^2(b_1a_2)(a_1c)(a_3c)(b_3c)b_{2x}$$
$$= ((a_1b_1)^2(a_3b_3)^2(b_1a_2)(a_1c)(a_3c)(b_3c)a_{2x}^2, b_{2x}^3)^2.$$

Now the left-hand member of this transvectant is a type of degree six and order two; looking back at the possible types of this order we see that it must be of the form

$$\lambda L_1L_2 + \mu(ab)^3K + \nu(H_1H_2)^2H_3.$$

The second and third of these terms contain invariant factors, and can therefore only lead to reducible terms in the above transvectant. Also

$$2L_1L_2 = 2(ab)^2(ac)(bc)c_x \,.\, (de)^2(df)(ef)f_x$$
$$= (ab)(de)c_xf_x \begin{vmatrix} (ad)^2 & (ae)^2 & (af)^2 \\ (bd)^2 & (be)^2 & (bf)^2 \\ (cd)^2 & (ce)^2 & (cf)^2 \end{vmatrix}$$
$$= \Sigma(H_1H_2)^2H_3 + \Sigma(H_1H_2)(H_2H_3)H_{1x}H_{3x}$$
$$= \Sigma(H_1H_2)^2H_3.$$

The type $(HK, F)^3$ is thus reducible.

The complete system of irreducible cubic types is then

$$(ab)^3,\ (ab)^2(bc)(cd)^2(da),\ (ab)^2(bc)(cd)^2(de)(ef)^2(fa),$$
$$(ab)^2(ac)(bc)c_x,\ (ab)^2(cd)^2(ae)(be)(ce)d_x,$$
$$(ab)^2a_xb_x,\ (ab)^2(bc)(cd)^2a_xd_x,$$
$$a_x^3,\ (ab)^2(bc)a_xc_x^2,$$
$$(ab)a_x^2b_x^2,$$

there being ten types in all.

The system of irreducible concomitants for two cubics may be obtained from the system of types or else directly, they will be found in the works of Clebsch and Gordan. The syzygies between them have been obtained by von Gall (*Math. Ann.* Bd. XXXI.).

258. Perpetuants. The irreducible seminvariants (§ 32) of the binary form of infinite order are called *perpetuants*. The complete system of perpetuants for one binary form of infinite order has been obtained by Macmahon* and Stroh†. The system is, of course, infinite in extent, but the individual members of it have all been identified.

The complete system of perpetuants for any simultaneous system of binary quantics of infinite order was obtained by Macmahon‡; and in particular the perpetuant types may be at once obtained from this paper.

The method by which these results were obtained, does not fall within the scope of this book. The results have been obtained more recently by means of the symbolical notation which has been here developed; and this investigation§ we shall follow.

259. A covariant is completely defined when the determinant factors in its symbolical expression are known; it will be convenient to use this part of the symbolical expression only. In dealing with forms of infinite order, it must be remembered that the complete expression for a covariant contains each of the factors a_x, b_x, ... raised to an indefinitely high power.

The identity

$$(bc)\, a_x + (ca)\, b_x + (ab)\, c_x = 0$$

may now be written

$$(bc) + (ca) + (ab) = 0.$$

By means of this identity any factor (bc), in a covariant, which does not contain a may be replaced by

$$(ac) - (ab),$$

i.e. by factors which do contain a. Thus all covariant types may be expressed in terms of those which are of the form

$$(ab)^\lambda (ac)^\mu (ad)^\nu \ldots$$

where a is any one of the letters chosen at will.

* *Proc. Lond. Math. Soc.*, vol. XXVI. See also *Am. Journal*, vols. VII. VIII.

† *Math. Ann.*, Bd. XXXVI.

‡ *Camb. Phil. Soc. Trans.*, vol. XIX. pp. 234—248.

§ Grace, *Proc. Lond. Math. Soc.*, vol. XXXV.

The types of this form are all linearly independent, for no linear algebraical identity can connect their symbolical expressions.

Hence if all reducible types were expressed in terms of types of this form, we should be able to write down the perpetuant types.

It must be remembered that a is a perfectly definite quantic of the system. Further that the remaining quantics concerned in any particular covariant will be considered in a particular order *determined beforehand.*

260. Consider the types of degree three; if w be the weight, we know that

$$\begin{aligned}(bc)^w &= \{(ac) - (ab)\}^w \\ &= (ac)^w - w\,(ab)\,(ac)^{w-1} + \ldots\ldots + (-1)^w\,(ab)^w.\end{aligned}$$

Hence the covariant $(ab)\,(ac)^{w-1}$ is expressible in terms of reducible covariants and of covariants in which the index of (ab) is greater than unity. Hence all perpetuant types of degree three are expressible in terms of the types

$$(ab)^\lambda\,(ac)^\mu, \quad \lambda \not< 2, \quad \mu \not< 1.$$

It should be noticed that of the three quantics concerned any one may be chosen to correspond to a, b or c respectively.

Further, the only reducible covariants of degree three and weight w are represented by

$$(bc)^w,\ (ab)^w,\ (ac)^w,$$

and hence the seminvariants $(ab)^\lambda\,(ac)^\mu\ (\lambda \not< 2,\ \mu \not< 1)$ are both independent and irreducible.

261. Types of degree four may all be expressed in terms of the independent forms

$$(ab)^\lambda\,(ac)^\mu\,(ad)^\nu.$$

If λ or μ be less than 2, then as in the previous paragraph the index of (ab) or (ac), as the case may be, can be increased at the expense of the index of (ad).

Thus, since the reducible covariant

$$(ab)^\lambda (cd)^{w-\lambda} = (ab)^\lambda \{(ad) - (ac)\}^{w-\lambda}$$

$$= (ab)^\lambda (ad)^{w-\lambda} - \binom{w-\lambda}{1} (ab)^\lambda (ac)(ad)^{w-\lambda-1}$$

$$+ \binom{w-\lambda}{2} (ab)^\lambda (ac)^2 (ad)^{w-\lambda-2} - \dots,$$

the covariant $(ab)^\lambda (ac)(ad)^{w-\lambda-1}$ can be expressed in terms of reducible forms and of covariants

$$(ab)^\lambda (ac)^\mu (ad)^{w-\mu-\lambda} \qquad (\mu > 1).$$

When both λ and μ are greater than unity, say $\lambda + \mu = M$, we may express, by means of Stroh's series § 64, the products

$$(ab)^\lambda (ac)^\mu$$

in terms of the following three sets:

(i) $(ab)^M, (ab)^{M-1}(ac), \dots (ab)^4 (ac)^{M-4}$;

(ii) $(ac)^M, (ac)^{M-1}(ab)$;

(iii) $(bc)^M, (bc)^{M-1}(ab)$.

The products contained in (ii) and (iii) need not be considered, for the corresponding covariants can be expressed linearly in terms of reducible covariants and of covariants in which the number of factors involving a, b, c only is greater than $\lambda + \mu$. These latter forms can be dealt with in the same way.

Thus we see that ultimately we can express all the covariants of degree four in terms of reducible covariants and of such as have the factor $(ab)^4$. Further we have seen that we may suppose the coefficient of (ac) to be greater than unity: hence all covariants of degree four can be expressed in terms of reducible covariants and of covariants of the form

$$(ab)^\lambda (ac)^\mu (ad)^\nu$$

where $\lambda \nless 4$, $\mu \nless 2$, $\nu \nless 1$, and the arrangement of the letters a, b, c, d has been fixed beforehand.

262. The theorem can now be proved in general by induction. We shall assume that all covariant types, of a system of binary forms of infinite order, which are of degree $n+1$ or less, can be expressed linearly in terms of reducible covariants and of covariants of the form

$$(aa_1)^{\lambda_1} (aa_2)^{\lambda_2} \dots (aa_n)^{\lambda_n},$$

where $\qquad \lambda_1 \nleqslant 2^{n-1}, \quad \lambda_2 \nleqslant 2^{n-2}, \quad \dots \lambda_n \nleqslant 1,$

and the arrangement of the letters $a, a_1, a_2, \dots a_n$ is fixed.

For degree $n + 2$ we have only to consider the covariants of the form

$$(aa_1)^{\lambda_1} (aa_2)^{\lambda_2} \dots (aa_{n+1})^{\lambda_{n+1}} = (aa_1)^{\lambda_1} (aa_2)^{\lambda_2} R.$$

Now $(aa_2)^{\lambda_2} R$ is of the same symbolical form as a covariant of degree $n + 1$ of the system; hence, using the result for that degree, we may, if $\lambda_2 < 2^{n-1}$, express it in terms of covariants of the same form but for which the index of (aa_2) is not less than 2^{n-1}, and of reducible covariants. In the same way, if $\lambda_1 < 2^{n-1}$, the index of (aa_1) in the product $(aa_1)^{\lambda_1} R$ can be increased.

Let $\qquad \lambda_1 + \lambda_2 = M \nleqslant 2 \cdot 2^{n-1}, \quad N \equiv 2^{n-1};$

then, as before, by means of Stroh's series all products $(aa_1)^{\lambda} (aa_2)^{\mu}$ can be expressed in terms of the $M + 1$ following products:

(i) $\quad (aa_1)^M, \; (aa_1)^{M-1}(aa_2), \; \dots (aa_1)^{2N}(aa_2)^{M-2N},$

(ii) $\quad (aa_2)^M, \; (aa_2)^{M-1}(aa_1), \; \dots (aa_2)^{M-N+1}(aa_1)^{N-1},$

(iii) $\quad (a_1a_2)^M, \; (a_1a_2)^{M-1}(aa_1), \; \dots (a_1a_2)^{M-N+1}(aa_1)^{N-1}.$

The products contained in (ii) and (iii) need not be considered, for the corresponding covariants have factors of the form $(aa_1)^{\rho}R$ where $\rho < 2^{n-1}$; hence these covariants can be expressed in terms of reducible forms and of products which contain a greater number of factors involving a, a_1, a_2 only.

The products contained in (i) all contain the factor $(aa_1)^{2N}$. Hence all covariants of degree $n + 2$ are expressible in terms of reducible forms and of covariants which have the symbolical factor $(aa_1)^{2N}$. But these can, by an application of the assumed result for degree $n + 1$, be expressed in terms of reducible forms and of the covariants

$$(aa_1)^{\lambda_1} (aa_2)^{\lambda_2} \dots (aa_{n+1})^{\lambda_{n+1}},$$

where $\qquad \lambda_1 \nleqslant 2^n, \quad \lambda_2 \nleqslant 2^{n-1}, \quad \dots \lambda_{n+1} \nleqslant 1.$

The theorem is then true for degree $n+2$ if it is true for degree $n+1$; it has been proved for degrees three and four and is therefore true for all degrees.

263. It should be noticed that it has not been proved that the covariants retained are irreducible. It is practically certain that this is so, but no rigorous proof has yet been given. The number of covariants retained which are of degree $n+1$ and weight w may be found as follows. If

$$w < 2^{n-1} + 2^{n-2} + \ldots + 1,$$

i.e. if $w < 2^n - 1$, all the covariants are reducible. If $w \nless 2^n - 1$, the covariants retained are of the form

$$(a\alpha_1)^{2^{n-1}} (a\alpha_2)^{2^{n-2}} \ldots (a\alpha_n) \,.\, R,$$

where R is any product

$$(a\alpha_1)^{\lambda_1} (a\alpha_2)^{\lambda_2} \ldots (a\alpha_n)^{\lambda_n},$$

and $$\lambda_1 + \lambda_2 + \ldots + \lambda_n = w - 2^n + 1.$$

Hence the number required is the coefficient of x^{w-2^n+1} in the expansion of $(1-x)^{-n}$—this being the number of homogeneous products of dimensions $w - 2^n + 1$ of n letters. This is equal to the coefficient of x^w in the expansion of

$$\frac{x^{2^n-1}}{(1-x)^n}.$$

This *generating function* for perpetuant types is the same as that obtained by Macmahon's methods.

264. The results thus obtained for perpetuant types are of great use in obtaining either the types or the ordinary covariants of a binary form of finite order. All that was required in the course of the argument was that the weight of the covariant under consideration should not exceed the order of the quantic—or quantics. Thus any covariant of weight w and of degree δ of the binary n-ic which is such that

$$2^{\delta-1} - 1 > w \ngtr n,$$

is reducible.

In § 114 the system of forms A_8 for a single binary form of order $\geqslant 12$ was discussed. The above considerations of weight, alone shew that the following forms which were there retained are reducible

$$(ab)^6(bc)^2(cd)^2(de), \quad (ab)^6(bc)^3(cd)^2(de), \quad (ab)^6(bc)(cd)^4(de)^2;$$

the argument applies to the last of these three covariants only if the order of the binary quantic a_x^n is greater than 12, it will be seen later that this is indeed reducible for the 12-ic.

The remaining forms of the system would not be reducible as perpetuants and hence we cannot hope to reduce them for forms of finite order; the two forms

$$(ab)^6(bc)^3(cd)^2, \quad (ab)^6(bc)(cd)^4,$$

however, are congruent mod. $(ab)^8$, as will be presently proved.

265. The theorem for covariants of forms of finite order corresponding to that which has been proved for perpetuant types is the following*.

All covariants which are of the first degree in the coefficients of each of the quantics

$$a_{1x}^{n_1}, \quad a_{2x}^{n_2}, \ldots a_{\delta x}^{n_\delta}$$

can be expressed linearly in terms of

(i) *covariants of the form*

$$(a_1a_2)^{\lambda_1}(a_2a_3)^{\lambda_2}\ldots(a_{\delta-1}a_\delta)^{\lambda_\delta},$$

where $\lambda_1 \not< 2^{\delta-2}, \quad \lambda_2 \not< 2^{\delta-3}, \ldots \lambda_\delta \not< 1,$

and the arrangement of the letters $a_1, a_2, \ldots a_\delta$ *is fixed;*

(ii) *covariants which have a symbolical factor*

$$(a_ha_k)^\lambda(a_ka_l)^{n_k-\lambda};$$

(iii) *products of covariants of lower total degree.*

The proof of this theorem follows that for perpetuants very closely. We first assume that it is true when the total degree of the covariant considered is not greater than $\delta-1$, and prove it when this total degree is δ.

Now the covariants to be considered can be expressed in terms of transvectants

$$(a_{1x}^{n_1}, C_{\delta-1})^\mu \ \ldots\ldots\ldots\ldots\ldots\ldots\ldots\ldots \text{(I)},$$

* Young, *Proc. Lond. Math. Soc.* 1903.

where $C_{\delta-1}$ is a covariant of the first degree in the coefficients of each of the quantics $a_{1_x}^{n_1}$, $a_{2_x}^{n_2}$, ... $a_{\delta_x}^{n_\delta}$.

On the assumption made $C_{\delta-1}$ can be expressed linearly in terms of covariants of the second class, of covariants of the form

$$(a_2 a_3)^{\lambda_2} (a_3 a_4)^{\lambda_3} \dots (a_{\delta-1} a_\delta)^{\lambda_\delta},$$

and of products of covariants of lower total degree.

If $C_{\delta-1}$ is of the second class the transvectant (I), and each of its terms, must be of the second class.

If $C_{\delta-1}$ is a product of two covariants P, Q, then the total degree of P being less than δ we may express it in terms of covariants of the form

$$(ab)^{\mu_1} (bc)^{\mu_2} \dots (fg)^{\mu_i},$$

and of covariants of the second and third classes.

If $\mu + \mu_1$ be greater than the order of the form a the transvectant contains a term of the second class; if it be not greater than this order the transvectant contains a reducible term.

If P is of the second class the transvectant itself belongs to the second class; and if P is of the third class, we may take one of its factors and proceed as before.

If $C_{\delta-1}$ belongs to the first class, we may take

$$C_{\delta-1} = (a_2 a_3)^{\lambda_2} (a_3 a_4)^{\lambda_3} \dots (a_{\delta-1} a_\delta)^{\lambda_\delta},$$

then when $\lambda_2 + \mu \nless n_2$, the transvectant contains a term belonging to the second class; but when $\lambda_2 + \mu < n_2$ it contains the term

$$(a_1 a_2)^{\mu} (a_2 a_3)^{\lambda_2} (a_3 a_4)^{\lambda_3} \dots (a_{\delta-1} a_\delta)^{\lambda_\delta},$$

and hence covariants of this form alone need be considered.

266. Let us now, for the sake of shortness, write

$$a_{4_x}^{n_4-\lambda_4} q_y^{\rho}$$
$$\equiv (a_4 a_5)^{\lambda_4} (a_5 a_6)^{\lambda_5} \dots (a_{\delta-1} a_\delta)^{\lambda_{\delta-1}} a_{4_x}^{n_4-\lambda_4} a_{5_y}^{n_5-\lambda_4-\lambda_5} \dots a_{\delta_y}^{n_\delta-\lambda_{\delta-1}}.$$

Then we shall proceed to shew that, if $\kappa < 2^{\delta-3}$, the transvectant

$$((a_1 a_2)^{\mu} (a_2 a_3)^{\kappa}, a_{4_x}^{n_4-\lambda_4} q_y^{\rho})^r{}_{y=x}$$

can be linearly expressed in terms of covariants of the second and third classes, and of covariants which contain a greater number of factors involving a_1, a_2, a_3 only.

The two sets of covariants $(a_1a_2)^\mu (a_2a_3)^\kappa$, and $((a_1a_2)^\mu, a_{3_x}^{n_3})^\kappa$, where $\mu+\kappa$ has a fixed value and κ takes all possible values less than $2^{\delta-3}$, are equivalent.

Hence if $$(a_1a_2)^\mu \equiv \alpha_x^{n_1+n_2-2\mu},$$

the above transvectants may be linearly expressed in terms of the following,

$$((\alpha a_3)^\kappa, a_{4_x}^{n_4-\lambda_4} q_y^\rho)^\tau.$$

Any one of these transvectants is a covariant of unit degree in the coefficients of each of the $\delta-1$ quantics

$$\alpha_x^{n_1+n_2-2\mu}, \quad a_{3_x}^{n_3}, \quad a_{4_x}^{n_4}, \; \ldots \; a_{\delta_x}^{n_\delta};$$

it can therefore, by hypothesis, be expressed in terms of covariants of the form

$$(\alpha a_3)^{\mu_1} (a_3a_4)^{\mu_2} \ldots (a_{\delta-1}a_\delta)^{\mu_{\delta-2}},$$

where $$\mu_1 \not< 2^{\delta-3}, \quad \mu_2 \not< 2^{\delta-4}, \; \ldots \; \mu_{\delta-2} \not< 1,$$

and of covariants belonging to the second and third classes. Thus the number of factors involving a_1, a_2, a_3 only, can be increased when $\kappa < 2^{\delta-3}$.

It should be noticed that covariants of the second class here include those which contain the factor

$$(a_h\alpha)^\nu (\alpha a_k)^{n_1+n_2-2\mu-\nu}.$$

It is easy to see that such a covariant belongs to the second class in the enunciation; for, we may suppose

$$\nu \geqslant n_1+n_2-2\mu-\nu, \quad n_1 \leqslant n_2,$$

and therefore $$\nu \geqslant n_1-\mu,$$

the covariant considered then contains the factor

$$(a_ha_1)^{n_1-\mu} (a_1a_2)^\mu.$$

267. The covariant

$$(a_1a_2)^\mu (a_2a_3)^{\lambda_2} (a_3a_4)^{\lambda_3} \ldots (a_{\delta-1}a_\delta)^{\lambda_{\delta-1}}$$

is a term of the transvectant

$$((a_1a_2)^\mu (a_2a_3)^{\lambda_2}, a_{4_x}^{n_4-\lambda_4} q_y^\rho)^{\lambda_3}{}_{y=x},$$

and hence differs from the whole transvectant or from any one of its terms by covariants in which the number of factors involving a_1, a_2, a_3 only is greater than $\lambda_2 + \mu$.

By § 266, we see that we may suppose that neither λ_2 nor μ is less than $2^{\delta-3}$; and hence that $\lambda_2 + \mu \nless 2^{\delta-2}$. The covariant

$$(a_1 a_2)^{\mu} (a_2 a_3)^{\lambda_2}$$

can be linearly expressed in terms of the covariants

(i) $(a_1 a_2)^{\lambda_2+\mu}, (a_1 a_2)^{\lambda_2+\mu-1} (a_2 a_3), \ldots (a_1 a_2)^{2^{\delta-2}} (a_2 a_3)^{\lambda_2+\mu-2^{\delta-2}}$,

(ii) $(a_1 a_3)^{\lambda_2+\mu}, (a_1 a_3)^{\lambda_2+\mu-1} (a_3 a_2), \ldots (a_1 a_3)^{\lambda_2+\mu-2^{\delta-3}+1} (a_3 a_2)^{2^{\delta-3}-1}$,

(iii) $(a_2 a_3)^{\lambda_2+\mu}, (a_2 a_3)^{\lambda_2+\mu-1} (a_3 a_1), \ldots (a_2 a_3)^{\lambda_2+\mu-2^{\delta-3}+1} (a_3 a_1)^{2^{\delta-3}-1}$.

Transvectants of a covariant from one of the last two rows with $a_{4x}^{n_4-\lambda_4} q_y^{\rho}$ can be expressed in terms of covariants which contain a greater number of factors involving a_1, a_2, a_3 only. Hence we may ultimately express all covariants in question linearly in terms of covariants having a factor $(a_1 a_2)^{\lambda_1}$, where $\lambda_1 \nless 2^{\delta-2}$, and of covariants belonging to the second and third classes. Proceeding, as in § 266, with the covariants which have a factor $(a_1 a_2)^{\lambda_1}$ where $\lambda_1 \nless 2^{\delta-2}$, we see that all covariants may be expressed linearly in terms of covariants of the form

$$(a_1 a_2)^{\lambda_1} (a_2 a_3)^{\lambda_2} \ldots (a_{\delta-1} a_{\delta})^{\lambda_{\delta}},$$

where $$\lambda_1 \nless 2^{\delta-2},\ \lambda_2 \nless 2^{\delta-3},\ \ldots\ \lambda_{\delta} \nless 1,$$

and of covariants of the second and third classes.

Thus the theorem is true when the total degree of the covariant is δ, provided that it is true when this total degree is less than δ.

268. It remains to shew that this theorem is true when the total degree is three.

The covariants to be considered are

$$(a_1 a_2)^{\lambda_1} (a_2 a_3)^{\lambda_2} (a_3 a_1)^{\lambda_3}.$$

Unless $\lambda_1 + \lambda_2 + \lambda_3$ is less than each of the numbers n_1, n_2, n_3, this covariant belongs to the second class. For let

$$\lambda_1 + \lambda_2 + \lambda_3 \geqslant n_2,$$

then by means of the identity

$$(a_3 a_1) = -(a_1 a_2) - (a_2 a_3)$$

the above covariant can be expressed in terms of the forms

$$(a_1 a_2)^r (a_2 a_3)^{n_2 - r} (a_3 a_1)^{\lambda_1 + \lambda_2 + \lambda_3 - n_2},$$

and belongs to the second class.

If $\lambda_1 + \lambda_2 + \lambda_3$ is less than each of n_1, n_2, n_3 the argument used for perpetuants may be repeated here word for word. The theorem is then true for total degree three, and therefore for any total degree.

269. If all the quantics are of the same order n we obtain a theorem concerning covariant types of a simultaneous system of binary n-ics.

In this case the covariants of the second class contain a factor of the form $(ab)^\lambda (bc)^{n-\lambda}$, and hence a factor of the form

$$(ab)^\mu (bc)^{n-\mu} (ca)^\rho,$$

where $$\mu \nless \frac{n}{2}, \quad \rho \nless 2n - 3\mu,$$

(see § 68).

Hence *all covariant types of a system of binary n-ics can be expressed linearly in terms of*

(i) *Covariants of the form*

$$(a_1 a_2)^{\lambda_1} (a_2 a_3)^{\lambda_2} \dots (a_{\delta-1} a_\delta)^{\lambda_\delta},$$

where $$\lambda_1 \nless 2^{\delta-2}, \; \lambda_2 \nless 2^{\delta-3}, \; \dots \; \lambda_\delta \nless 1,$$

and the order of the letters is fixed beforehand.

(ii) *Covariants which have a factor of the form*

$$(ab)^\lambda (bc)^{n-\lambda} (ca)^\rho,$$

where $$\lambda \nless \frac{n}{2}, \quad \rho \nless 2n - 3\lambda.$$

(iii) *Products of covariants of lower total degree.*

270. The theorem just proved expresses all irreducible covariants, of grade $\leqslant \frac{n}{2}$, in terms of a certain number of forms, which, there is good reason to believe, are irreducible when n is infinite. If this be so, these forms are certainly irreducible for finite values of n.

However, it does not follow that we cannot express them in terms of covariants of higher grade or else of covariants of the second class.

In this connection we shall prove the following: *all covariant types of the binary n-ic can be expressed linearly in terms of covariants of the form*

$$(a_1 a_2)^{\lambda_1} (a_2 a_3)^{\lambda_2} \dots (a_{\delta-1} a_\delta)^{\lambda_{\delta-1}},$$

where $\lambda_1 \geqslant 2\lambda_2,\ \lambda_2 > \lambda_3,\ \lambda_3 > \lambda_4,\ \dots\ \lambda_{\delta-2} > \lambda_{\delta-1},$

and of covariants belonging to the second and third classes.

Using the previous theorem we see that covariants of the form

$$C = (a_1 a_2)^{\mu_1} (a_2 a_3)^{\mu_2} \dots (a_{\delta-1} a_\delta)^{\mu_{\delta-1}}$$

alone need be considered.

Let $\mu_1 \not> 2\mu_2$, then C is a term of

$$((a_1 a_2)^{\mu_1} (a_2 a_3)^{\mu_2}, (a_4 a_5)^{\mu_4} \dots (a_{\delta-1} a_\delta)^{\mu_{\delta-1}} a_{4_x}^{\ n-\delta_4} a_{5_y}^{\ n-\delta_4-\delta_5} \dots)^{\mu_3}{}_{y=x},$$

and differs from any other term by covariants which involve a greater number of factors containing a_1, a_2, a_3 only.

Then by Stroh's theorem, we may express $(a_1 a_2)^{\mu_1} (a_2 a_3)^{\mu_2}$ in terms of covariants of the form

$$(ab)^{\lambda_1} (bc)^{\lambda_2},$$

where $\lambda_1 \not< 2\lambda_2,\quad \lambda_1 + \lambda_2 = \mu_1 + \mu_2,$

and a, b, c are the letters a_1, a_2, a_3 in some order.

Let $(ab)^{\lambda_1} \equiv \alpha_x^{2n-2\lambda_1},$

we have then to consider covariants of the form

$$(\alpha a_3)^{\lambda_2} (a_3 a_4)^{\mu_3} \dots (a_{\delta-1} a_\delta)^{\mu_{\delta-1}}.$$

If $\lambda_2 < \mu_3$ we may consider the transvectant

$$((\alpha a_3)^{\lambda_2} (a_3 a_4)^{\mu_3}, (a_5 a_6)^{\mu_5} \dots (a_{\delta-1} a_\delta)^{\mu_{\delta-1}} a_{5_x}^{\ n-\mu_5} a_6^{\ n-\mu_5-\mu_6})^{\mu_4}{}_{y=x}.$$

Then $(\alpha a_3)^{\lambda_2} (a_3 a_4)^{\mu_3}$ can be expressed linearly in terms of $(a_3 a_4)^{\lambda_2+\mu_3}$, and of members of the sets

$$(\alpha a_3)^{\kappa} (a_3 a_4)^{\lambda_2+\mu_3-\kappa},\quad (\alpha a_4)^{\kappa} (a_4 a_3)^{\lambda_2+\mu_3-\kappa},$$

where $\kappa > \dfrac{\lambda_2 + \mu_3}{2}.$

We may proceed in exactly the same way at every step, and so prove the theorem.

It will be noticed here that the order of the letters in the covariants

$$(a_1 a_2)^{\lambda_1} (a_2 a_3)^{\lambda_2} \dots (a_{\delta-1} a_\delta)^{\lambda_{\delta-1}}$$

is not fixed.

271. The Maximum Order of a covariant. Let us consider the covariant types of a system of quantics of which none of the orders exceeds n. By § 265, these types can all be expressed in terms of covariants of three kinds. Consider the covariants of the second kind. These contain a factor of the form

$$(a_1 a_2)^{\lambda} (a_2 a_3)^{n_2-\lambda}.$$

We may suppose that $\lambda \nless n_2 - \lambda$; the order of the covariant

$$(a_1 a_2)^{\lambda} a_{1_x}{}^{n_1-\lambda} a_{2_x}{}^{n_2-\lambda}$$

is then $$n_1 + n_2 - 2\lambda \ngtr n_1 \leqslant n.$$

Now let us introduce a new symbol, for each covariant whose order does not exceed n. Covariants of the second kind are thus at once reduced in degree. Covariants thus reduced may themselves be expressed in terms of covariants of the three different kinds. The covariants of the second kind may again be reduced, and so on. Hence finally we have expressed the system of covariants in terms of covariants of the form

$$(a_1 a_2)^{\lambda_1} (a_2 a_3)^{\lambda_2} \dots (a_{\delta-1} a_\delta)^{\lambda_{\delta-1}},$$

where $$\lambda_1 \nless 2^{\delta-2},\ \lambda_2 \nless 2^{\delta-3},\ \dots\ \lambda_{\delta-1} \nless 1,$$

—$a_{1_x}{}^{n_1}$, $a_{2_x}{}^{n_2}$, ... $a_{\delta_x}{}^{n_\delta}$ being either members of the original system of quantics or covariants of that system whose order does not exceed n—; and of products of covariants of lower degree.

The covariant of maximum order must then be of the form

$$(a_1 a_2)^{\lambda_1} (a_2 a_3)^{\lambda_2} \dots (a_{\delta-1} a_\delta)^{\lambda_{\delta-1}} a_{1_x}{}^{n-\lambda_1} a_{2_x}{}^{n_2-\lambda_1-\lambda_2} \dots a_{\delta_x}{}^{n_\delta-\lambda_{\delta-1}},$$

where $$\lambda_1 \nless 2^{\delta-2},\ \lambda_2 \nless 2^{\delta-3},\ \dots\ \lambda_{\delta-1} \nless 1,$$

and $n_1, n_2, \dots n_\delta$ are all equal to or less than n.

The order of this for a given value of δ is a maximum when

$$\lambda_1 = 2^{\delta-2},\ \lambda_2 = 2^{\delta-3},\ \dots\ \lambda_{\delta-1} = 1,$$

$$n_1 = n_2 = \dots = n_\delta = n.$$

In this case the order is

$$n\delta - 2(1 + 2 + \ldots + 2^{\delta-2}) = n\delta - 2^{\delta} + 2.$$

The maximum order is then the greatest of the numbers

$$n,\ 2n - 2,\ 3n - 6,\ \ldots\ n\delta - 2^{\delta} + 2,\ \ldots.$$

It is easy to see that if $n = 2^{\lambda} + n_1$, where $n_1 < 2^{\lambda}$, then this maximum order is

$$(\lambda + 1)(2^{\lambda} + n_1) - 2^{\lambda+1} + 2 = (\lambda - 1)\,2^{\lambda} + n_1(\lambda + 1) + 2.$$

Comparison with perpetuants shews at once, that if the results for these are absolutely accurate, then the maximum order just obtained is always reached—even for a single quantic of order n except for the case $n = 3$.

Ex. (i). The covariant $(ab)^6(bc)(cd)^4(de)^2$ of the twelvic referred to in § 264 is of weight 13 and hence must be reducible to covariants of the second and third classes; it is evidently reducible in the usual sense of the word.

Ex. (ii). Shew that the following covariants of the ten-ic can be expressed in terms of reducible covariants and of covariants of higher grade:

$$(ab)^4(bc)^3(cd),$$
$$(ab)^6(bc)^3(cd)^3,$$
$$(ab)^6(bc)^3(cd)^2(de).$$

References to papers by Jordan and Sylvester on the problem of § 271 and allied problems regarding weight and degree will be found in Meyer. The limits hitherto given are much too high for large values of n.

CHAPTER XVI.

GENERAL THEOREMS ON QUANTICS.

272. In this chapter certain results are obtained by an application of the theory, or rather the notation of the theory, of finite substitution groups. So little knowledge of this subject is required, that, for the sake of readers unacquainted with it, we shall start from the commencement, and prove the few well-known theorems required.

In the first place a function of n variables

$$f(x_1, x_2, x_3, \dots x_n)$$

is under consideration; the function

$$f(x_2, x_1, x_3, \dots x_n)$$

is derived from this by the interchange of the two variables x_1 and x_2. The operation by which the latter function is obtained from the former is called a substitution, it is usually denoted by the symbol $(x_1 x_2)$. Thus we may write

$$f(x_2, x_1, x_3, \dots x_n) = (x_1 x_2) f(x_1, x_2, x_3, \dots x_n).$$

A more general example of a substitution is the operation by which the arrangement of variables

$$x_1, x_2, \dots x_n$$

is changed to

$$y_1, y_2, \dots y_n,$$

the y's being the variables $x_1, x_2, \dots x_n$ arranged in some order. This substitution is often written

$$\begin{pmatrix} x_1 & x_2 \dots & x_n \\ y_1 & y_2 \dots & y_n \end{pmatrix};$$

thus

$$\begin{pmatrix} x_1 & x_2 \dots & x_n \\ y_1 & y_2 \dots & y_n \end{pmatrix} f(x_1, x_2, \dots x_n) = f(y_1, y_2, \dots y_n).$$

Substitutions which represent merely the interchange of two variables are called transpositions; thus the substitution (x_1x_2) introduced above is a transposition.

The product of two substitutions. Let s_1, s_2 be any two substitutions of the letters $x_1, x_2, \dots x_n$, the meaning here attached to the product s_1s_2 is that it is an operation which when applied to a function of $x_1, x_2, \dots x_n$ is equivalent to the operation first of s_2 on this function and then of s_1 on the resulting function. (The usual convention is that the substitution on the left is the first to operate, but the above is more convenient for our present purpose.) Thus

$$s_1s_2f(x_1, x_2, \dots x_n) = s_1[s_2f(x_1, x_2, \dots x_n)].$$

The effect of s_2 is merely to produce a rearrangement of the variables, the effect of s_1 on the resulting function is to produce a fresh rearrangement; thus the product of two substitutions is a substitution.

It will be seen at once that substitutions obey the distributive law, for

$$\begin{aligned} s_1[s_2s_3]f(x_1, x_2, \dots x_n) &= s_1[s_2s_3f(x_1, x_2, \dots x_n)] \\ &= s_1[s_2\{s_3f(x_1, x_2, \dots x_n)\}] \\ &= [s_1s_2]s_3f(x_1, x_2, \dots x_n). \end{aligned}$$

On the other hand substitutions are not in general commutative, for example:

$$\begin{aligned} (x_1x_2)(x_1x_3)f(x_1, x_2, x_3) &= (x_1x_2)f(x_3, x_2, x_1) \\ &= f(x_3, x_1, x_2), \end{aligned}$$

but

$$\begin{aligned} (x_1x_3)(x_1x_2)f(x_1, x_2, x_3) &= (x_1x_3)f(x_2, x_1, x_3) \\ &= f(x_2, x_3, x_1). \end{aligned}$$

Any substitution can be represented as a product of transpositions.

For any rearrangement of the letters $x_1, x_2, \dots x_n$ can be produced, first by an interchange of x_1 and one other letter by which x_1 takes its new position, next by an interchange of x_2 and another letter by which x_2 is brought to its new position, and so on.

It will be found that a substitution can be represented as a product of transpositions in a great number of ways, *e.g.*

$$(x_1x_2) = (x_2x_3)(x_1x_3)(x_2x_3);$$

but *the number of transpositions in a product which represents a given substitution is always even or always odd.* For consider the function

$$\Delta = \begin{vmatrix} x_1^{n-1} & x_1^{n-2} \dots 1 \\ x_2^{n-1} & x_2^{n-2} \dots 1 \\ \dots\dots & \dots\dots \\ x_n^{n-1} & x_n^{n-2} \dots 1 \end{vmatrix} = \prod_{r,s} (x_r - x_s);$$

the effect of any transposition operating on Δ is merely to change its sign. Hence

$$s\Delta = \pm \Delta$$

according as s is a product of an even or odd number of substitutions. Substitutions will be called even or odd according as the number of transpositions of which they are composed is even or odd.

Consider any rearrangement of the letters $x_1, x_2, \dots x_n$; let x_r be the letter which takes the place of x_1; x_s that which takes the place of x_r; x_t that which takes the place of x_s, and so on; we must sooner or later arrive at a stage when x_1 is the letter which takes the place of x_u the last of the series. The substitution which replaces x_1 by x_r, x_r by x_s, x_s by x_t and so on, and finally x_u by x_1 is usually written $(x_1 x_r x_s x_t \dots x_u)$ and is called a *cycle.* It is evident from the definition that

$$(x_1 x_r x_s x_t \dots x_u) = (x_r x_s x_t \dots x_u x_1).$$

The rearrangement considered may be produced so far as the letters in the cycle are concerned by operating with $(x_1 x_r x_s x_t \dots x_u)$. Let x_α be one of the letters not contained in this cycle, then we may suppose that x_β in the new arrangement takes the place of x_α and proceed as before. Thus we see that the rearrangement may be produced by operating with a number of independent cycles, *i.e.* cycles such that no two contain a common letter. *Hence any substitution is equal to a product of a number of independent cycles.*

That operation which leaves any function operated on unaltered is called the identical substitution and is written 1. The product

$$(x_1 x_2) . (x_1 x_2) = (x_1 x_2)^2$$

leaves every function unaltered, hence

$$(x_1 x_2)^2 = 1.$$

Consider the rearrangement of the letters $x_1, x_2, \ldots x_n$ produced by any substitution s; there is a perfectly definite substitution which will change this new arrangement back to the old arrangement. This is called the *inverse* substitution of s and is written s^{-1}. In virtue of the definition of s^{-1} we see that

$$s^{-1}sf(x_1, x_2, \ldots x_n) = f(x_1, x_2, \ldots x_n),$$

and hence $$s^{-1}s = 1.$$

Again it is to be observed that the result of operating on the new arrangement of the letters with ss^{-1} is to leave it unaltered, hence

$$ss^{-1} = 1.$$

Consider the powers of any substitution,

$$s, s^2, s^3, \ldots;$$

they are all substitutions, and since the number of different substitutions of n letters is finite—in fact $n!$, the number of possible arrangements of those letters—these powers cannot be all different. Hence for some values of h, k,

$$s^h = s^k.$$

In virtue of the associative law, we can write

$$s^l s^m = s^{l+m}.$$

Hence $$s^{h+l} = s^{k+l}.$$

Further, whatever substitution σ may be,

$$\sigma . s^h = \sigma s^k;$$

hence if $\sigma = (s^{-1})^h$, we see that

$$(s^{-1})^h . s^h = (s^{-1})^{h-1} s^{-1} s s^{h-1} = (s^{-1})^{h-1} s^{h-1} = \ldots = 1;$$

and hence $$1 = s^{k-h}.$$

We may suppose that $k > h$, hence among the positive powers of s we must find the identical substitution. Let p be the smallest positive index for which $s^p = 1$, then p is called the *order* of the substitution.

The substitution $$\sigma s \sigma^{-1}$$

is said to be conjugate to s.

In σ let the letter which replaces x_r be denoted by x_r'; let

$$s = \begin{pmatrix} x_1 & x_2 \dots x_n \\ y_1 & y_2 \dots y_n \end{pmatrix},$$

where $y_1, y_2, \dots y_n$ are the letters $x_1, x_2, \dots x_n$ arranged in some order.

Now $$\sigma = \begin{pmatrix} x_1 & x_2 \dots x_n \\ x_1' & x_2' \dots x_n' \end{pmatrix} = \begin{pmatrix} y_1 & y_2 \dots y_n \\ y_1' & y_2' \dots y_n' \end{pmatrix},$$

and $$\sigma^{-1} = \begin{pmatrix} x_1' & x_2' \dots x_n' \\ x_1 & x_2 \dots x_n \end{pmatrix}.$$

Hence

$$\sigma s \sigma^{-1} = \begin{pmatrix} y_1 & y_2 \dots y_n \\ y_1' & y_2' \dots y_n' \end{pmatrix} \begin{pmatrix} x_1 & x_2 \dots x_n \\ y_1 & y_2 \dots y_n \end{pmatrix} \begin{pmatrix} x_1' & x_2' \dots x_n' \\ x_1 & x_2 \dots x_n \end{pmatrix}$$

$$= \begin{pmatrix} y_1 & y_2 \dots y_n \\ y_1' & y_2' \dots y_n' \end{pmatrix} \begin{pmatrix} x_1' & x_2' \dots x_n' \\ y_1 & y_2 \dots y_n \end{pmatrix} = \begin{pmatrix} x_1' & x_2' \dots x_n' \\ y_1' & y_2' \dots y_n' \end{pmatrix}.$$

Therefore $\sigma s \sigma^{-1}$ is the substitution which would be obtained by operating on the expression for s with the substitution σ. It must then be a product of cycles each having the same number of letters as the cycles of s; and is obtained from s by permuting the letters. Such substitutions are called *similar*. Every substitution similar to s is obtained from s by a suitable permutation of the letters, and is therefore of the form $\sigma s \sigma^{-1}$.

Now if s be any cycle $(x_1 x_2 \dots x_k)$ then

$$s^{-1} = (x_k \dots x_2 x_1)$$

as may easily be verified; and hence s^{-1} is similar to s. But every substitution is a product of a number of independent cycles, the inverse substitution is then the product of the inverse cycles; hence any substitution is similar to its inverse.

273. If m substitutions $s_1, s_2, \dots s_m$ are such that the product of any two of them is itself one of the m substitutions, these m substitutions are said to form a group.

Thus as may be at once verified

$$1, (x_1 x_2);\ 1, (x_1 x_2 x_3), (x_1 x_3 x_2)$$

are groups.

The number of substitutions included in a group is called the *order*, the number of letters affected is called the *degree* of the group.

Thus the two groups written down are of order 2 degree 2, and of order 3 degree 3 respectively.

Now the n letters $x_1, x_2, \dots x_n$ can be arranged in $n!$ ways, hence the total number of substitutions affecting n letters is $n!$. These substitutions obviously form a group, it is of degree n and of order $n!$. This group is called the symmetric group for the n letters $x_1, x_2, \dots x_n$.

It is useful to have a symbol by which to denote this group, the symmetric group for the letters $x_1, x_2, \dots x_n$ will be written

$$\{x_1 x_2 \dots x_n\}.$$

More particularly this symbol will be used to denote the sum of all the substitutions of the symmetric group.

Again the product of two even substitutions is obviously an even substitution, hence the even substitutions which affect n letters form a group. This group is called the alternating group. Let

$$s_1, s_2, \dots s_m$$

be the members of the alternating group, and

$$\sigma_1, \sigma_2, \dots \sigma_{m'}$$

the remaining substitutions which affect the letters $x_1, x_2, \dots x_n$.

Then these latter substitutions are all odd; and hence the product of any two of them is an even substitution.

Now if t_1, t_2, t_3 be any three substitutions and

$$t_1 t_2 = t_1 t_3,$$

then

$$t_1^{-1} t_1 t_2 = t_1^{-1} t_1 t_3,$$

and hence

$$t_2 = t_3.$$

By hypothesis the substitutions $s_1, s_2, \dots s_m, \sigma_1, \sigma_2, \dots \sigma_{m'}$ are all different, hence the substitutions

$$\sigma_1 s_1, \sigma_1 s_2, \dots \sigma_1 s_m, \sigma_1^2, \sigma_1 \sigma_2, \dots \sigma_1 \sigma_{m'}$$

are all different. But the former set include all the substitutions of the n letters, hence the latter must do so too. Hence the even substitutions

$$\sigma_1^2, \sigma_1 \sigma_2, \dots \sigma_1 \sigma_{m'}$$

form the alternating group. And therefore

$$m = m' = \tfrac{1}{2} n!.$$

The symbol $$\{x_1x_2 \dots x_n\}'$$
will be used to denote the sum of the even substitutions minus the sum of the odd substitutions of the letters

$$x_1, x_2, \dots x_n.$$

This will be called the negative symmetric group, and on the other hand
$$\{x_1x_2 \dots x_n\}$$
will be called the positive symmetric group of the n letters.

For example

$$\{x_1x_2x_3\} = 1 + (x_1x_2x_3) + (x_1x_3x_2) + (x_1x_2) + (x_2x_3) + (x_3x_1),$$
$$\{x_1x_2x_3\}' = 1 + (x_1x_2x_3) + (x_1x_3x_2) - (x_1x_2) - (x_2x_3) - (x_3x_1).$$

274. As an illustration of the notation just introduced we remark that the determinant

$$\begin{vmatrix} a_1 & a_2 & \dots & a_n \\ b_1 & b_2 & \dots & b_n \\ \dots & \dots & \dots & \dots \\ \dots & \dots & \dots & \dots \\ k_1 & k_2 & \dots & k_n \end{vmatrix}$$

may be written
$$\{ab \dots k\}' a_1b_2 \dots k_n,$$
the substitutions being supposed to affect the letters and not the suffixes. Or adopting a double suffix notation we may write

$$\begin{vmatrix} a_{1,1} & a_{1,2} & \dots & a_{1,n} \\ a_{2,1} & a_{2,2} & \dots & a_{2,n} \\ \dots & \dots & \dots & \dots \\ \dots & \dots & \dots & \dots \\ a_{n,1} & a_{n,2} & \dots & a_{n,n} \end{vmatrix} = \{a_1a_2 \dots a_n\}' a_{1,1}a_{2,2} \dots a_{n,n},$$

where the first suffix only appears in the substitutions and is alone affected by them.

Or again

$$\begin{vmatrix} a^p & b^p & c^p \\ a^q & b^q & c^q \\ a^r & b^r & c^r \end{vmatrix} = \{abc\}' a^p b^q c^r.$$

As an example of the use of the positive symmetric group, referring to § 44, we observe that the rth polar of the form

$$a_x{}^n = a_{1_x} a_{2_x} \dots a_{n_x}$$

may be conveniently written

$$a_x^{n-r} a_y^r = \frac{1}{n!} \{\alpha_1 \alpha_2 \dots \alpha_n\} \alpha_{1_y} \alpha_{2_y} \dots \alpha_{r_y} \alpha_{r+1_x} \dots \alpha_{n_x}.$$

275. Consider any function F of the coefficients of certain linear binary forms α_x, β_x, γ_x, which is homogeneous and linear in the coefficients of each form separately. We may write

$$F = \alpha_1 \beta_1 \phi_1 + \alpha_1 \beta_2 \phi_2 + \alpha_2 \beta_1 \phi_3 + \alpha_2 \beta_2 \phi_4,$$

where the ϕ's are functions of the coefficients of the linear forms γ_x, δ_x, ... of the same character as F.

Then $$\{\alpha\beta\}' F = (\alpha\beta) [\phi_2 - \phi_3];$$

i.e. $\{\alpha\beta\}' F$ is the product of $(\alpha\beta)$ and a function which does not contain the coefficients of α_x or β_x.

Again we may write

$$F = \Sigma \alpha_r \beta_s \gamma_t \phi_{r,s,t}, \quad (r, s, t = 1, 2),$$

but here the suffixes r, s, t can never be all different.

Now

$$\{\alpha\beta\gamma\}' \alpha_r \beta_s \gamma_t = \begin{vmatrix} \alpha_r & \alpha_s & \alpha_t \\ \beta_r & \beta_s & \beta_t \\ \gamma_r & \gamma_s & \gamma_t \end{vmatrix},$$

hence $$\{\alpha\beta\gamma\}' F = 0.$$

In the same way if α_{1_x}, α_{2_x}, ... be any p-ary linear forms, where

$$\alpha_{r_x} = \alpha_{r,1} x_1 + \alpha_{r,2} x_2 + \dots + \alpha_{r,p} x_p,$$

and F be any function homogeneous and linear in the coefficients of each,

$$\{\alpha_1 \alpha_2 \dots \alpha_{p+1}\}' F = 0,$$

$$\{\alpha_1 \alpha_2 \dots \alpha_p\}' F = \begin{vmatrix} \alpha_{1,1} & \alpha_{1,2} & \dots & \alpha_{1,p} \\ \alpha_{2,1} & \alpha_{2,2} & \dots & \alpha_{2,p} \\ \dots & \dots & \dots & \dots \\ \dots & \dots & \dots & \dots \\ \alpha_{p,1} & \alpha_{p,2} & \dots & \alpha_{p,p} \end{vmatrix} . \psi,$$

where the substitutions affect the first suffixes only, and ψ is a function of the coefficients of

$$\alpha_{p+1_x}, \alpha_{p+2_x}, \dots.$$

The expression $$| \alpha_1 \alpha_2 \dots \alpha_p |$$
will be used as an abbreviation for the determinant just written down.

Again, if $r > p$,
$$\{\alpha_1 \alpha_2 \dots \alpha_r\}' F = 0;$$
and if $r < p$, it is easy to see that
$$\{\alpha_1 \alpha_2 \dots \alpha_r\}' F = \sum_\lambda \Delta_\lambda \psi_\lambda,$$
where Δ_λ is one of the determinants of the matrix
$$\left| \begin{matrix} \alpha_{1,1} & \alpha_{1,2} & \dots & \alpha_{1,p} \\ \alpha_{2,1} & \alpha_{2,2} & \dots & \alpha_{2,p} \\ \dots & \dots & \dots & \dots \\ \alpha_{r,1} & \alpha_{r,2} & \dots & \alpha_{r,p} \end{matrix} \right|.$$

It is unnecessary to suppose that $\alpha_{r,1}, \alpha_{r,2}, \dots \alpha_{r,p}$ are the coefficients of a linear p-ary form α_{r_x}. The facts just established are true if F is homogeneous and linear in each of any m sets of quantities
$$\begin{matrix} \alpha_{1,1} & \alpha_{1,2} & \dots & \alpha_{1,p} \\ \alpha_{2,1} & \alpha_{2,2} & \dots & \alpha_{2,p} \\ \dots & \dots & \dots & \dots \\ \alpha_{m,1} & \alpha_{m,2} & \dots & \alpha_{m,p} \end{matrix}$$
there being p quantities in each set; and a one-to-one correspondence between the members of any two sets.

Thus in particular: *If F is a function homogeneous and linear in the coefficients of each of m binary n-ics*
$$(\alpha_{1,0}, \alpha_{1,1}, \dots \alpha_{1,n} \between x_1, x_2)^n, \dots (\alpha_{m,0}, \alpha_{m,1}, \dots \alpha_{m,n} \between x_1, x_2)^n,$$
m being greater than $n+1$, then
$$\{\alpha_1 \alpha_2 \dots \alpha_{n+2}\}' F = 0 \quad \dots\dots\dots\dots\text{(i)},$$
$$\{\alpha_1 \alpha_2 \dots \alpha_{n+1}\}' F = | \alpha_1 \alpha_2 \dots \alpha_{n+1} | . F_1 \quad \dots\dots\text{(ii)},$$
$$\{\alpha_1 \alpha_2 \dots \alpha_n\}' F = | \alpha_1 \alpha_2 \dots \alpha_n \beta | \quad \dots\dots\dots\text{(iii)},$$
where the substitutions affect the first suffixes only of the coefficients $\alpha_{p,q}$; $| \alpha_1 \alpha_2 \dots \alpha_{n+1} |$ is the determinant of $n+1$ rows and columns formed by the coefficients of the $n+1$ quantics concerned;
$$| \alpha_1 \alpha_2 \dots \alpha_n \beta |$$
is the same determinant with quantities $\beta_0, \beta_1, \dots \beta_n$ replacing
$$\alpha_{n+1,0}, \alpha_{n+1,1}, \dots \alpha_{n+1,n},$$

these quantities being homogeneous and linear in the coefficients of the quantics represented by $\alpha_{n+1}, \alpha_{n+2}, \dots \alpha_m$; *and where* F_1 *is homogeneous and linear in the coefficients of the quantics represented by* $\alpha_{n+2}, \dots \alpha_m$.

The first two of the above results are sufficiently clear from what has already been said. As regards the last we observe that

$$\{\alpha_1\alpha_2 \dots \alpha_n\}' F = \sum_\lambda \Delta_\lambda \beta_\lambda,$$

where Δ_λ is one of the determinants of the matrix

$$\begin{vmatrix} \alpha_{1,0} & \alpha_{1,1} & \dots & \alpha_{1,n} \\ \alpha_{2,0} & \alpha_{2,1} & \dots & \alpha_{2,n} \\ \dots & \dots & \dots & \dots \\ \alpha_{n,0} & \alpha_{n,1} & \dots & \alpha_{n,n} \end{vmatrix}.$$

That is Δ_λ may be taken to be the minor of $\alpha_{n+1,\lambda}$ in the determinant

$$|\, \alpha_1\alpha_2 \dots \alpha_{n+1} \,|,$$

and hence $$\{\alpha_1\alpha_2 \dots \alpha_n\}' F = |\, \alpha_1\alpha_2 \dots \alpha_n\beta \,|.$$

276. Let the quantics of the last paragraph be represented symbolically thus

$$(\alpha_{r,0}, \alpha_{r,1}, \dots \alpha_{r,n} \between x_1, x_2)^n = [a_x^{(r)}]^n, \quad (r = 1, 2, \dots m).$$

Then the determinant $|\, \alpha_1\alpha_2 \dots \alpha_{n+1} \,|$ is an invariant, for it may be written symbolically

$$\prod_{r,s} (a^{(r)}a^{(s)}) \qquad (r, s = 1, 2, \dots m;\ r \neq s).$$

Hence if F is an invariant, then F_1 is also an invariant.

Again we can shew that if F is an invariant, then

$$(\beta_0, \beta_1, \dots \beta_n \between x_1, x_2)^n$$

is a covariant. For let A_r be the minor of $\alpha_{n+1,r}$ in the determinant

$$|\, \alpha_1\alpha_2 \dots \alpha_{n+1} \,|.$$

Then since F is an invariant

$$\Sigma\beta_r A_r$$

is also an invariant.

Now we know that $\Sigma a_{n+1,\,r} A_r$ and $\Sigma a_{n+1,\,r} \binom{n}{r} x_1^{n-r} x_2^r$ are both invariantive, and hence that the quantities

$$A_0,\ A_1,\ \ldots\ A_n$$

and

$$x_1^n,\ \binom{n}{1} x_1^{n-1} x_2,\ \ldots\ \binom{n}{n} x_2^n$$

form two cogredient sets. Therefore, since $\Sigma \beta_r A_r$ is an invariant, $\Sigma \beta_r \binom{n}{r} x_1^{n-r} x_2^r$ must also be invariantive. In other words

$$(\beta_0,\ \beta_1,\ \ldots\ \beta_n \between x_1,\ x_2)^n$$

is a covariant.

If F were a covariant, and contained the variables x_1, x_2, then β_0, β_1, ... β_n would also contain these quantities; in this case the form

$$(\beta_0,\ \beta_1,\ \ldots\ \beta_n \between y_1,\ y_2)^n$$

is a covariant in two sets of variables.

277. As examples of the use of the results just established we may instance the fundamental identity for binary forms

$$\{abc\}'\,(ab)\,(cd) = 0\,;$$

that for ternary forms

$$\{abcd\}'\,(abc)\,(def) = 0,$$

or

$$\{abcd\}'\,(abe)\,(cdf) = 0\,;$$

those for quaternary forms

$$\{abcde\}'\,(abcd)\,(efgh) = 0,$$

$$\{abcde\}'\,(abfg)\,(cdhi)\,(ejkl) = 0,$$

and so on.

Let I be an invariant linear in the coefficients of each of $n+1$ binary n-ics; then with the notation of § 276 we have

$$\{\alpha_1 \alpha_2 \ldots \alpha_{n+1}\}'\, I = \lambda\, |\, \alpha_1 \alpha_2 \ldots \alpha_{n+1}\, |$$

where λ is some constant, possibly zero.

If n is odd we may take

$$I = (a^{(1)}\, a^{(2)})^n\, (a^{(3)}\, a^{(4)})^n\, \ldots\, (a^{(n)}\, a^{(n+1)})^n.$$

In this case, provided the $n+1$ quantics are all linearly independent,

$$\{\alpha_1 \alpha_2 \ldots \alpha_{n+1}\}'\, I$$

is different from zero, for all terms of I in which two coefficients have the same suffixes are destroyed by the operator, and the rest are obtained from

$$[\alpha_{1,0}\,\alpha_{2,n}]\left[-\binom{n}{1}\alpha_{3,1}\,\alpha_{4,n-1}\right]\left[\binom{n}{2}\alpha_{5,2}\,\alpha_{6,n-2}\right]\ldots\ldots$$

$$\left[(-1)^{\frac{n-1}{2}}\binom{n}{\frac{n-1}{2}}\alpha_{n,\frac{n-1}{2}}\,\alpha_{n+1,\frac{n+1}{2}}\right]$$

by interchanging both the first suffixes of pairs of brackets in all possible ways keeping the order of the three first suffixes unaltered—so that the first suffixes of any one bracket are always of the form $(2r-1)$, $(2r)$ and in this order—and by then interchanging the first suffixes inside individual brackets, each such interchange being accompanied by a change of sign. But both these operations are effected by

$$\{\alpha_1\alpha_2\ldots\alpha_{n+1}\}'$$

—since the first only requires even substitutions. Hence

$$\{\alpha_1\alpha_2\ldots\alpha_{n+1}\}'\,I=\lambda\,\{\alpha_1\alpha_2\ldots\alpha_{n+1}\}'\,\alpha_{1,0}\,\alpha_{2,n}\,\alpha_{3,1}\,\alpha_{4,n-1}\ldots\alpha_{n,\frac{n-1}{2}}\,\alpha_{n+1,\frac{n+1}{2}}$$
$$=\pm\lambda\,|\,\alpha_1\alpha_2\ldots\alpha_{n+1}\,|$$

where λ is not zero. Hence when n is odd

$$|\,\alpha_1\alpha_2\ldots\alpha_{n+1}\,|$$

is reducible.

Again, if $n=4$, it is easy to see that this invariant is irreducible. Let us suppose that it be reducible, then

$$|\,\alpha_1\alpha_2\alpha_3\alpha_4\alpha_5\,|=\Sigma I_2I_3$$

where I_r is an invariant of degree r.

Now I_2 must be of the form $(a^{(r)}a^{(s)})^4$, where r, s are two of the numbers 1, 2, 3, 4, 5; hence

$$\{\alpha_1\alpha_2\alpha_3\alpha_4\alpha_5\}'\,I_2I_3=0,$$

and therefore

$$\{\alpha_1\alpha_2\alpha_3\alpha_4\alpha_5\}'\,|\,\alpha_1\alpha_2\alpha_3\alpha_4\alpha_5\,|$$
$$=5\,!\,|\,\alpha_1\alpha_2\alpha_3\alpha_4\alpha_5\,|=0.$$

This we know to be untrue in general, hence the hypothesis, that the invariant in question is reducible, is false.

278. Let s be any substitution of the symmetric group

$$\{a_1 a_2 \ldots a_n\},$$

then $$\{a_1 a_2 \ldots a_n\} s = \{a_1 a_2 \ldots a_n\}.$$

For $\{a_1 a_2 \ldots a_n\} s$ contains $n!$ terms, which are all different (if σ and σ' are different substitutions, then σs and $\sigma' s$ are also different), and are all members of the positive symmetric group

$$\{a_1 a_2 \ldots a_n\}.$$

Similarly $$s \{a_1 a_2 \ldots a_n\} = \{a_1 a_2 \ldots a_n\}.$$

Now any purely formal relation between substitutions will still hold good if the sign of every transposition be changed; hence

$$s \{a_1 a_2 \ldots a_n\}' = \pm \{a_1 a_2 \ldots a_n\}'$$

according as s is an even or odd substitution, in particular

$$(a_1 a_2) \{a_1 a_2 \ldots a_n\}' = - \{a_1 a_2 \ldots a_n\}'.$$

Again if $\{a_1 a_2 \ldots a_n\}$ be any positive symmetric group and

$$\{a_1 a_2 b_3 \ldots b_m\}'$$

be a negative symmetric group, the two groups having a pair of common letters, then

$$\begin{aligned} &\{a_1 a_2 \ldots a_n\} \{a_1 a_2 b_3 \ldots b_m\}' \\ &\qquad = \{a_1 a_2 \ldots a_n\} (a_1 a_2) [- (a_1 a_2) \{a_1 a_2 b_3 \ldots b_m\}'] \\ &\qquad = - \{a_1 a_2 \ldots a_n\} \{a_1 a_2 b_3 \ldots b_m\}' = 0. \end{aligned}$$

Similarly $$\{a_1 a_2 b_3 \ldots b_m\}' \{a_1 a_2 \ldots a_n\} = 0.$$

Thus if two symmetric groups, one positive the other negative, have a pair of common letters, their product is always zero.

279. The following purely formal theorem enables us to establish various results relating to invariants.

Let the letters $a_1, a_2, \ldots a_n$ be arranged in any manner in horizontal rows, so that each row has its first letter in the same vertical column, its second letter in a second vertical column, and so on; and so that no row contains more letters than any row above it.

Thus for four letters $a_1a_2a_3a_4$ the five possible kinds of arrangement of the tableau would be

$$a_1a_2a_3a_4; \qquad \begin{matrix} a_1a_2a_3; \\ a_4 \end{matrix} \qquad \begin{matrix} a_1a_2; \\ a_3a_4 \end{matrix} \qquad \begin{matrix} a_2a_1; \\ a_3 \\ a_4 \end{matrix} \qquad \begin{matrix} a_1 \\ a_2 \\ a_3 \\ a_4 \end{matrix}$$

Then form the substitutional expression

$$S = \Sigma G_1 G_2 \ldots G_h \Gamma_1' \Gamma_2' \ldots \Gamma_k'$$

such that G_1 is the positive symmetric group of the letters of the first row, G_2 that of the letters of the second row, and so on, G_h being that of the letters of the last row; and that Γ_1' is the negative symmetric group of the letters of the first column, Γ_2' that of the letters of the second column, and so on, Γ_k' being that of the letters of the last column (in case a row or column contains only one letter, it is understood that the positive or negative symmetric group of a single letter is unity).

Let us suppose that in the tableau considered there are α_1 letters in the first row, α_2 in the second, and so on; where owing to the conditions laid down

$$\left.\begin{matrix} \alpha_1 + \alpha_2 + \ldots + \alpha_h = n \\ \alpha_1 \nless \alpha_2 \nless \alpha_3 \ldots \nless \alpha_h, \; \alpha_1 = k \end{matrix}\right\} \ldots\ldots\ldots\ldots\ldots(\text{I});$$

let $T_{\alpha_1, \alpha_2, \ldots \alpha_h}$ be the sum of the $n!$ expressions S obtained by permuting the letters in the tableau in all possible ways, the numbers $\alpha_1, \alpha_2, \ldots \alpha_h$ of letters in the various rows remaining fixed.

Then $$\Sigma A_{\alpha_1, \alpha_2, \ldots \alpha_h} T_{\alpha_1, \alpha_2, \ldots \alpha_h} = 1$$

where the summation extends to all possible values of the numbers $\alpha_1, \alpha_2, \ldots \alpha_h$ which satisfy the conditions (I); and $A_{\alpha_1, \alpha_2, \ldots \alpha_h}$ is a numerical coefficient which can be uniquely determined.

For two letters we have

$$\begin{aligned} 1 &= \tfrac{1}{4} T_2 + \tfrac{1}{4} T_{1,1} \\ &= \tfrac{1}{2}\{a_1a_2\} + \tfrac{1}{2}\{a_1a_2\}'. \end{aligned}$$

For three letters we have

$$\begin{aligned} 1 &= \tfrac{1}{36} T_3 + \tfrac{1}{9} T_{2,1} + \tfrac{1}{36} T_{1,1,1} \\ &= \tfrac{1}{6}\{a_1a_2a_3\} + \tfrac{1}{9}\Sigma\{a_1a_2\}'\{a_1a_3\} + \tfrac{1}{6}\{a_1a_2a_3\}', \end{aligned}$$

as can easily be verified.

280. Let the T's be arranged in the order defined by the convention that $T_{\alpha_1, \alpha_2, \dots \alpha_h}$ comes before $T_{\beta_1, \beta_2, \dots \beta_{h'}}$ when the first of the differences

$$\alpha_1 - \beta_1,\ \alpha_2 - \beta_2,\ \dots$$

which does not vanish is positive.

Now if S be one of the $n!$ expressions of which $T_{\alpha_1, \alpha_2, \dots \alpha_h}$ is the sum, then $T_{\alpha_1, \alpha_2, \dots \alpha_h}$ is obtained from S by permuting the letters in all possible ways and taking the sum. Hence if when S is expanded as a sum of substitutions, any particular substitution s occurs in it, then in $T_{\alpha_1, \alpha_2, \dots \alpha_h}$ the sum of all the substitutions of the symmetric group of the letters $a_1, a_2, \dots a_n$ similar to s must occur. Hence defining $t_{\beta_1, \beta_2, \dots \beta_k}$ to be the sum of all those substitutions which are formed of k cycles of orders $\beta_1, \beta_2, \dots \beta_k$ respectively, it follows that

$$T_{\alpha_1, \alpha_2, \dots \alpha_h} = \Sigma \lambda_{\beta_1, \beta_2, \dots \beta_k} t_{\beta_1, \beta_2, \dots \beta_k} \quad \dots\dots\dots\dots(\text{II}),$$

where $\lambda_{\beta_1, \beta_2, \dots \beta_k}$ is a numerical coefficient.

If cycles of order unity, which are equivalent to the identical substitution, be introduced, we may suppose that the suffixes of $t_{\beta_1, \beta_2, \dots \beta_k}$ satisfy the conditions

$$\beta_1 + \beta_2 + \dots + \beta_k = n$$

$$\beta_1 \not< \beta_2 \not< \beta_3 \dots \not< \beta_k.$$

The t's are now defined by numbers which obey exactly the same conditions as those which define the T's. The number of t's must then be equal to the number of T's, let us say equal to M. Now the equations (II) may be regarded as a system of linear equations expressing the t's in terms of the T's. Hence if these equations are all independent

$$t_{\beta_1, \beta_2, \dots \beta_k} = \Sigma \mu_{\alpha_1, \alpha_2, \dots \alpha_h} T_{\alpha_1, \alpha_2, \dots \alpha_h}$$

and in particular

$$1 = t_{1, 1, \dots 1} = \Sigma A_{\alpha_1, \alpha_2, \dots \alpha_h} T_{\alpha_1, \alpha_2, \dots \alpha_h}.$$

If these equations are not all linearly independent, there must be a relation of the form

$$\Sigma BT = 0.$$

In order to prove the impossibility of such a relation, it will be shewn that

(i) $$T_{\alpha_1, \alpha_2, \ldots \alpha_h} T_{\beta_1, \beta_2, \ldots \beta_{h'}} = 0$$

when $T_{\alpha_1, \alpha_2, \ldots \alpha_h}$ comes after $T_{\beta_1, \beta_2, \ldots \beta_{h'}}$, the T's being arranged in the order defined above; and that

(ii) $$T^2_{\alpha_1, \alpha_2, \ldots \alpha_h} \neq 0.$$

Let $S = PN$ be one of the $n!$ expressions of which $T_{\alpha_1, \alpha_2, \ldots \alpha_h}$ is the sum, where P denotes the product of the positive symmetric groups, and N that of the negative symmetric groups. Similarly let $S' = P'N'$ be one of the expressions of which $T_{\beta_1, \beta_2, \ldots \beta_{h'}}$ is the sum. If one of the groups of P' contains a pair of letters contained in any one group of N then, § 278,

$$NP' = 0.$$

Consider the tableau by means of which S is formed, α_1 is the number of letters in the top row, it is also the number of columns, and consequently it is the number of the positive symmetric groups in P. Again α_2 is the number of letters in the second row, and hence $\alpha_1 - \alpha_2$ is the number of columns which contain one letter only. Similarly $\alpha_2 - \alpha_3$ is the number of columns containing exactly two letters, and in general $\alpha_i - \alpha_{i+1}$ is the number of columns containing exactly i letters.

If $\beta_1 > \alpha_1$ there are more letters in the first row of the tableau for S' than there are columns in the tableau for S. Hence one group at least of the product P' contains a pair of letters belonging to the same group of N; and therefore

$$NP' = 0.$$

In order that NP' may be other than zero, we must then have $\beta_1 \ngtr \alpha_1$. If this condition be satisfied but

$$\beta_1 + \beta_2 > \alpha_1 + \alpha_2,$$

we see that there are more letters in the first two rows of the tableau for S' than can be arranged in the tableau for S with the condition that no three occur in the same column. In this case some group of N must contain three of the letters of the first two groups of P', and therefore two of the letters belonging to one of these groups. Hence if

$$\beta_1 + \beta_2 > \alpha_1 + \alpha_2,$$

then
$$NP' = 0.$$

Again if
$$\beta_1 + \beta_2 + \beta_3 > \alpha_1 + \alpha_2 + \alpha_3,$$

some one group of N must contain four of the letters belonging to the first three groups of P', and again NP' vanishes.

Proceeding thus we see that NP' is always zero unless

$$\alpha_1 \not< \beta_1,\ \alpha_1 + \alpha_2 \not< \beta_1 + \beta_2,\ \ldots\ \alpha_1 + \alpha_2 + \ldots + \alpha_i \not< \beta_1 + \beta_2 + \ldots + \beta_i.$$

We deduce that if the first of the differences

$$\alpha_1 - \beta_1,\ \alpha_2 - \beta_2,\ \ldots$$

which is other than zero is negative, then NP' is zero, *i.e.* if $T_{\alpha_1, \alpha_2, \ldots \alpha_h}$ comes after $T_{\beta_1, \beta_2, \ldots \beta_{h'}}$ then

$$NP' = 0;$$

and hence
$$T_{\alpha_1, \alpha_2, \ldots \alpha_h} T_{\beta_1, \beta_2, \ldots \beta_{h'}} = (\Sigma PN)(\Sigma P'N') = 0.$$

Next we must prove that

$$T_{\alpha_1, \alpha_2, \ldots \alpha_h} T_{\alpha_1, \alpha_2, \ldots \alpha_h} = T^2_{\alpha_1, \alpha_2, \ldots \alpha_h} \neq 0.$$

For this purpose it is only necessary to shew that the coefficient of the identical substitution is other than zero. Now

$$T_{\alpha_1, \alpha_2, \ldots \alpha_h} = \Sigma \lambda_{\beta_1, \beta_2, \ldots \beta_k} t_{\beta_1, \beta_2, \ldots \beta_k}$$

but every substitution is similar to its reciprocal substitution; hence if s be any substitution contained in $t_{\beta_1, \beta_2, \ldots \beta_k}$, s^{-1} must also be contained in this expression. It follows that in $T_{\alpha_1, \alpha_2, \ldots \alpha_h}$ both s and s^{-1} have the same numerical coefficient. In $T^2_{\alpha_1, \alpha_2, \ldots \alpha_h}$ the only products which produce the identical substitution are those obtained when a substitution s is taken in the first T, and s^{-1} in the second. Hence the required coefficient of the identical substitution is of the form $\Sigma\lambda^2$, which, being the sum of a number of positive terms, cannot be zero.

Let us now suppose that there is a relation of the form

$$\Sigma BT = 0,$$

and let $\lambda T_{\alpha_1, \alpha_2, \dots \alpha_h}$ be the first term of this relation, the terms being arranged in the order defined above. Then

$$[\Sigma BT]\, T_{\alpha_1, \alpha_2, \dots \alpha_h} = 0.$$

And hence by the relation (i)

$$\lambda T^2_{\alpha_1, \alpha_2, \dots \alpha_h} = 0;$$

therefore by (ii) $\qquad \lambda = 0.$

Thus no T can be the first term in such a relation; in other words the equations (II) are linearly independent. Hence

$$1 = \Sigma A_{\alpha_1, \alpha_2, \dots \alpha_h} T_{\alpha_1, \alpha_2, \dots \alpha_h} \quad \text{...............(III).}$$

Q.E.D.

281. The coefficients in this series have been calculated* but as their values are not of importance for our present purpose we merely quote the formula

$$A_{\alpha_1, \alpha_2, \dots \alpha_h} = \left(\frac{\prod\limits_{r, s} (\alpha_r - \alpha_s - r + s)}{\prod\limits_{r} (\alpha_r + h - r)!} \right)^2.$$

The substitutions $s_1 s_2$ and $s_2 s_1$ are similar, for

$$s_1 s_2 = s_2^{-1} (s_2 s_1)\, s_2,$$

their coefficients in the expansion of $T_{\alpha_1, \alpha_2, \dots \alpha_h}$ are therefore the same. Hence since

$$T_{\alpha_1, \alpha_2, \dots \alpha_h} = \Sigma PN,$$

it follows that also

$$T_{\alpha_1, \alpha_2, \dots \alpha_h} = \Sigma NP.$$

If $T_{\beta_1, \beta_2 \dots \beta_{h'}} = \Sigma P'N'$ comes before $T_{\alpha_1, \alpha_2 \dots \alpha_h}$ it has been shewn that

$$NP' = 0.$$

Hence also $\qquad P'N = 0,$

and $\qquad N'P'NP = 0,$

and $\qquad [\Sigma N'P'][\Sigma NP] = 0,$

and therefore $\qquad T_{\beta_1, \beta_2, \dots \beta_{h'}} T_{\alpha_1, \alpha_2, \dots \alpha_h} = 0.$

That is, the product of any two different T's is zero.

* Young, *Proc. Lond. Math. Soc.* vol. XXXIV. p. 361.

Now multiply the relation (III) by $T_{\alpha_1, \alpha_2, \dots \alpha_h}$, we obtain at once

$$T_{\alpha_1, \alpha_2, \dots \alpha_h} = A_{\alpha_1, \alpha_2, \dots \alpha_h} T^2_{\alpha_1, \alpha_2, \dots \alpha_h}.$$

282. By polarizing the form

$$a_x^{(1)} a_x^{(2)} \dots a_x^{(m)} = F$$

once with respect to each of the sets of variables

$$\begin{array}{cccc} x_1^{(1)} & x_2^{(1)} & \dots & x_n^{(1)} \\ x_1^{(2)} & x_2^{(2)} & \dots & x_n^{(2)} \\ \dots & \dots & \dots & \dots \\ x_1^{(m)} & x_2^{(m)} & \dots & x_n^{(m)} \end{array}$$

we obtain an expression which may be written

$$F_1 = \frac{1}{n!} \{x^{(1)} x^{(2)} \dots x^{(m)}\} a^{(1)}{}_{x^{(1)}} a^{(2)}{}_{x^{(2)}} \dots a^{(m)}{}_{x^{(m)}}$$

$$= \frac{1}{n!} \{a^{(1)} a^{(2)} \dots a^{(m)}\} a^{(1)}{}_{x^{(1)}} a^{(2)}{}_{x^{(2)}} \dots a^{(m)}{}_{x^{(m)}}.$$

If in F_1 each of the sets of variables is replaced by the original set $x_1, x_2, \dots x_n$, we obtain the form F from which we started. Neither the passage from F to F_1 nor that from F_1 to F affects the invariant properties of F, so that we may regard F and F_1 as equivalent.

Similarly if $f(a^{(1)}, a^{(2)}, \dots a^{(m)})$ be an invariant linear in the coefficients of each of m quantics of the same order, whose coefficients are

$$a_0^{(r)}, a_1^{(r)}, \dots \qquad (r = 1, 2, \dots m),$$

then $$\{a^{(1)} a^{(2)} \dots a^{(m)}\} f(a^{(1)}, a^{(2)}, \dots a^{(m)})$$

is an invariant which may be obtained by means of Aronhold operators from an invariant of a single quantic.

Again if P be the product of h positive symmetric groups, which between them contain all the letters $a^{(1)}, a^{(2)}, \dots a^{(m)}$, but no two contain the same letter, then

$$Pf(a^{(1)}, a^{(2)}, \dots a^{(m)})$$

is an invariant which may be obtained by means of Aronhold operators from an invariant of h quantics.

283. Peano's Theorem. Let $F(a^{(1)}, a^{(2)}, \dots a^{(m)})$ be a covariant linear in the coefficients of each of m binary n-ics

$$(a_0^{(r)}, a_1^{(r)}, \dots a_n^{(r)} \between x_1, x_2)^n; \quad r = 1, 2, \dots m.$$

Operate on F with the two sides of the identity § 280 (III.)

$$1 = \Sigma A_{\alpha_1, \alpha_2, \dots \alpha_h} T_{\alpha_1, \alpha_2, \dots \alpha_h},$$

then

$$F = \Sigma A_{\alpha_1, \alpha_2, \dots \alpha_h} T_{\alpha_1, \alpha_2, \dots \alpha_h} F.$$

Now

$$T_{\alpha_1, \alpha_2, \dots \alpha_h} = \Sigma PN,$$

where N has a factor of the form

$$\{a^{(1)} a^{(2)} \dots a^{(h)}\}'.$$

If $h > n + 1$, then by § 275,

$$\{a^{(1)} a^{(2)} \dots a^{(h)}\}' F = 0,$$

if $h = n + 1$,

$$\{a^{(1)} a^{(2)} \dots a^{(h)}\}' F = | a^{(1)} a^{(2)} \dots a^{(h)} | . R,$$

if $h = n$,

$$\{a^{(1)} a^{(2)} \dots a^{(h)}\}' F = | a^{(1)} a^{(2)} \dots a^{(h)} Q |,$$

where $(Q_0, Q_1, \dots Q_n \between x_1, x_2)^n$ is a covariant of the forms considered.

Hence when $h > n + 1$,

$$T_{\alpha_1, \alpha_2, \dots \alpha_h} F = 0;$$

when $h = n + 1$, $T_{\alpha_1, \alpha_2, \dots \alpha_h} F$ is a sum of terms each of which contains a factor of the form $| a^{(1)} a^{(2)} \dots a^{(n+1)} |$; when $h = n$, $T_{\alpha_1, \alpha_2, \dots \alpha_h} F$ is a sum of terms of the form $| a^{(1)} a^{(2)} \dots a^{(n)} Q |$.

Now the number of positive symmetric groups in P is h; hence by § 282, $T_{\alpha_1, \alpha_2, \dots \alpha_h} F$ is a sum of terms each of which is obtainable by means of Aronhold operators from a covariant of only h different n-ics.

Let us suppose that F is a covariant type which does not give any irreducible covariant, unless we are considering a system of more than n n-ics. Then

$$F = \Sigma A_{\alpha_1, \alpha_2, \dots \alpha_h} T_{\alpha_1, \alpha_2, \dots \alpha_h} F,$$

if $h < n + 1$, $T_{\alpha_1, \alpha_2, \dots \alpha_h} F$ is reducible for it is obtained by Aronhold operators from reducible forms: if $h > n + 1$

$$T_{\alpha_1, \alpha_2, \dots \alpha_h} F = 0;$$

and if $h = n + 1$, $T_{a_1, a_2, \ldots a_h} F$ is a sum of terms each of which contains a factor of the form $| a^{(1)} a^{(2)} \ldots a^{(n+1)} |$, and is thus reducible unless $m = n + 1$.

Hence *every type of the binary n-ic which does not furnish an irreducible covariant for a system of n n-ics is reducible, with the possible exception of the invariant type*

$$| a^{(1)} a^{(2)} \ldots a^{(n+1)} |.$$

Further it has been shewn that, if $h = n$, $T_{a_1, a_2, \ldots a_h} F$ is equal to a sum of terms of the form $| a^{(1)} a^{(2)} \ldots a^{(n)} Q |$. But if the invariant $| a^{(1)} a^{(2)} \ldots a^{(n+1)} |$ is reducible, the invariant

$$| a^{(1)} a^{(2)} \ldots a^{(n)} Q |$$

is reducible in the same way. Hence when the above invariant is reducible every irreducible type furnishes an irreducible covariant for a system of $n - 1$ n-ics.

It has been shewn § 277 that this invariant is reducible when n is odd; thus the covariants of any number of cubics can be obtained by means of Aronhold operators from those for two cubics.

Similar results may be obtained in exactly the same way for ternary forms, or for forms involving any number of variables.

Thus all covariants of any number of ternary n-ics may be obtained by means of Aronhold operators from the system for $\binom{n+2}{2} - 1$ ternary n-ics with the possible exception of the invariant determinant of $\binom{n+2}{2}$ rows and columns, each row being formed by the coefficients of one n-ic.

Ex. Shew that the determinant formed by the coefficients of six conics is an irreducible invariant of the system.

284. Peano's theorem was first proved by means of an expansion due to Capelli*, which is virtually the same as the expansion used here, but is expressed in terms of polar operators.

* "Sur les Opérations dans la Théorie des formes Algébriques," *Math. Ann.* Bd. 37. See also "Lezioni sulla Teoria delle Forme Algebriche," Ch. I. § XXIII. by the same writer.

It is sufficient, in order to make the comparison clear, to point out that if f be a function linear in each of n sets of q-ary variables, a_1, a_2, ... a_n; then

$$\{a_1 a_2 \dots a_h\}' f = \Sigma \Delta_\lambda f_\lambda$$

where Δ_λ is one of the determinants of the matrix

$$\left| \begin{array}{cccc} a_{1,1} & a_{1,2} & \cdots\cdots & a_{1,q} \\ a_{2,1} & a_{2,2} & \cdots\cdots & a_{2,q} \\ \cdots & \cdots & \cdots & \cdots \\ a_{h,1} & a_{h,2} & \cdots\cdots & a_{h,q} \end{array} \right|,$$

and in fact is equal to $H_{a_1, a_2, \dots a_h} f$ where

$$H_{a_1, a_2, \dots a_h} = \Sigma \Delta_\lambda D_\lambda,$$

D_λ being the determinant of the matrix

$$\left| \begin{array}{cccc} \frac{\partial}{\partial a_{1,1}} & \frac{\partial}{\partial a_{1,2}} & \cdots\cdots & \frac{\partial}{\partial a_{1,q}} \\ \frac{\partial}{\partial a_{2,1}} & \frac{\partial}{\partial a_{2,2}} & \cdots\cdots & \frac{\partial}{\partial a_{2,q}} \\ \cdots & \cdots & \cdots & \cdots \\ \frac{\partial}{\partial a_{h,1}} & \frac{\partial}{\partial a_{h,2}} & \cdots\cdots & \frac{\partial}{\partial a_{h,q}} \end{array} \right|$$

obtained by writing $\frac{\partial}{\partial a_{r,s}}$ for $a_{r,s}$ in each element of Δ_λ.

In the case of binary forms, if $a_x{}^m b_y{}^n$ be polarized so as to obtain a function of $m+n$ sets of binary variables, and then the identity § 280 (III.) be applied, we obtain a proof of Gordan's series, § 52, which is a particular case of the series of Capelli. The coefficients in Gordan's series are not apparent, but it is possible to obtain them by these methods*.

Thus if $$f = a_{x^{(1)}} a_{x^{(2)}} \dots a_{x^{(m)}} b_{y^{(1)}} b_{y^{(2)}} \dots b_{y^{(n)}},$$
then $$T_{a_1, a_2, \dots a_h} f = 0,$$
when $h > 2$.

If $h = 2$, we need only consider those expressions PN which are of the form

$$PN = \{x^{(1)} x^{(2)} \dots x^{(m)} y^{(1)} \dots y^{(i)}\} \{y^{(i+1)} y^{(i+2)} \dots y^{(n)}\} \{x^{(1)} y^{(i+1)}\}' \dots \{x^{(n-i)} y^{(n)}\}';$$

* See Young, *Proc. Lond. Math. Soc.* vol. XXXIII.

and in this case

$$PN.f = P.(x^{(1)} y^{(i+1)}) \dots (x^{(n-i)} y^{(n)}) (ab)^{n-i} a_{x^{(n-i+1)}} \dots a_{x^{(m)}} b_{y^{(1)}} \dots b_{y^{(i)}},$$

which may be obtained by polarization from

$$(xy)^{n-i} (ab)^{n-i} a_x^{m+i-n} b_x^i.$$

In just the same way we see that, if x, y, z be ternary variables, $a_x^m b_y^n c_z^p$ can be expanded in a series each term of which may be obtained by polarization from a term of the form

$$\lambda (xyz)^i (abc)^i (ab\widehat{xy})^{j_1} (bc\widehat{xy})^{j_2} (ca\widehat{xy})^{j_3} a_x^{k_1} b_x^{k_2} c_x^{k_3},$$

where

$$(ab\widehat{xy}) = \begin{vmatrix} a_1 & b_1 & x_2y_3 - x_3y_2 \\ a_2 & b_2 & x_3y_1 - x_1y_3 \\ a_3 & b_3 & x_1y_2 - x_2y_1 \end{vmatrix}.$$

Let

$$u_1 = x_2y_3 - x_3y_2$$
$$u_2 = x_3y_1 - x_1y_3$$
$$u_3 = x_1y_2 - x_2y_1,$$

then

$$u_z = 0$$

is the condition that the point z lies on the straight line joining the points x and y. Hence u_1, u_2, u_3 are really the coordinates of a straight line.

We see then that all covariants of ternary forms which contain any number of variables can be expressed in terms of polars of covariants which contain the variables x and u only. The u's were introduced for geometrical reasons § 207, but it is now apparent that they are necessary to make the analytical theory complete.

285. Let $x^{(1)}, x^{(2)}, \dots x^{(n)}$ represent n sets of q-ary variables

$$x_1^{(r)}, x_2^{(r)}, \dots x_q^{(r)}; \quad r = 1, 2, \dots n.$$

Then if F be a function linear in each of these sets, we see, in the same way as before, that PNF is a function obtainable by polarization from a function, not necessarily linear, of

1st the single set $\quad x_1^{(1)}, x_2^{(1)}, \dots x_q^{(1)},$

2nd the determinants of the matrix

$$\begin{vmatrix} x_1^{(1)} & x_2^{(1)} \dots\dots x_q^{(1)} \\ x_1^{(2)} & x_2^{(2)} \dots\dots x_q^{(2)} \end{vmatrix},$$

3rd the determinants of the matrix

$$\begin{vmatrix} x_1^{(1)} & x_2^{(1)} \ldots\ldots & x_q^{(1)} \\ x_1^{(2)} & x_2^{(2)} \ldots\ldots & x_q^{(2)} \\ x_1^{(3)} & x_2^{(3)} \ldots\ldots & x_q^{(3)} \end{vmatrix},$$

and so on, and lastly of the determinant

$$\begin{vmatrix} x_1^{(1)} & x_2^{(1)} \ldots\ldots & x_q^{(1)} \\ x_1^{(2)} & x_2^{(2)} \ldots\ldots & x_q^{(2)} \\ \ldots & \ldots & \ldots \\ x_1^{(q)} & x_2^{(q)} \ldots\ldots & x_q^{(q)} \end{vmatrix}.$$

These may be all regarded as auxiliary variables. It will be useful to denote the variables of the 2nd set, viz. the determinants of the matrix

$$\begin{vmatrix} x_1^{(1)} & x_2^{(1)} \ldots\ldots & x_q^{(1)} \\ x_1^{(2)} & x_2^{(2)} \ldots\ldots & x_q^{(2)} \end{vmatrix}$$

by the letter ${}_2x$, with appropriate suffixes. Similarly the third set will be denoted by ${}_3x$ and so on.

Then in taking the complete system of concomitants of q-ary forms we have $q-1$ different kinds of variables which may appear.

The geometric meaning of these variables is easy to obtain. A space of $q-1$ dimensions being under consideration, the variables x or ${}_1x$ represent point coordinates: the variables ${}_2x$ are line coordinates; the variables ${}_3x$ are plane coordinates, and so on.

The linear substitutions by which these auxiliary variables are transformed, when any linear transformation of the point coordinates is made, are easy to find. The variables ${}_ix$ and ${}_{q-i}x$ are contragredient, and in fact

$$\Sigma\, {}_ix\, {}_{q-i}x$$

is an absolute concomitant. As a particular case, when q is even*, this leads to a relation between the variables ${}_{\frac{q}{2}}x$. This remark has already been illustrated for quaternary forms § 220.

* The case of binary forms is of course an exception.

286. Let $a_x^{(1)}, a_x^{(2)}, \ldots a_x^{(n)}$ be any n linear q-ary forms; and F a function linear in the coefficients of each. Then apply the formula

$$F = \Sigma A_{\alpha_1, \alpha_2, \ldots \alpha_h} T_{\alpha_1, \alpha_2, \ldots \alpha_h} F,$$

where substitutions interchange the sets of coefficients

$$a^{(1)},\ a^{(2)},\ \ldots\ a^{(n)}.$$

Consider the form PNF, it is a sum of terms each of which is a product of determinants of matrices of the form

$$\begin{vmatrix} a_1^{(1)} & a_2^{(1)} & \ldots\ldots & a_q^{(1)} \\ a_1^{(2)} & a_2^{(2)} & \ldots\ldots & a_q^{(2)} \\ \ldots & \ldots & \ldots & \ldots \\ a_1^{(i)} & a_2^{(i)} & \ldots\ldots & a_q^{(i)} \end{vmatrix}.$$

Moreover if N has two negative symmetric groups of degree i and $A_{i,\lambda}$, $B_{i,\mu}$ represent determinants from each of the corresponding matrices (the particular determinant being defined by the second suffix), then every term of NF has a factor of the form

$$A_{i,\lambda} B_{i,\mu}.$$

Hence every term of PNF has a factor of the form

$$A_{i,\lambda} B_{i,\mu} + A_{i,\mu} B_{i,\lambda}$$

as is evident when the tableau from which PN is constructed is considered. But this is a coefficient of the expression

$$[\Sigma A_{i\ i}x]\,[\Sigma B_{i\ i}x]$$

which is a covariant since the a's are contragredient to the x's.

Now if PN is a term of $T_{\alpha_1, \alpha_2, \ldots \alpha_h}$, N contains α_h groups of degree h, $\alpha_{h-1} - \alpha_h$ groups of degree $h-1$, and so on; hence PNF is a linear function of the coefficients of concomitants of the form

$$_hA^{(1)}{}_{_hx}\ {}_hA^{(2)}{}_{_hx} \cdots {}_hA^{(\alpha_h)}{}_{_hx}\ {}_{h-1}A^{(1)}{}_{_{h-1}x} \cdots {}_1A^{(\alpha_1-\alpha_2)}{}_{_1x}.$$

Thus every function of the coefficients of certain linear forms can be expressed in terms of coefficients of concomitants of those forms.

It is unnecessary to assume that the function is linear in the coefficients of each of the forms, in order that the above theorem may be true. For the function can be made linear by means of

Aronhold operators, and after the above process the original coefficients can be restored without affecting the invariantive properties in question.

Further the forms considered may be symbolical, and we at once deduce that *every integral function homogeneous in the coefficients of each of certain q-ary forms can be expressed as a linear function of the coefficients of the concomitants of those forms.*

APPENDIX I.

NOTE ON THE SYMBOLICAL NOTATION.

As we have said in § 82 the notation used in this work is really equivalent to Cayley's hyperdeterminants. The great advance made by the German school lies in the possibility of transforming symbolical expressions, and, of course, in the proof that every invariant form can be represented as a combination of hyperdeterminants. The reader may feel the need of justifying directly the results obtained by manipulating umbral expressions and accordingly we shall indicate how the whole theory can be made to rest on differential operators.

There are different ways of doing this. Salmon has remarked that, f being a binary form of order n, since

$$f = \frac{1}{n!}\left(x_1 \frac{\partial}{\partial y_1} + x_2 \frac{\partial}{\partial y_2}\right)^n f_y,$$

we may regard f as being equal to

$$(\alpha_1 x_1 + \alpha_2 x_2)^n \cdot \frac{f_y}{n!},$$

where $$\alpha_1^p \alpha_2^q = \left(\frac{\partial}{\partial y_1}\right)^p \left(\frac{\partial}{\partial y_2}\right)^q.$$

Hence we may suppose that

$$\alpha_1 = \frac{\partial}{\partial y_1}, \quad \alpha_2 = \frac{\partial}{\partial y_2},$$

and that the final operation is on

$$\frac{f_y}{n!}.$$

Any symbolical expression can be thus at once transformed into one exactly like it but involving only differential operators, *e.g.* if

$$f = \alpha_x^m, \ \phi = \beta_x^n,$$

then

$$(\alpha\beta)^r \alpha_x^{m-r} \beta_x^{n-r}$$

is $$\left(\frac{\partial^2}{\partial y_1 \partial z_2} - \frac{\partial^2}{\partial z_1 \partial y_2}\right)^r \left(x_1 \frac{\partial}{\partial y_1} + x_2 \frac{\partial}{\partial y_2}\right)^{m-r} \left(x_1 \frac{\partial}{\partial z_1} + x_2 \frac{\partial}{\partial z_2}\right)^{n-r} \frac{f_y}{m\,!} \frac{\phi_z}{n\,!}$$

(see § 82).

In any calculation we may omit the operand while we transform the operator.

After this the reader who wishes to do so will have no difficulty in developing the theory of the symbols when they are regarded as differential operators.

For another method see Kempe, *Proc. L.M.S.* vol. XXIV. p. 102 and Elliott, *Proc. L.M.S.* vol. XXXIII. p. 231.

Some interesting general remarks on the underlying principles of the symbolical notation will be found in Study, *Methoden Ternäre Formen*; and some very curious remarks in Lie-Scheffers, *Vorlesungen über Continuierlichen Grüppen*, p. 720.

This is the most convenient place to give a brief explanation of the so-called Chemico-Algebraic theory—an idea originally due to Sylvester which has perhaps attracted more attention than its intrinsic merits deserve.

In this theory an atom in chemistry corresponds to a binary form in algebra, and the valency of the atom to the order of the form. To each unit in the valency of an atom, in the chemical theory, a bond is supposed to correspond, and each such bond can connect the atom in question with an atom of valency one such as Hydrogen. Thus Oxygen is of valency two, and there exists a compound OH_2 which is written graphically

H—O—H,

there being two bonds proceeding from O and one from each H.

Since each unity in the order of a form gives rise to one possibility of transvection with another form, the analogy is evident

—if we have a binary quadratic o_x^2 and two linear forms h_x, h'_x the formula OH_2 corresponds to the algebraic expression

$$(oh)(oh')$$

an invariant of the three forms.

Then Carbon being of valency four we have the compound (Marsh Gas)

$$CH_4 \text{ or } \begin{array}{c} \text{H} \\ | \\ \text{H—C—H} \\ | \\ \text{H} \end{array}$$

and this corresponds to the invariant

$$(ch)(ch')(ch'')(ch''')$$

of a binary quartic and four linear forms.

The four hydrogen atoms in CH_4 are supposed on chemical grounds to occupy similar positions* in the structure of the compound and hence CH_4 is more naturally like

$$(ch)^4$$

an invariant of a quartic and a single linear form h_x.

Then the compounds CH_3Cl, CH_2Cl_2 etc. may be supposed like the invariants

$$(ch)^3(ck), \quad (ch)^2(ck)^2$$

where k is written for Cl and we see in the chemistry an analogue to polarizing in algebra.

Guided by the above the reader will have no difficulty in writing down an invariant corresponding to any graphical formula however complicated—in fact the algebraic form of the invariant is only a different (perhaps a more concise) way of writing down the chemical formula.

Difficulties arise when we recollect that some atoms have apparently different valencies illustrated by S in SO_2 and H_2S, for of course a binary form can have only one order. Gordan and Alexeleff suppose that the corresponding algebraic form is then polarized. Thus S in SO_2 would correspond to S_x^4 and S in H_2S to $S_x^2 S_y^2$ and now the degree available for transvection is two.

* For an explanation of this and the other chemical facts we have referred to see Scott, *Chemical Theory*, chap. VI.

The idea has been developed by various writers in the direction of making the algebraic methods graphical (references in Meyer) and lately Gordan and Alexeleff * have written several papers in which the algebra is applied to chemistry; the papers were criticised by Study † and from the objections and the replies the reader may be able to form his own opinion. We venture only to remark that the wonderful feature of the algebra is the capacity for reduction, and that, unless there is something corresponding in chemistry, the whole theory seems to be no more than a superficial analogy. It is of course certain that a reducible invariant often corresponds to a stable compound—moreover the *general* features which lead an algebraist to suppose a form reducible and a chemist to suppose a compound unstable are as nearly opposite in character as they can be.

It has been stated in Chapter I. § 21 that there are two ways of obtaining relations between concomitants symbolically expressed, viz.:

(i) By means of the fundamental identities

$$(bc)\, a_x + (ca)\, b_x + (ab)\, c_x = 0,$$

$$(bc)(ad) + (ca)(bd) + (ab)(cd) = 0\,;$$

(ii) By means of the fact that a concomitant is left unaltered when a pair of letters which refer to the same quantic is interchanged.

We give here a demonstration of the fact that all relations may be thus obtained ‡.

Let $F(C_i) = 0$ be any identical relation (supposed rational integral and homogeneous) between concomitants C_i of any system of binary forms.

Let
$$F(C_i) = \Sigma P_j$$
each term P_j being itself a concomitant. Also let

$$(A_0, A_1, \ldots A_n \between x_1, x_2)^n = a_x^n = b_x^n = \ldots$$

be one of the quantics of the system. The introduction of the symbolical notation may be effected by operators like

$$a_1^n \frac{\partial}{\partial A_0} + a_1^{n-1} a_2 \frac{\partial}{\partial A_1} + \ldots + a_2^n \frac{\partial}{\partial A_n}$$

* *Wiedemann's Annalen der Physik*, 1899, 1900. † In *Wiedemann*.

‡ Cf. Gordan, *Invariantentheorie*, Bd. II. § 117.

operating on $F(C_i)$. These operators do not destroy the property that $F(C_i)$ considered as a function of the coefficients of the quantics is identically zero.

Hence if $F(C_i) = \Sigma P_j$ becomes $\Sigma \Pi_k$, where each term Π_k is a symbolical product of factors of the forms (ab), a_x,

$$\Sigma \Pi_k \equiv 0$$

considered as a function of the symbolical letters.

Owing to the manner in which the symbolical letters were introduced no distinction was possible between two letters which refer to the same quantic, hence $\Sigma \Pi_k$ is unaltered by any interchange of two such letters, and therefore the second method of obtaining relations between forms will have no effect here.

Let $a_1, a_2, a_3, \dots a_r$ be the symbolical letters, and $n_1, n_2, \dots n_r$ the orders of the quantics to which they refer—if the forms Π_k are covariants we shall suppose that $a_{r,2} = x_1$, $a_{r,1} = -x_2$ and that n_r is the order of Π_k. Then applying the theorem of § 47, x being replaced by a_1 and y by a_2, we obtain

$$\Sigma \Pi_k = \sum_{j=0}^{n_2} \lambda_j (a_1 a_2)^j \left[a_2 \frac{\partial}{\partial a_1} \right]^{n_2 - j} [\Sigma \Pi_k^{(j)}],$$

where $\Sigma \Pi_k^{(j)}$ does not contain a_2.

Now $\Sigma \Pi_k \equiv 0$, hence every term of this series is identically zero. For if the jth term be the first which does not vanish, we may divide by $(a_1 a_2)^j$, and then put $a_2 = a_1$, whence

$$\lambda_j [\Sigma \Pi_k^{(j)}] = 0.$$

But this series is merely obtained by repeated use of the fundamental identities (see § 46), hence the reductions used so far belong entirely to the two classes mentioned.

It may happen that $\Sigma \Pi_k^{(j)}$ is not identically zero considered as a function of symbolical factors. In this case we may apply the same process again.

Each time we do this the number of letters in the function under consideration is reduced. Hence in $(r-2)$ steps at most the expressions are reduced to expressions which are identically zero when considered as functions of the symbolical factors—for when only two letters are left there is only one possible symbolical factor.

Thus the identity $F(C_i) = 0$ is made to depend entirely on the two fundamental methods of reduction.

APPENDIX II.

ON WRONSKI'S THEOREM AND THE APPLICATION OF TRANSVECTANTS TO DIFFERENTIAL EQUATIONS.

THE form of Wronski's theorem used in § 189 is not the usual one, but the determinant, which there vanishes, can be transformed in the following manner.

If
$$f = a_x^n,$$

then
$$\frac{\partial^r f}{\partial x_1^\lambda \partial x_2^\mu} = \frac{n\,!}{(n-\lambda-\mu)\,!}\, a_x^{n-\lambda-\mu}\, a_1^\lambda a_2^\mu,$$

and
$$x_2^\mu \frac{\partial^r f}{\partial x_1^\lambda \partial x_2^\mu} = \frac{n\,!}{(n-\lambda-\mu)\,!}\, a_x^{n-\lambda-\mu}\, a_1^\lambda\,(a_x - a_1 x_1)^\mu$$

$$= \frac{n\,!}{(n-\lambda-\mu)\,!}\, a_x^{n-\lambda-\mu} a_1^\lambda \left\{a_x^\mu - \mu a_x^{\mu-1}\, a_1 x_1 + \ldots\right\}.$$

It follows that
$$x_2^\mu \frac{\partial^r f}{\partial x_1^\lambda \partial x_2^\mu} = A_0^{(\mu)} \frac{\partial^\lambda f}{\partial x_1^\lambda} + A_1^{(\mu)}\, x_1 \frac{\partial^{\lambda+1} f}{\partial x_1^{\lambda+1}} + \ldots + A_\mu^{(\mu)} x_1^\mu \frac{\partial^{\lambda+\mu} f}{\partial x_1^{\lambda+\mu}},$$
where the A's are numbers depending on n, r, λ and μ.

Now in § 189 the determinant whose typical row is
$$\frac{\partial^r f'}{\partial x_1^r},\ \frac{\partial^r f'}{\partial x_1^{r-1}\partial x_2}, \ldots \frac{\partial^r f'}{\partial x_2^r},$$
vanishes, hence so also does that whose typical row is
$$\frac{\partial^r f'}{\partial x_1^r},\ x_2 \frac{\partial^r f'}{\partial x_1^{r-1}\partial x_2}, \ldots x_2^r \frac{\partial^r f'}{\partial x_2^r}.$$

On using the value found above for

$$x_2^{\mu}\frac{\partial^r f'}{\partial x_1^{\lambda}\partial x_2^{\mu}}$$

the typical row of the determinant becomes

$$\frac{\partial^r f'}{\partial x_1^r},\quad A_0^{(1)}\frac{\partial^{r-1}f'}{\partial x_1^{r-1}}+A_1^{(1)}x_1\frac{\partial^r f'}{\partial x_1^r},\ \ldots$$

$$A_0^{(r)}f'+A_1^{(r)}x_1\frac{\partial f'}{\partial x_1}+\ldots+A_r^{(r)}x_1^r\frac{\partial^r f'}{\partial x_1^r},$$

and since this determinant vanishes we infer, on modification of the columns, that the determinant whose typical row is

$$\frac{\partial^r f'}{\partial x_1^r},\ \frac{\partial^{r-1}f'}{\partial x_1^{r-1}},\ldots f'$$

vanishes.

If we replace x_2 by unity in the f's, this is the usual form of Wronski's determinant.

For the sake of completeness, we shall give an easy proof of Wronski's theorem.

If $u_1, u_2, \ldots u_{n+1}$ be $n+1$ functions of a single variable x, and the determinant whose rth row is

$$u_r,\ \frac{du_r}{dx},\ \frac{d^2u_r}{dx^2},\ \ldots\frac{d^nu_r}{dx^n}$$

vanish, then there is an identical relation of the form

$$\lambda_1u_1+\lambda_2u_2+\ldots+\lambda_{n+1}u_{n+1}=0,$$

where the λ's are constants.

In fact the vanishing of the determinant is the condition that the u's should be solutions of the same linear differential equation of order n, say

$$p_0\frac{d^ny}{dx^n}+p_1\frac{d^{n-1}y}{dx^{n-1}}+\ldots+p_ny=0.$$

We have therefore to prove that such an equation cannot have $(n+1)$ linearly independent integrals.

The theorem is easy to establish when n is unity, so we assume it true for $n-1$ and proceed inductively.

Now u_1 being a solution of the equation of order n, write

$$y=u_1\int w\,dx.$$

It is quickly seen that w is given by an equation of order $n-1$, say

$$q_0 \frac{d^{n-1} w}{dx^{n-1}} + q_1 \frac{d^{n-2} w}{dx^{n-2}} + \dots + q_n w = 0,$$

and this equation is satisfied by

$$\frac{d}{dx}\left(\frac{u_2}{u_1}\right), \quad \frac{d}{dx}\left(\frac{u_3}{u_1}\right), \dots \frac{d}{dx}\left(\frac{u_{n+1}}{u_1}\right),$$

hence by hypothesis there is a relation of the type

$$\lambda_2 \frac{d}{dx}\left(\frac{u_2}{u_1}\right) + \lambda_3 \frac{d}{dx}\left(\frac{u_3}{u_1}\right) + \dots + \lambda_{n+1} \frac{d}{dx}\left(\frac{u_{n+1}}{u_1}\right) = 0.$$

Integrating and multiplying up by u_1 we obtain a relation of the form

$$\lambda_1 u_1 + \lambda_2 u_2 + \dots + \lambda_{n+1} u_{n+1} = 0$$

between the $(n+1)$ functions u, so Wronski's theorem is completely established.

The application of the result to the $(r+1)$ functions f shews that if x_2 be replaced by unity, there is a relation of the type

$$\lambda_1 f_1 + \lambda_2 f_2 + \dots + \lambda_{r+1} f_{r+1} = 0,$$

and then making each f homogeneous again by the introduction of x_2, the result quoted in § 189 follows at once.

The device of changing from differential coefficients with respect to two variables to differentiation with respect to a single variable is often useful.

For example, the rth transvectant of

$$f = a_x^m, \quad \phi = b_x^n$$

is

$$\psi = (ab)^r a_x^{m-r} b_x^{n-r}.$$

Thus

$$x_2^r \psi = (a_1 b_2 x_2 - a_2 b_1 x_2)^r a_x^{m-r} b_x^{n-r}$$

$$= (a_1 b_x - a_x b_1)^r a_x^{m-r} b_x^{n-r}$$

$$= a_x^{m-r} a_1^r b_x^n - r a_x^{m-r+1} a_1^{r-1} b_x^{n-1} b_1 + \dots + (-1)^r a_x^m b_x^{n-r} b_1^r$$

$$= \frac{(n-r)!}{n!} \frac{\partial^r f}{\partial x_1^r} \phi - r \frac{(n-r+1)!}{n!} \frac{(m-1)!}{m!} \frac{\partial^{r-1} f}{\partial x_1^{r-1}} \frac{\partial \phi}{\partial x_1} + \dots$$

$$+ (-1)^r \frac{(m-r)!}{m!} \frac{\partial^r \phi}{\partial x_1^r} f.$$

Applying this to the nth transvectant of two forms f and ϕ, each of order n, we can reduce the problem of finding a form f apolar to ϕ to the solution of a linear differential equation of order n; it follows at once that there are not more than n linearly independent forms, and moreover from the algebraic theory we infer that all integrals of the equation are polynomials in x_1.

Conversely a differential equation whose coefficients are polynomials can be reduced to a relation between transvectants. To give a simple example, consider the equation

$$P_0 \frac{d^2y}{dx^2} + P_1 \frac{dy}{dx} + P_2 y = 0,$$

the P's being polynomials in x.

If ϕ_1, ϕ_2, ϕ_3 be three forms of orders r_1, r_2, r_3 respectively, and f be a form of order n,

$$(f\phi_1)^0 = f\phi_1,$$

$$x_2 (f\phi_2)^1 = \frac{1}{n} \phi_2 \frac{\partial f}{\partial x_1} - \frac{1}{r_2} f \frac{\partial \phi_2}{\partial x_1},$$

$$x_2^2 (f\phi_3)^2 = \frac{1}{n(n-1)} \phi_3 \frac{\partial^2 f}{\partial x_1^2} - \frac{2}{nr_3} \frac{\partial \phi_3}{\partial x_1} \frac{\partial f}{\partial x_1} + \frac{1}{r_3(r_3-1)} f \frac{\partial^2 \phi_3}{\partial x_1^2}.$$

Now replacing x_2 by unity as usual, we can choose ϕ_1, ϕ_2, ϕ_3, so that

$$x_2^2 (f\phi_3)^2 + x_2 (f\phi_2)^1 + (f\phi_1)^0 = 0$$

is the same as

$$P_0 \frac{d^2 f}{dx_1^2} + P_1 \frac{df}{dx_1} + P_2 f = 0,$$

for the transvectant relation is

$$\frac{1}{n(n-1)} \phi_3 \frac{d^2 f}{dx_1^2} - \frac{2}{nr_3} \frac{df}{dx_1} \frac{d\phi_3}{dx_1} + \frac{1}{r_3(r_3-1)} f \frac{d^2\phi_3}{dx_1^2}$$
$$+ \frac{1}{n} \phi_2 \frac{df}{dx_1} - \frac{1}{r_2} f \frac{d\phi_2}{dx_1} + f\phi_1 = 0,$$

and we must have

$$\frac{1}{n(n-1)} \phi_3 = P_0,$$

$$\frac{1}{n} \phi_2 - \frac{2}{nr_3} \frac{d\phi_3}{dx_1} = P_1,$$

$$\phi_1 - \frac{1}{r_2} \frac{d\phi_2}{dx_1} + \frac{1}{r_3(r_3-1)} \frac{d^2\phi_3}{dx_1^2} = P_2.$$

The first equation gives ϕ_3 and its order is that of P_0, the next equation gives ϕ_2 and then the last gives ϕ_1; hence the transformation is always possible, but of course the coefficients depend on the order of the form f that is chosen to represent y.

References to further developments in connection with Differential Equations will be found in Meyer's *Berichte* and in Klein's lithographed lectures on Linear Differential Equations of the Second Order.

APPENDIX III.

JORDAN'S LEMMA.

IN Chapter IV. great use was made of this theorem:—*If* $x+y+z=0$, *then any product of powers of* x, y, z *of order* n *can be expressed linearly in terms of such products as contain one exponent equal to or greater than* $\frac{2n}{3}$.

In § 64 a proof due to Stroh is given. We shall now give a simpler proof in which a much more general theorem is incidentally established.

The general theorem may be stated as follows:—

If
$$a_x, \; b_x, \; c_x, \; \ldots$$
be a system of r distinct linear forms and
$$\alpha, \; \beta, \; \gamma, \ldots$$
be r positive integers satisfying the relation
$$\alpha+\beta+\gamma+\ldots=n-r+1,$$
then it is impossible to find binary forms
$$A, \; B, \; C, \; \ldots$$
of orders $\alpha, \beta, \gamma, \ldots$
respectively such that
$$a_x^{n-\alpha} A + b_x^{n-\beta} B + c_x^{n-\gamma} C + \ldots = 0 \;\ldots\ldots\ldots\ldots (\text{I}).$$

In fact, suppose that such an identical relation exists and operate $\alpha+1$ times with
$$a_2 \frac{\partial}{\partial x_1} - a_1 \frac{\partial}{\partial x_2} = D.$$

It is easily seen, by making a_2 zero, that D^{a+1} annihilates a binary form of order n only when that form contains the factor a_x^{n-a}; hence since $a_x, b_x, c_x, \ldots$ are all different, we have

$$b_x^{n-\beta-a-1} B' + c_x^{n-\gamma-a-1} C' + \ldots = 0 \qquad \text{(II)},$$

where $B', C', \ldots$

are of orders $\beta, \gamma, \ldots$

respectively and do not vanish identically unless

$$B, C, \ldots$$

do also.

The relation (II) is of the same form as (I) except that r is changed into $r-1$ and n is changed into $n-\alpha-1$ for

$$\beta+\gamma+\ldots = n-r-\alpha+1 = (n-\alpha-1)-(r-1)+1.$$

Now a relation of the type (I) is impossible when $r=1$ for any value of n unless the form A (the only one occurring) vanishes identically, hence by induction it is impossible for all values of n and the theorem is established.

The forms $A, B, C, \ldots$

involve $\alpha+1, \beta+1, \gamma+1, \ldots$

arbitrary coefficients respectively, or in all

$$\alpha+\beta+\gamma+\ldots+r = (n+1).$$

Any binary form can be expressed linearly in terms of $(n+1)$ linearly independent forms of the same order, and since there is no identical relation of the form

$$a_x^{n-a} A + b_x^{n-\beta} B + c_x^{n-\gamma} C + \ldots = 0,$$

it follows that any binary form of order n can be expressed uniquely in the form

$$a_x^{n-a} A + b_x^{n-\beta} B + c_x^{n-\gamma} C + \ldots,$$

where $a_x, b_x, c_x, \ldots$ are all different, $A, B, C, \ldots$ are of orders $\alpha, \beta, \gamma, \ldots$ respectively, and

$$\alpha+\beta+\gamma+\ldots = n-r+1.$$

Consider now three linear forms

$$x, y, z,$$

where $z = -(x+y)$ and x, y are the variables.

It follows from the above that if

$$\alpha+\beta+\gamma=n-2$$

then any homogeneous expression of x, y, z of order n can be expressed in the form

$$x^{n-\alpha} P + y^{n-\beta} Q + z^{n-\gamma} R,$$

where P, Q, R are of order α, β, γ respectively.

In other words, changing α into $n-\lambda$, β into $n-\mu$ and γ into $n-\nu$, any homogeneous product of order n of x, y, z can be expressed in the form

$$x^{\lambda} P + y^{\mu} Q + z^{\nu} R,$$

where $$\lambda + \mu + \nu = 2n + 2,$$

and the expression is unique.

This is Stroh's generalized form of Jordan's lemma given in § 64, and the lemma itself follows at once since we can always choose integers λ, μ, ν satisfying the relation

$$\lambda+\mu+\nu=2n+2$$

and $$\lambda \geqslant \frac{2n}{3}, \quad \mu \geqslant \frac{2n}{3}, \quad \nu \geqslant \frac{2n}{3}.$$

On expressing P in terms of x and y, Q in terms of y and z, and R in terms of z and x, it follows that any homogeneous product of order n of x, y, z $(x+y+z=0)$ can be expressed linearly in terms of

$$\begin{matrix} x^n, & x^{n-1} y, & x^{n-2} y^2, \ldots & x^{\lambda} y^{n-\lambda} \\ y^n, & y^{n-1} z, & y^{n-2} z^2, \ldots & y^{\mu} z^{n-\mu} \\ z^n, & z^{n-1} x, & z^{n-2} x^2, \ldots & z^{\nu} x^{n-\nu} \end{matrix}$$

and the same would still be true if z were changed into x in the second row.

The reader will have no difficulty in modifying the above so as to obtain another proof of the fact that the general system of apolar forms constructed in § 178 contains n linearly independent forms.

APPENDIX IV.

FURTHER RESULTS ON COVARIANT TYPES.

THE expression given in § 262 for perpetuant types, may be used to determine the perpetuants when the forms are supposed not all different.

The result (using the notation of the paragraph referred to) for one quantic is

$$\lambda_{\delta-1} = 1 + \xi_{\delta-1}$$
$$\lambda_{\delta-2} = 2 + \xi_{\delta-1} + \xi_{\delta-2}$$
$$\ldots\ldots\ldots\ldots\ldots\ldots\ldots\ldots$$
$$\ldots\ldots\ldots\ldots\ldots\ldots\ldots\ldots$$
$$\lambda_2 = 2^{\delta-3} + \xi_{\delta-1} + \xi_{\delta-2} + \ldots + \xi_2$$
$$\lambda_1 = 2^{\delta-2} + 2(\xi_{\delta-1} + \xi_{\delta-2} + \ldots + \xi_2 + \xi_1),$$

where all the ξ's are zero or positive integers.

For two quantics the covariant can be written in the form

$$(a_1 a_2)^{2\alpha_1} (a_2 a_3)^{\alpha_2} \ldots (a_{i-1} a_i)^{\alpha_{i-1}} (a_i b_1)^{\alpha_i} (b_1 b_2)^{\beta_1} (b_2 b_3)^{\beta_2} \ldots (b_{j-1} b_j)^{\beta_{j-1}},$$

where the a's refer to one quantic and the b's to the other; the indices satisfy the conditions

$$\beta_{j-1} = 1 + \zeta_j$$
$$\beta_{j-2} = 2 + \zeta_j + \zeta_{j-1}$$
$$\ldots\ldots\ldots\ldots\ldots\ldots\ldots\ldots$$
$$\beta_1 = 2^{j-2} + \zeta_j + \zeta_{j-1} + \ldots + \zeta_2$$
$$\alpha_i = 2^{j-1} + \zeta_j + \zeta_{j-1} + \ldots + \zeta_1$$
$$\alpha_{i-1} = 2^j + \xi_{i-1}$$
$$\alpha_{i-2} = 2^{j+1} + \xi_{i-1} + \xi_{i-2}$$
$$\ldots\ldots\ldots\ldots\ldots\ldots\ldots\ldots$$
$$\alpha_2 = 2^{j+i-3} + \xi_{i-1} + \xi_{i-2} + \ldots + \xi_2$$
$$2\alpha_1 = 2^{j+i-2} + 2(\xi_{i-1} + \xi_{i-2} + \ldots + \xi_1).$$

For the proof of these results see Grace, "On Perpetuants," *Proc. London Math. Soc.*, vol. XXXV., p. 219.

The reasoning of this paper for the case of a single quantic may be applied to types, with the result that all perpetuant types of degree δ may be expressed in terms of products of types and of types of the form

$$(a_1 a_2)^{\lambda_1} (a_2 a_3)^{\lambda_2} \dots (a_{\delta-1} a_\delta)^{\lambda_{\delta-1}},$$

where the λ's satisfy the conditions laid down above for types for a single quantic, except that

$$\lambda_1 = 2^{\delta-2} + 2(\xi_{\delta-1} + \xi_{\delta-2} + \dots + \xi_2) + \xi_1,$$

and where the order of the letters $a_1, a_2, \dots$ is no longer fixed.

The method of §§ 262, 265 can be extended to the case of any concomitant which is linear in the coefficients of each of certain binary forms

$$a_{1_x}^{\ n_1},\ a_{2_x}^{\ n_2},\ \dots\ a_{\delta_x}^{\ n_\delta}.$$

Thus any such concomitant can be expressed by repeated transvection in terms of

(i) forms of the type

$$((\dots(((a_1 a_2)^{\lambda_1} a_3)^{\lambda_2} a_4)^{\lambda_3} \dots a_{\delta-1})^{\lambda_{\delta-2}} a_\delta)^{\lambda_{\delta-1}},$$

(ii) reducible forms.

The relations satisfied by the indices are however somewhat complicated. They are given as follows:—

A reduction (in the sense of § 265) is always possible when $\lambda_1 < 2^{\delta-2}$ unless one of the following conditions is satisfied:

$$\lambda_i + \Sigma(2\lambda_j + 2^{\delta-j} - n_{j+1}) > n_1 \quad - (2^{\delta-i} - 1),$$
$$\lambda_i + \Sigma(2\lambda_j + 2^{\delta-j} - n_{j+1}) > n_2 \quad - (2^{\delta-i} - 1),$$
$$\lambda_i \qquad\qquad > n_{i+1} - (2^{\delta-i} - 1),$$

where $\delta > i > j > 1$, but i and j are otherwise unrestricted, as also is the number of terms under the sign of summation—in particular there may be none.

In general, the condition that $\lambda_1 = 2^{\delta-2} - \alpha$ may not mean a reduction is that certain positive integers,

$f_1(0), f_2(0), f_3(0)$;

$f_1(\xi_1), f_2(\xi_1), f_3(\xi_1), \quad \xi_1 = 1, 2$;

$f_1(\xi_1, \xi_2), f_2(\xi_1, \xi_2), f_3(\xi_1, \xi_2), \quad \xi_1 = 1, 2 : \xi_2 = 1, 2$;

....

$f_1(\xi_1, \xi_2, \dots \xi_r), f_2(\xi_1, \xi_2, \dots \xi_r), f_3(\xi_1, \xi_2, \dots \xi_r),$

$\xi_1 = 1, 2 : \xi_2 = 1, 2 : \dots \xi_r = 1, 2,$

can be found such that

$$\alpha = f_1(0) + f_2(0) + f_3(0) + f(1) + f(2)$$
$$f(\xi) = f_1(\xi) + f_2(\xi) + f_3(\xi) + f(\xi, 1) + f(\xi, 2)$$

....

$$f(\xi_1, \xi_2, \dots \xi_r) = f_1(\xi_1, \xi_2, \dots \xi_r) + f_2(\xi_1, \xi_2, \dots \xi_r)$$
$$+ f_3(\xi_1, \xi_2, \dots \xi_r) + f(\xi_1, \xi_2, \dots \xi_r, 1) + f(\xi_1, \xi_2, \dots \xi_r, 2);$$

(where the quantities $f(\xi_1, \xi_2, \dots \xi_r)$ are positive integers which satisfy

$$f(\xi_1, \xi_2, \dots \xi_r, \xi_{r+1}) \ngtr 2^{\delta-r-3} - f_1(\xi_1, \xi_2, \dots \xi_r) - f_3(\xi_1, \xi_2, \dots \xi_r)$$

when $\xi_1 + \xi_2 + \dots + \xi_r + \xi_{r+1} + r$ is odd, and

$$f(\xi_1, \xi_2, \dots \xi_r, \xi_{r+1}) \ngtr 2^{\delta-r-3} - f_2(\xi_1, \xi_2, \dots \xi_r) - f_3(\xi_1, \xi_2, \dots \xi_r)$$

when $\xi_1 + \xi_2 + \dots + \xi_r + \xi_{r+1} + r$ is even);

$$\lambda_{r+2} + 2^{\delta-r-2} + \phi_1(\xi_1, \xi_2, \dots \xi_r) > n_1 + f_1(\xi_1, \xi_2, \dots \xi_r) - 1^*,$$
$$\lambda_{r+2} + 2^{\delta-r-2} + \phi_2(\xi_1, \xi_2, \dots \xi_r) > n_2 + f_2(\xi_1, \xi_2, \dots \xi_r) - 1,$$
$$\lambda_{r+2} + 2^{\delta-r-2} - f(\xi_1, \xi_2, \dots \xi_r) > n_{r+3} + f_3(\xi_1, \xi_2, \dots \xi_r) - 1;$$

where, finally, the ϕ's are defined by the laws

$$\phi_\eta(\xi_1, \xi_2, \dots \xi_r) + f(\xi_1, \xi_2, \dots \xi_r) - \phi_\eta(\xi_1, \xi_2, \dots \xi_{r-1})$$
$$- f(\xi_1, \xi_2, \dots \xi_{r-1})$$

is zero if $\eta + \xi_1 + \xi_2 + \dots + \xi_r$ is even, but if this sum is odd, the expression considered

$$= (2\lambda_{r+1} + 2^{\delta-r-1} - n_{r+2}) - 2\{f(\xi_1, \xi_2, \dots \xi_{r-1}) - f(\xi_1, \xi_2, \dots \xi_r)\},$$

and
$$\phi_\eta(0) = -\alpha = -f(0),$$
$$\phi_1(1) + f(1) = \mu_2 - 2\{f(0) - f(1)\},$$
$$\phi_2(1) + f(1) = 0,$$
$$\phi_1(2) + f(2) = 0,$$
$$\phi_2(2) + f(2) = \mu_2 - 2\{f(0) - f(2)\}.$$

* In case $f_1(\xi_1, \xi_2, \dots \xi_r)$ is zero, this inequality need not be satisfied; this remark applies also to the other two inequalities.

INDEX.

CAMBRIDGE: PRINTED BY J. & C. F. CLAY, AT THE UNIVERSITY PRESS.

Printed in the United States
By Bookmasters